PRINCIPLES AND PRACTICE OF BIOLOGICAL MASS SPECTROMETRY

Wiley-Interscience Series on Mass Spectrometry

Series Editors

Dominic M. Desiderio
Departments of Neurology and Biochemistry
University of Tennessee Health Science Center

Nico M. M. Nibbering
University of Amsterdam

The aim of the series is to provide books written by experts in the various disciplines of mass spectrometry, including but not limited to basic and fundamental research, instrument and methodological developments, and applied research.

Books in the Series

Michael Kinter, *Protein Sequencing and Identification Using Tandem Mass Spectrometry*
                                                                0-471-32249-07

Chhabil Dass, *Principles and Practice of Biological Mass Spectrometry*
                                                                0-471-33053-1

PRINCIPLES AND PRACTICE OF BIOLOGICAL MASS SPECTROMETRY

Chhabil Dass

Department of Chemistry
University of Memphis
Memphis, Tennesse

A JOHN WILEY & SONS, INC., PUBLICATION

New York • Chichester • Weinheim • Brisbane • Singapore • Toronto

For ordering and customer service, call 1-800-CALL-WILEY.

Library of Congress Cataloging-in-Publication Data:

Dass, Chhabil.
  Principles and practice of biological mass spectrometry / Chhabil Dass.
    p. cm.
  ISBN 0-471-33053-1 (cloth: alk. paper)
    1. Mass spectrometry. 2. Biomolecules—Analysis. I. Title.

  QP519.9.M3 D33 2000
  572'.36—dc21   00-038129

Printed in the United States of America.

10 9 8 7 6 5 4 3 2

His Invisible Hand is behind this work.

CONTENTS

FOREWORD

The field of mass spectrometry is undergoing a renaissance. Exciting developments have occurred, for example, in time-of-flight (TOF) and quadropole ion-trap instruments. Increased levels of detection sensitivity, mass resolution, and instrument performance are available in today's commercial instruments. Concomitant with the instrumental developments, mass spectrometry is a major contributor to the developments in the fields of basic ion chemistry studies, environmental studies, and biological mass spectrometry. For example, a worldwide effort is occurring in genomics and proteomics; mass spectrometry plays a key role in both fields. At the time of writing, it is anticipated that the human genome will be sequenced within the near future, and that much research activity will shift to proteomics. The *proteome* is defined as the expression of all the proteins from a gene at any one time in a tissue, fluid, cell, or other biological component. High-resolution separation techniques such as two-dimensional gel electrophoresis are used to separate the thousands of proteins in a proteome, matrix-assisted laser desorption/ionization (MALDI), and TOF quadrupole ion traps are used to identify those proteins, and bioinformatics are used to compare the MS data (usually in the form of a tryptic map) to the data contained in protein databases.

With all this research activity proceeding at a very high rate, mass spectrometry is poised to play a significant role. This present volume addresses the various aspects of mass spectrometry that are pertinent for today's research, for example, in biological mass spectrometry. A well-balanced and in-depth approach describes the various ionization processes that are used. The description of current basic instrumentation will be appreciated by the newcomers that enter the field from so many other different backgrounds. The basic principles of protein characterization are presented.

This volume will serve well to introduce newcomers to the field of mass spectrometry, and to provide a good teaching tool for those interested in learning the basic principles and applications of mass spectrometry.

D. M. DESIDERIO
Memphis, Tennessee

SERIES PREFACE

This series provides books written by experts in every area of mass spectrometry, including basic and fundamental research, instrument and methodological developments, and applied research. The books in this series will be of use not only to researchers who use mass spectrometry and wish to focus on one particular area, but also to teachers in the classroom and newcomers to the field of mass spectrometry. Mass spectrometry is being used in a variety of rapidly developing disciplines, and this series will provide an effective way to collect pertinent information in each area. Finally, the sum total of the research collected within this book series will be of interest to researchers in related areas such as chemistry, physics, biology, medicine, and nutrition.

PREFACE

Since the early 1900s, mass spectrometry has played a prominent role in several areas of physics, chemistry, geology, cosmochemistry, nuclear science, material science, archeology, petroleum industry, forensic science, and environmental science. The ultrahigh detection sensitivity and unsurpassed molecular specificity have contributed to this prominence. In the past, the contribution of mass spectrometry to biological sciences, however, remained obscure. The older ionization techniques were inadequate for the analysis of biological compounds. Remarkable developments in gentler modes of ionization, such as fast-atom bombardment, electrospray ionization, and matrix-assisted laser desorption/ionization, have changed this situation. In addition, several innovative developments have taken place in mass analyzer technology; time-of-flight mass spectrometry is undergoing a renaissance; and Fourier transform ion cyclotron resonance technology has matured. The upper mass range that is amenable by mass spectrometry has been extended beyond imagination. The detection sensitivity has reached in the zeptomole range. Remarkably efficient new approaches and paradigms have evolved that can solve practical issues in biochemical sciences. As a consequence, the applications of mass spectrometry to biochemical fields have grown in astronomical proportion. A new branch of mass spectrometry called *biological mass spectrometry* has emerged.

At present a strong void exists for a compendium that provides an up-to-date treatment of basic principles of instrumentation, techniques, and applications of the expanding field of biological mass spectrometry in a single volume. The expressed purpose of this volume is to fulfill this need. The first half of this book contains a comprehensive discussion on various ionization techniques, mass analyzers, tandem mass spectrometry instrumentation, the concepts of molecular mass measurement and quantitative analysis, and coupling of high-resolution separation devices. The second half addresses mass spectrometry approaches to the analysis of a diverse variety of biological compounds, such as proteins, peptides, glycoproteins, oligosaccharides, lipids, glycolipids, and oligonucleotides. The subjects of folding and unfolding of proteins, noncovalent interactions of biomolecules, and screening of combinatorial libraries are also discussed at length. Several real-world issues are highlighted in the last chapter, which attests to the role that mass spectrometry is destined to play in exploring diverse biological phenomena. Each chapter provides insight on background information concerning a specific class of biological compound (structure, biological function, etc.), and also describes modern approaches to characterization of those compounds.

This book should appeal to beginners and experts alike. It can serve as a reference text in a mass spectrometry course and as a resource for professionals working in various fields of biological sciences. The search to understand the intricacies of and to improve human health will continue. Mass spectrometry will become an integral part of these studies, and this book will be a handy companion in this endeavor.

I would like to take this opportunity to express my gratitude to Professor Dominic M. Desiderio for his invaluable suggestions, editorial comments, and for writing a forward to this volume. I would also like to acknowledge the assistant of Dhammika Jayawardene in the preparation of this volume. My wife Asha also deserves appreciation for her patience and understanding for my extra preoccupation during the progress of this project.

CHHABIL DASS
Memphis, Tennessee

ABBREVIATIONS

ACE	affinity capillary electrophoresis
ADH	alcohol dehydrogenase
AMEL	amelogenin loci
amol	attomole
APA	3-aminopicolinic acid
APCI	atmospheric pressure chemical ionization
API	atmospheric pressure ionization
ATT	6-aza-2-thiothymine
ATTR	transthyretin (TTR) amyloidosis
BAC	bioaffinity characterization
bp	base pair
BSA	bovine serum albumin
CAD	computer-assisted design; computed-aided drafting
CCD	charge-coupled device
CCSD	Complex Carbohydrate Structure Database
CD	charge detection; circular dichroism
CE	capillary electrophoresis; charge exchange
CEC	capillary electrochromatography
CEM	channel electron multiplier
CF	continuous flow; cystic fibrosis
CGE	capillary gel electrophoresis
α-CHCA	α-cyano-4-hydroxycinnamic acid
CI	chemical ionization
CID	collision-induced dissociation
CIEF	capillary isoelectric focusing
CLND	chemiluminescent nitrogen detector
CNBr	cyanogen bromide
CoA	coenzyme A
CREMS	charge reduction electrospray mass spectrometry
CRF	charge-remote fragmentation
CRM	charge residue model

CrPV	cricket paralysis virus
CSF	cerebrospinal fluid
CSP	calf spleen phosphodiesterase
CTL	cytotoxic T lymphocytes
CX	charge exchange
CZE	capillary-zone electrophoresis
Da	dalton
dc	direct current
DCTA	*trans*-1,2-diaminocyclohexane-N,N,N',N'-tetracetic acid
DE	delayed extraction
DHB	2,5-dihydroxybenzoic acid
DIOS	desorption/ionization on silicone
DNA	deoxyribonucleic acid
DTE	dithioerythritol
DTT	dithiothreitol
ECD	electron-capture dissociation
ECF	extracellular fluid
EI	electron ionization
ELISA	enzyme-linked immunoassay
EM	electron multiplier
endo H	endo-β-N-acetyl-glucosaminidase H
ESA	electrostatic analyzer
ESI	electrospray ionization
EST	expressed sequence tags
eV	electronvolt
FAB	fast-atom bombardment
FD	field desorption
FFR	field-free region
FI	field ionization
FIA	flow-injection analysis
FIS	field ion spectrometry
fmol	femtomole
FPD	focal (or point) plane detector
FSOT	fused-silica open tubular
FT	Fourier transform
FWHM	full width at half maximum
GalNAc	N-acetylglucosamine
GC	gas chromatography
GHRF	growth-hormone-releasing factor
GLP	good laboratory practices
GPA	glycerophosphatidic acid
GPC	glycerophosphocholine; gel-permeation chromatography
GPE	glycerophosphoethanolamine
GPG	glycerophosphoglycerol
GPI	glycerophosphoinositol

GPS	glycerophosphoserine
GSL	glycosphingolipid
HABA	2-(4-hydroxyphenylazo)benzoic acid
HBsAg	hepatitis B surface antigen
HBV	hepatitis B virus
HCD	heated capillary dissociation
HEL	hen egg lysozyme
HGP	human genome project
HIV	human immunodeficiency virus
HPA	3-hydroxypicolinic acid
HPLC	high-performance liquid chromatography
IAE	immunoaffinity extraction
ICP	inductively coupled plasma
ICR	ion cyclotron resonance
i.d.	internal diameter
IDM	ion desorption model
IE	ionizing energy
IEF	isoelectric focusing
IKE	ion kinetic energy
IMAC	immobilized metal ion affinity chromatography
IMS	ion mobility spectrometry
IPD	ion–photon detector
IR	infrared
IRMPD	infrared multiphoton dissociation
KDO	2-keto-3-deoxymannooctulosonic acid
KER	kinetic energy release
LAP	leucyl aminopeptidase
LAS	library affinity selection
LC	liquid chromatography
LDI	laser desorption/ionization
LLE	liquid–liquid extraction
LOS	lipooligosaccharides
LPS	lipopolysaccharides
liquid–SIMS	liquid–secondary ionization mass spectrometry
LT	leukotriene
mAb	monoclonal antibody
MAGIC	monodisperse aerosol generation interface for chromatography
MALDI	matrix-assisted laser desorption/ionization
MCA	multichannel acquisition
MCP	microchannel (or multichannel) plate
MEKC	miceller electrokinetic chromatography
MHC	major histocompatibility complex
MIKES	mass-analyzed ion kinetic energy spectroscopy
MRM	multiple-reaction monitoring
MS	mass spectrometry

MS/MS	tandem mass spectrometry
m/z	mass-to-charge ratio
NADH	nicotinamide adenine dinucleotide (reduced form)
nanoES	nanoelectrospray
NC	nitrocellulose
Nd/YAG	neodynium/yttrium aluminum garnet
NMR	nuclear magnetic resonance
NOG	*n*-octyl pyranoglucoside
nt	nucleotide
oa	orthogonal acceleration
o.d.	outer diameter
PA	proton affinity; picolinic acid
PAD	postacceleration detector
PAGE	polyacrylamide gel electrophoresis
PAH	polycylic aromatic hydrocarbon
PATRIC	position- and time-resolved ion counting
PC	phenylcarbamyl
PCA	principal-component analysis
PCD	plasma-coupled device
PCR	polymerase chain reaction
PD	plasma desorption
PEG	polyethylene glycol
PEEK	polyetheretherketone
PFK	perfluorokerosene
PFTBA	perfluorotributylamine
PG	prostaglandin
PIC	phenylisocynate
PIF	proteolysis-inducing factor
PITC	phenylisothiocyanate
pmol	picomole
PNGase F	*N*-glycosidase F
POMC	proopiomelanocortin
PPG	polypropylene glycol
PSD	postsource decay
PSI	pulsed sample introduction
PTC	phenylthiocarbamyl
PTH	phenylthiohydantoin
PVDF	polyvinylidene difluoride
QIT	quadrupole ion trap
QUISTOR	quadrupole ion store
RE	recombination energy
RBC	resonance electron capture
rf	radiofrequency
RIA	radioimmunoassay
RNA	ribonucleic acid

RNase	ribonuclease
RP	reversed phase
RPMCs	rat peritoneal mast cells
RRA	radioreceptor assay
RRKM	Rice–Ramsperger–Kassel–Marus (theory)
RTOF	reflectron time of flight
SDS	sodium dodecyl sulfate
SFC	supercritical fluid chromatography
SH2	src homology 2
SID	surface-induced dissociation
SIM	selected-ion monitoring
SIMS	secondary ionization mass spectrometry
SIS	superconductor–insulator–superconductor
S/N	signal-to-noise ratio
SORI	sustained off-resonance irradiation
SPE	solid-phase extraction
SPPS	solid-phase peptide synthesis
SRM	selected-reaction monitoring
SRY	sex-determining region Y
SSI	sonic spray interface
STAT	saccharide topology analysis tool
SVP	snake venom phosphodiesterase
SWIFT	stored waveform inverse Fourier transform
TBDMS	t-butyldimethylsilyl
TDC	time-to-digital converter
TEA	triethylamine
TFA	trifluoroacetic acid
THAP	$2',4',6'$-trihydroxyacetophenone
TIC	total ion current
TLC	thin-layer chromatography
TMA	trimethylamine
TOF	time of flight
TRF	thyrotropin-releasing factor
TSQ	triple-sector quadrupole
UV	ultraviolet
VEE	Venezuelan equine encephalitis
zmol	zeptomole

1

INTRODUCTION TO MASS SPECTROMETRY

Mass spectrometry probably is the most versatile and comprehensive analytic technique currently at the disposal of chemists and biochemists. Since the early 1900s, it has enjoyed prominence in several areas of physics, chemistry, geology, cosmochemistry, nuclear science, material science, archeology, petroleum industry, forensic science, and environmental science. The ultrahigh detection sensitivity and high molecular specificity are the hallmarks of this technique. Molecular mass determination, structure elucidation, quantification at trace levels, and mixture analysis are some of the major applications of mass spectrometry. In addition, the technique has been used to study ion chemistry and ion–molecule reaction dynamics; to provide data on physical properties such as ionizing energy, appearance energy, enthalpy of a reaction, and proton affinities; and to verify theoretical predictions that are based on molecular orbital calculations.

1.1. HISTORICAL PERSPECTIVE

Sir Joseph J. Thomson (1856–1940) conceptualized the idea of mass spectrometry (MS) in 1897 through his cathode ray tube experiments [1]. He measured the mass-to-charge ratio (m/z) of the negatively charged cathode ray particles by passing the collimated beam through crossed electric and magnetic fields. The birth of mass spectrometry is credited to his work on the analysis of positive rays with a parabola mass spectrograph in the early part of the twentieth century [2]. At that time, it was prophesized by Thomson that this new technique would be of immense use to the chemical analysis. However, chemists failed to realize the implication of this development for the next two decades thereafter. After Thomson, the developments of mass spectrometry continued in the hands of Aston, Dempster, Bainbridge, and Nier, who developed its applications to the discovery of new isotopes and to determine their relative abundances and exact masses.

Since the original work of Thomson, mass spectrometry has undergone extensive innovative developments, and a wide range of applications have evolved. In the 1940s, mass spectrometry was embraced by the petroleum industry. The applications to organic chemistry began in the 1950s and exploded in the 1960s and 1970s. High-resolution mass spectrometry became available in the 1950s, and paved the way for exact mass measurements. The development of gas chromatography (GC)MS in the 1960s marked the beginning of the analysis of complex mixtures by mass spectro-metry [3,4]. The 1980s and 1990s witnessed a rapid pace of developments in instrumentation and ionization techniques.

In the past, the applications of mass spectrometry to biological fields remained obscure, primarily because of the lack of suitable ionization techniques for compounds of biological origin. That situation has changed. Several unique developments in gentler modes of ionization have allowed the production of ions from nonpolar compounds, compounds of large molecular mass, and compounds of biological relevance. These methods include fast-atom bombardment (FAB) [5], electrospray ionization (ESI) [6,7], and matrix-assisted laser desorption/ionization (MALDI) [8]. The last two methods have extended the upper mass range that is amenable by mass spectrometry. Currently, several biopolymers with molecular mass over 100 kilodaltons (kDa) are analyzed routinely. Concurrent with these develop-ments, several innovations have taken place in mass analyzer technology; high-field and superfast magnets have been introduced, time-of-flight (TOF)MS is undergoing a renaissance, and Fourier transform (FT) ion cyclotron resonance–mass spectro-metry (ICRMS) has matured. Other unique developments in the field of mass spectrometry are tandem mass spectrometry (in 1970s), also known as MS/MS, the interfacing of high-resolution separation devices such as GC, high-performance liquid chromatography (HPLC), and capillary electrophoresis (CE) with mass spectrometry, the improvements in detection devices, and the introduction of fast data processing systems. The coupling of HPLC and CE to mass spectrometry has provided one of the most useful instruments to biochemists. The emergence of tandem mass spectrometry (MS/MS) is one of the high-points in the field of structure analysis by mass spectrometry.

These developments have all elevated mass spectrometry to a level where it has become an indispensable component in the arsenal of biomedical research. A new branch of mass spectrometry called *biological mass spectrometry* has emerged, and is constantly evolving.

1.2. WHY MASS SPECTROMETRY?

The wide popularity of mass spectrometry is the result of its following unique capabilities:

- It provides unsurpassed molecular specificity, which is the result of its ability to provide molecular mass and structurally diagnostic fragment ions of the analyte.
- It provides ultrahigh detection sensitivity. It has the ability to detect a single molecule; sensitivities in the attomole (amol) and zeptomole (zmol) ranges have been reported [9].
- It has unparalleled versatility to determine the structures of most classes of unknown compounds.
- It is applicable to all elements.
- It is applicable to all kinds of samples: volatile or nonvolatile, polar or nonpolar, and solid, liquid, or gaseous materials.
- In combination with high-resolution separation devices, it is uniquely qualified to analyze real-world complex samples.

In the past, the exorbitant cost and need for skilled operators of state-of-the-art magnetic sector mass spectrometers restricted mass spectrometry to the domain of a very few select laboratories. Currently, GCMS, LCMS, CEMS, ESIMS, and MALDIMS instruments are commercially available, reasonably priced, and user-friendly. As a result, mass spectrometry has become a much more accessible and an essential component of any contemporary chemical and biochemical research laboratory. Almost every chemical and biotechnical industrial establishment and higher academic institution in the United States has the distinction of having some sort of mass spectrometry facility.

1.3. BASIC CONCEPTS OF MASS SPECTROMETRY

Mass spectrometry is an analytic technique that measures the masses of individual molecules and atoms. As conceptualized in Figure 1.1, the first essential step in mass spectrometry analysis is to convert the analyte molecules into gas-phase ionic species because one can experimentally manipulate the motion of ions, and to detect them (which is not possible with neutral species). The excess energy transferred to the molecule during the ionization event leads to fragmentation. Next, a mass analyzer separates these molecular ions and their charged fragments according to their m/z (mass/charge) ratio. The ion current due to these mass-separated ions is detected by a suitable detector and displayed in the form of a mass spectrum. To enable the ions to move freely in space without colliding or interacting with other species, each of these steps is carried out under high vacuum (10^{-4}–10^{-8} torr).

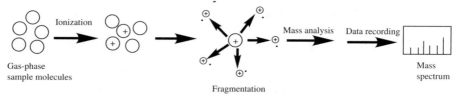

Figure 1.1. Basic concept of mass spectrometry analysis.

Thus, a mass spectrometer consists of several essential functional units; they are depicted in Figure 1.2 in the form of a block diagram. These units are

- An inlet system to transfer a sample to the ion source
- A vacuum system to maintain a very low pressure in the mass spectrometer
- An ion source to convert the neutral sample molecules into gas-phase ions
- A mass analyzer to separate and mass-analyze ionic species
- A detector to measure the relative abundance of the mass-resolved ions
- Electronics to control the operation of various units
- A data system to record, process, store, and display the data

The overall analytic capability of a mass spectrometry system depends on the combined performance of these individual units. Several ionization techniques have emerged, each with special purpose. A detailed account of the ionization methods that are applicable to compounds of biological relevance is presented in Chapter 2. These methods include electron ionization (EI), chemical ionization (CI), FAB [5], ^{252}Cf plasma desorption (PD) [10], field desorption (FD) [11,12], ESI [6,7], and MALDI [8]. Several different types of mass spectrometers are in common use [13], and have been categorized on the basis of a mass analyzer system. Chapter 3 discusses the basic principles of a variety of commonly used mass analyzers, which include double-focusing magnetic sector, quadrupole mass filters, TOFMS, quadrupole ion traps, and Fourier transform (FT)-ICRMS. Some of the common detector systems used in these mass analyzers are also described in this chapter.

1.4. THE NATURE OF MASS SPECTROMETRY DATA

The most important form of data that a mass spectrometer provides is a mass spectrum, which is a plot of m/z values of all ions that reach the detector versus their abundance. A typical computer-generated (bar-graph plot) positive-ion EI mass spectrum is shown in Figure 1.3. This is the spectrum of 10-[3′-(N-bishydroxyethyl)-amino]propyl-2-trifluoromethylphenoxazine (396 Da), a putative chemosensitizer [14]. It is a general practice that the most abundant ion in the spectrum is designated as the base peak (here, m/z 88), and is arbitrarily given a relative height of 100. The

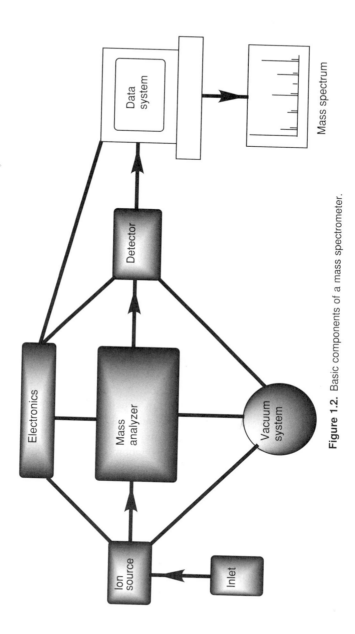

Figure 1.2. Basic components of a mass spectrometer.

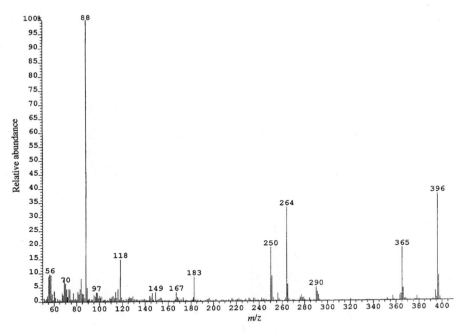

Figure 1.3. Electron ionization mass spectrum of 10-[3′-(N-bishydroxyethyl)amino]propyl-2-trifluoromethylphenoxazine.

abundances of all other ions in the spectrum are reported as a percentage abundance relative to this base peak.

A useful feature of a mass spectrum is that it contains a wealth of structure-specific information, and the most important of which is the molecular mass of the analyte. Most of the EI-formed ions are singly charged ions. Therefore, the molecular mass can be readily deduced from the identify of the molecular ion (designated as $M^{+ \cdot}$) because it represents the intact molecule. This ion usually is the most abundant peak among the high-mass cluster of peaks in the spectrum (e.g., the m/z 396 in Figure 1.3). Chapter 5 provides a more detailed discussion on the subject of the molecular mass determination with mass spectrometry. From the m/z values of the fragment ions, the structure of the analyte can be deduced. Precisely how that interpretation is done is beyond the scope of this book. A treatise by McLafferty and Turecek contains a wealth of information on this subject [15]. A text by Watson is also a useful reading material [16]. In some other ionization techniques (e.g., CI, FAB, MALDI), the molecular ion is obtained as a protonated molecule ($[M + H]^{+}$). Figure 1.4 is a typical example of this type of mass spectrum, which is the spectrum of 10-[3′(N-bishydroxyethyl)amino]propyl-2-trifluoromethylphenoxazine that was acquired by the liquid-secondary ionization mass spectrometry (SIMS) technique, which is a variation of FAB. The molecular mass from this spectrum is obtained by subtracting the mass of a proton from the m/z value of the $[M + H]^{+}$ ion. The

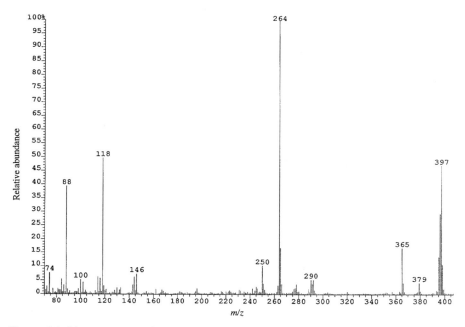

Figure 1.4. Mass spectrum of 10-[3'-(N-bishydroxyethyl)amino]propyl-2-trifluoromethylphenox-azine obtained with the liquid–SIMS technique.

molecular mass can also be obtained from a negative-ion spectrum. However, the structural information is very sparse in negative-ion spectra because negative ions are more stable. The structural information is also sparse in a mass spectrum that is acquired by ionizing the molecule with a gentler mode of ionization such as FAB, ESI, or MALDI. The structural information from such techniques is obtained by coupling them to tandem mass spectrometry, and subjecting the molecular ion to collision-induced dissociation (CID) (see Chapter 4 for more details).

1.5. APPLICATIONS

A survey of the current literature provides an ample testament to the potential and practical usefulness of mass spectrometry in the field of biochemistry and medicine. This development has become possible with the introduction of ESI and MALDI ionization techniques. Astonishing features of these techniques are sensitivity and high-mass capability; thus, they have permitted the molecular mass measurement of large biological compounds with unprecedented accuracy, speed, and low sample amounts. In addition, the coupling of chromatography and electrophoretic techniques with mass spectrometry has produced powerful tools for the simultaneous separation and detection of complex mixtures of biological compounds. Equipped with these developments, mass spectrometry has become an integral part of biological research. Most significantly, the use of CID and tandem mass spectrometry has allowed structural details to be unraveled.

A high level of ionization efficiency of the ESI and APCI modes of ionization provides an opportunity for the ultra-high-sensitivity detection of several analytes. Furthermore, the coupling of these techniques with HPLC and other separation devices provides an opportunity for the automation and high-throughput analysis. These developments have pushed mass spectrometry to the forefront of analytic techniques for the quantitative analysis of a variety of compount types in biological fluids. These methods have found a unique niche in the pharmacodynamic and pharmacokinetic evaluation of new drugs, in the management of various forms of illnesses, and in the study of the functional role of bioactive compounds in various neurological and pathophysiological events. A more detailed description of the field of quantitative analysis can be found in Chapter 6.

Mass spectrometry now appears to be the primary means of protein identification. A combination of techniques, such as gel electrophoresis, peptide mapping, and database searching with the molecular mass and tandem mass spectrometry data of the peptide fragments of a protein digest, has contributed to this success. With these advancements, proteomic analysis has become a relatively simpler task. These developments are discussed in Chapters 8 and 9. Chapter 10 deals with a related topic of the characterization of glycoproteins and oligosaccharides.

The continued developments in ESI and MALDI modes of ionization have propelled mass spectrometry to a still higher level. Besides obtaining the molecular mass, amino acid sequence, and nature and site of covalent modifications of proteins and other biomolecules, the study of supramolecules by mass spectrometry has expanded. ESIMS and MALDIMS are increasingly used to investigate the higher-order structures and the folding–unfolding dynamics of proteins and noncovalent complexes of biomolecules. The details of these studies are discussed in Chapters 11 and 12.

Mass spectrometry has played a pivotal role in the structural characterization of fatty acids, phospholipids, glycolipids, lipopolysaccharides that are derived from different organisms, and lipid mediators. A variety of mass spectrometry techniques that include FABMS, ESIMS, and MALDIMS have been used for this objective. This field is discussed in Chapter 13.

Combinatorial chemistry has the potential to simultaneously synthesize and screen a large number of drug candidates. This enormity poses a challenge for an analytic technique to rapidly analyze a large number of diverse library components. Because of its potential in distinguishing closely related compounds on the basis of molecular mass and fragmentation pattern, mass spectrometry is an ideal technique for these applications. In response to the need for rapid screening of diverse types of libraries, several mass spectrometry-based methods have been developed. These techniques are discussed in Chapter 14.

The analysis of megadalton ions has become a reality. The mass spectrometry-based techniques to sequence DNA have matured (see Chapter 15). These developments are destined to contribute to the human genome project.

Mass spectrometry now offers new perspectives on solving real-world problems. With sensitive and faster analysis methods at hand, its role has expanded in clinical, immunologic, and cancer-related studies, in the profiling of bacteria and viruses, and in the characterization of peptides, proteins, and other biomolecules at the level of a single cell and single neuron. The role of mass spectrometry in a few selected real-world problems is demonstrated in Chapter 16.

The search to understand the intricacies of human health and to improve it will continue. A concerted role of several biological sciences has contributed and will continue to contribute to this understanding. Mass spectrometry will become an integral part of these studies, and in the future, we will witness the expanded role of mass spectrometry in biomedical research.

1.6. LITERATURE

Mass spectrometry research on the topics of instrumental developments, new techniques, and applications are published in several specialized mass spectrometry journals such as *Journal of the American Society for Mass Spectrometry, Rapid Communications in Mass Spectrometry, Journal of Mass Spectrometry, International Journal of Mass Spectrometry and Ion Processes, European Mass Spectrometry*, and *Journal of the Mass Spectrometry Society of Japan. Mass Spectrometry Reviews* publishes critical reviews on topics of current interest. In addition, readers are advised to look at articles in *Analytical Chemistry, Journal of American Chemical Society, Journal of Chromatography, Journal of Biological Chemistry, Protein Science*, and *Proceedings of the National Academic of Sciences* (USA). Occasionally, high-impact articles appear in *Science* and *Nature*. Biennial reviews on mass spectrometry and the A-page articles in *Analytical Chemistry* are also a source of important information in mass spectrometry.

REFERENCES

1. J. J. Thomson, *Philos. Mag. V* **44**, 293 (1897).
2. J. J. Thomson, *Rays of Positive Electricity and the Application to Chemical Analyses*, Longmans Green, London, 1913.
3. J. T. Watson and K. Biemann, *Anal. Chem.* **36**, 1135–1137 (1964).
4. R. Ryhage, *Anal. Chem.* **36**, 759–764 (1964).
5. M. Barber, R. S. Bordoli, R. D. Sedgwick, and A. N. Tyler, *J. Chem. Soc. Chem. Commun.* 325–327 (1981).
6. J. B. Fenn, M. Mann, C. K. Meng, S. F. Wong, and C. M. Whitehouse, *Science* **246**, 64–71 (1989).
7. R. D. Smith, J. A. Loo, R. R. Ogorzalek Loo, M. Busman, and H. R. Udseth, *Mass Spectrom. Rev.* **10**, 359–451 (1991).

8. M. Karas and F. Hillenkamp, *Anal. Chem.* **60**, 2299–2301 (1988).

9. M. E. Belov, M. V. Groshkov, H. R. Udseth, G. A. Anderson, and R. D. Smith, *Anal. Chem.* **72**, 2271–2279 (2000).

10. B. Sundqvist and R. D. Macfarlane, *Mass Spectrom. Rev.* **2**, 421–460 (1985).

11. H. D. Beckey, *Principles of Field Ionization and Field Desorption Mass Spectrometry*, Pergamon, Oxford, England, 1977.

12. L. Prokai, *Field Desorption Mass Spectrometry*, Marcel Dekker, New York, 1990.

13. C. Dass, in D. M. Desiderio, ed., *Mass Spectrometry: Clinical and Biomedical Applications*, Plenum Press, New York, Vol. 2, 1994, pp. 1–52.

14. C. Dass, K. N. Thimmaiah, B. S. Jayashree, R. Seshadiri, M. Israel, and P. J. Houghton, *Biol. Mass Spectrom.* **23**, 140–146 (1994).

15. F. W. McLafferty and F. Turecek, *Interpretation of Mass Spectra*, University Science Books, Mill Valley, CA, 1993.

16. J. T. Watson, *Introduction to Mass Spectrometry*, Lippincott-Raven, New York, 1997.

2

IONIZATION METHODS

The objective of this chapter is to introduce various ionization methods currently in use for the analysis of biomolecules. The success of a mass spectrometry experiment relies to a large extent on the way that a neutral compound is transferred into gas-phase ionic species. The choice of a method for the ionization of a compound is contingent on the nature of the sample under investigation. Because the central theme of this book is the characterization of biomolecules by mass spectrometry, the most widely applied ionization techniques in this field, such as fast-atom bombardment (FAB) [1], electrospray ionization (ESI) [2], and matrix-assisted laser desorption/ionization (MALDI) [3] will be described in more detail. For the sake of completeness of discussion, a few other common ionization techniques will also be discussed briefly. A monograph written on this subject is worth reading [4].

2.1. ELECTRON IONIZATION

Historically, electron ionization (EI) has been the most popular mode of ionization for organic compounds. This method is applicable to thermally stable and relatively volatile compounds, or to those compounds that can be converted to the gaseous state at the prevailing vacuum and temperature of the ion source. Many compounds of biological interest are also accessible to this technique. The upper mass limit of compounds that are amenable to EI is roughly 1000 Da.

2.1.1. Basic Principles of Electron Ionization

In this mode, the sample is first converted to the vapor phase, and then bombarded with a beam of energetic electrons at low-pressure conditions ($\sim 10^{-5}$–10^{-6} torr). This collision process displaces an electron from a molecule (M) of the target compound to convert it to a positive ion that contains an odd number of electrons. The positive ions thus formed are usually referred to as odd-electron *molecular ions* ($M^{+\cdot}$) or radical cations (Eq. 2.1). In order for this process to occur, the energy of the bombarding electrons must be greater than the ionization energy (IE) of the sample molecule. Because the mass of an electron is negligible, the m/z value of the molecular ion is a direct measure of its molecular mass. The negative ions can be formed via the capture of an electron by a neutral molecule. Although this process is less probable in an EI source, abundant negative ions are produced under chemical ionization (CI) conditions (see Section 2.2.3).

$$M + e^- \rightarrow M^{+\cdot} + 2e^- \tag{2.1}$$

The energy gained in excess of IE causes primary ions to promptly fragment into structurally diagnostic smaller mass fragment ions. Often, these fragment ions have sufficient energy to fragment further. The fragmentation pattern thus obtained is an indication of the structure of the sample molecule. The fragmentation of the molecular ions occurs mostly within the ion-source region. The efficiency of ionization and of subsequent fragmentation increases with electron energy, and reaches a plateau at 50–100 eV. At these electron energies, the EI spectrum becomes a "fingerprint" of the compound being analyzed. For this reason, a large majority of the mass spectra that are reported in the literature and stored in spectral libraries have been acquired at 70 eV.

Discussion of the fragmentation mechanisms of small ions is beyond the scope of this volume. However, readers are advised to consult a well-written volume on this subject [5]. The relative abundance of an ion depends not only on its rate of formation but also on the rates of the subsequent fragmentation reactions. The driving force for a reaction to occur is governed by the stability of the resulting fragments. Thus, the final appearance of a mass spectrum of a compound is the result of a series of competitive and consecutive unimolecular reactions, which have been described in terms of quasiequilibrium theory [6] and Rice–Ramsperger–Kassel–Marcus (RRKM) theory [7].

The construction of a typical EI source is illustrated in Figure 2.1. Heating to an incandescent temperature a thin filament of rhenium wire creates an electron beam, which is collected at the anode (trap) after traveling through the ionization chamber. To increase the probability of ionization, electrons are made to travel in narrow, helical trajectories by applying a weak magnetic field in parallel with the direction of the electron beam. The positive ions thus formed are pushed into the accelerating region by applying a positive potential to a repeller electrode. Before entering the mass analyzer, all ions are accelerated to a certain kinetic energy, the value of which is determined by the potential difference between the source block and the exit slit

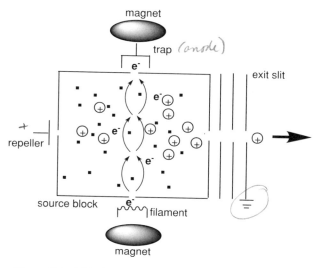

Figure 2.1. A block diagram of an electron ionization source.

held at ground potential. In the case of magnetic sector and time-of-flight instruments, the value of this accelerating potential is several kilovolts. In contrast, only a few volts of accelerating potential is used with quadrupole-based mass spectrometers. The source block is usually heated up to 300°C to prevent any condensation of the sample.

The sample ion current, I^+, is given by Eq. 2.2, where β is the ion extraction efficiency, Q_i is the total ionizing cross section, L is the effective ionizing pathlength, $[N]$ is the concentration of the sample molecules, and I_e is the ionizing current. Thus, the sample ion current can be augmented by increasing the ionizing current, the sample pressure, ionizing cross section, ionizing pathlength, and ion extraction efficiency. There is an upper limit to the sample pressure because, beyond a certain value, mass resolution is degraded and the probability of ion–molecule reactions becomes higher. The ion extraction efficiency is improved by increasing the repeller voltage and acceleration voltage. The chemical nature of the sample molecules and the energy of the ionizing electrons are the factors that can increase the ionizing cross section. However, the most feasible approach to increase the sample ion current is to use higher values of the ionizing current:

$$I^+ = \beta Q_i L[N]I_e \tag{2.2}$$

Electron ionization is a popular technique in conjunction with gas chromatography (GC). These two techniques are perfectly matched because they deal with relatively volatile samples. GC also acts as a mode of sample introduction into the EI source. Highly volatile samples that do not require any prior chromatography can be introduced via a sample reservoir. Solid and liquid samples are introduced via a direct insertion probe.

2.1.2. Limitations of Electron Ionization

Although EI is a simple and sensitive technique for the analysis of small molecules, it suffers from a serious drawback that many compounds are not stable under EI conditions. During ionization, 2–8 eV energy is transferred to a molecule. Therefore, with certain types of molecules, fragmentation is so pervasive that the molecular ion signal is either absent or of very little significance. In the absence of a molecular ion, it is an uphill task to determine the molecular mass of a compound. A further pitfall of EI is its incompatibility with thermally labile and nonvolatile compounds. Thermal energy supplied to break the intermolecular bonds in such polar compounds may also cause their decomposition. The volatility requirement, however, can be circumvented in some cases if an analyte is derivatized with specific chemical reagents. This process blocks those polar functional groups from participating in intermolecular hydrogen bonding. However, derivatization is not an attractive proposition because it requires additional sample-handling steps and because it creates uncertainty in the determination of the molecular mass of the original compound when incomplete derivatization occurs and when the exact number of derivatization sites is not known.

2.2. CHEMICAL IONIZATION

In view of the limitations of the EI procedure in producing stable molecular ion species, a compelling need to develop less energetic ionization methods was felt. Chemical ionization (CI) was the first serious attempt in this endeavor [8]. Although this technique uses the same (or slightly modified) ion source and the same sample introduction techniques as those used in EI, the ionization process is drastically different. Ionization process in CI involves ion–molecule reactions of the sample molecules with reagent gas ions. A complete process requires several steps. A reagent gas, the partial pressure of which is 10–100 times greater than the sample pressure, is introduced into a modified EI source. In the first step, this reagent gas is ionized at somewhat higher pressures (between 0.1 and 1 torr) by bombardment with a beam of 200–500-eV electrons. In the second step, one or more stable reagent ions are produced by ion–molecule reactions. In some cases, the initially formed reagent ions are not stable; they undergo simple collisions with neutral reagent gas molecules to produce thermally cool stable reagent ions. The sample molecules are finally ionized by ion–molecule reaction with those stable reagent ions. For further reading, readers are referred to a volume entirely devoted to this subject [9].

The principle of CI is illustrated with respect to methane as the reagent gas in the following equations:

Step 1:
$$CH_4 + e^- \rightarrow CH_4^{+\cdot} + 2e^- + CH_3^+ + \cdots \qquad (2.3)$$

Step 2:
$$CH_4 + CH_4^{+\cdot} \rightarrow CH_5^+ + CH_3^{\cdot} \qquad (2.4)$$

$$CH_4 + CH_3^+ \rightarrow C_2H_5^+ + H_2 \qquad (2.5)$$

Step 3:
$$M + CH_5^+ \rightarrow [M + H]^+ + CH_4 \qquad (2.6)$$

$$M + CH_5^+ \rightarrow [M + CH_5]^+ \qquad (2.7)$$

$$M + C_2H_5^+ \rightarrow [M + C_2H_5]^+ \qquad (2.8)$$

$$M + C_2H_5^+ \rightarrow [M - H]^+ + C_2H_6 \qquad (2.9)$$

The primary ions ($CH_4^{+\cdot}$ and CH_3^+) that are formed in Eq. 2.3 undergo further reaction with neutral methane to produce stable secondary reagent ions (Eqs. 2.4 and 2.5), which behave as Brønsted acids. The ionization of the sample molecules (M) takes place through the proton transfer reaction, provided the proton affinity of M is greater than methane (Eq. 2.6). The formed molecular ions are the adducts of the sample molecule with a proton or the reagent ions (e.g., $[M + H]^+$, $[M + CH_5]^+$, or $[M + C_2H_5]^+$). The hydride abstraction to form $[M - H]^+$ (Eq. 2.9) is also observed for certain compounds. The ions formed by CI are called _even-electron quasimo-lecular ions_ and, unlike EI-formed molecular ions, their mass is different from that of the neutral molecule. The molecular mass of the analyte can be deduced from the _m/z_ values of these adducts after accounting for the mass of the attached species. The energy of the CI-formed ions is quickly attenuated via collisions with the neutral reagent gas molecules. As a consequence, the CI-formed ions are generally more stable than the odd-electron ions that are formed in EI. A simple spectrum is produced via CI, and contains primarily the molecular ion signal and a few fragment ions. Thus, the primary use of CI is to confirm or determine the molecular mass of volatile compounds. The rules that govern the fragmentation of even-electron ions are different from those of the odd-electron ions [10,11]. Therefore, the structural information, if available from the CI process, is complementary to that obtained via the EI technique.

Various reagent gases have been discovered for analytical advantages, and are listed in Table 2.1 along with the corresponding reagent ions and their proton affinities. Several unusual positive-ion CI reagents have been reviewed [13]. The literature also sites references to the use of metal ions (e.g., Cu^+ and Fe^{3+}) for CI of organic compounds [14,15].

A key feature of CI is that the extent of fragmentation can be controlled by a proper choice of a reagent gas. The amount of energy transferred to the sample ion under CI conditions is a function of the exothermicity, $\Delta H°$ of the acid–base proton transfer reaction. With respect to the reaction shown in Eq. 2.10, the energy transferred during CI is given by Eq. 2.11. Thus, the reaction is more efficient when the proton affinity (PA) of the sample molecule, M, is greater than that of the

Table 2.1. Reagents for positive-ion chemical ionization

Reagent Gas	Reagent ion	Proton Affinity[a] (kcal/mol)
H_2	H_3^+	101
CH_4	CH_5^+	132
H_2O	H_3O^+	167
CH_3OH	$CH_3OH_2^+$	182
C_2H_5OH	$C_2H_5OH_2^+$	186
i-C_4H_{10}	i-$C_4H_9^+$	196
$(CH_3)_2CO$	$(CH_3)_2COH^+$	197
NH_3	NH_4^+	204
CH_3NH_2	$CH_3NH_3^+$	211

[a] Proton affinities are from Ref. 12.

reagent gas molecule, B. The PA is measured as the heat that is liberated during the protonation of a molecule.

$$M + [B + H]^+ \rightarrow [M + H]^+ + B \tag{2.10}$$

$$\Delta H° = -[PA(B) - PA(M)] \tag{2.11}$$

Thus, for a given sample, the $[M + H]^+$ ions that are produced via CI with H_3^+ ions would be more energetic than those produced by CI with i-$C_4H_9^+$ ions.

Another unique feature of CI is the feasibility of the selective ionization of a specific compound [16]. By a proper choice of a reagent gas, only those compounds with a PA value greater than that of the reagent gas are ionized. As an example, a reagent gas with a high PA, such as ammonia, is an ineffective CI reagent for many organic compounds, but an efficient reagent for organic amines [17].

An electron ionization source can also be adapted for CI experiments, but after certain modifications. First, electrons are produced from a heated metal filament or a ribbon. Second, to maintain a higher pressure, the ionization chamber is made as gastight as possible by reducing the apertures of the electron beam entry and ion exit slits. Third, the permanent magnet used to collimate the electron beam and the anode used to trap the electrons are eliminated because there is no need for the electron beam to travel all the way to the other end of the gas chamber. However, electrons of much higher energies (≤ 500 eV) are needed so that they can penetrate to a reasonable length in the ion chamber. Finally, a more efficient pumping system is employed to maintain the source pressure to $< 10^{-4}$ torr, and the analyzer region is differentially pumped as the result of a higher pressure in the CI source.

2.2.1. Charge Exchange Chemical Ionization (CECI)

Charge exchange (CE) is a special mode of CI that is explicitly designed to produce odd-electron molecular ions instead of even-electron quasimolecular ions [18]. The basic concept of ionization is similar to the normal mode of CI. That is, a reagent gas is first ionized via EI to produce reagent gas ions, $G^{+\cdot}$. The abstraction of an electron from a neutral analyte molecule converts it into a radical cation (Eq. 2.12). The internal energy, ε, of $M^{+\cdot}$ that are produced in this reaction is given by the difference between the recombination energy (RE) of $G^{+\cdot}$ and ionizing energy (IE) of the reagent gas molecules (Eq. 2.13). The RE of $G^{+\cdot}$ is usually approximated to the negative value of the vertical IE of gas molecules provided $G^{+\cdot}$ is formed in the ground electronic state:

$$G^{+\cdot} + M \rightarrow M^{+\cdot} + G \tag{2.12}$$

$$\varepsilon(M^{+\cdot}) = RE(G^{+\cdot}) - IE(M) \tag{2.13}$$

Similar to the normal mode of CI, ions of a well-defined internal energy can be produced by carefully selecting a reagent gas. A variety of CE gases has been used for specific applications. These gases include (with increasing RE) toluene, benzene, NO, CS_2, COS, Xe, CO_2, CO, N_2, Ar, and He. The greater the difference in $RE(G^{+\cdot})$ and IE(M), the more extensive is the fragmentation. Thus, an EI-like spectrum can be generated with a reagent gas of a higher recombination energy. By using CECI with some of the abovementioned reagent gases, ions with energies below their threshold for isomerization have been generated [19,20].

2.2.2. Mixed Chemical Ionization

In some applications, it is beneficial to acquire a spectrum that has the features of EI and CI both. This objective is achieved when a mixture of reagent gases such as Ar/H_2O or $Ar/i\text{-}C_4H_{10}$ is employed. For example, with a mixture of $Ar/i\text{-}C_4H_{10}$, the $i\text{-}C_4H_9^+$ reagent ions will produce mainly the $[M + H]^+$ ions and almost no fragment ions, whereas CECI with $Ar^{+\cdot}$ ions being highly exothermic will produce EI-like spectra. Thus, a suitable reagent gas mixture provides a mixed spectrum in a single scan, from which the molecular mass and structural information can be derived simultaneously. $EICI$?

2.2.3. Negative-Ion Chemical Ionization

Although a majority of applications of mass spectrometry are concerned with the analysis of positive ions, attention has also been focused on negative ions. The chemistry of gas-phase negative ions has fascinated many researchers because negative ions are more stable than their positively charged counterparts. Also, quantitative measurements can benefit from the increased detection sensitivity of negative ions.

Two different reaction schemes are used to produce negative ions of volatile compounds. The first scheme involves a direct interaction of the sample molecules with electrons. Depending on the energy of the electron involved, the formation of negative ions can proceed via one of the following processes:

$$AB + e^-(\approx 0 \text{ eV}) \rightarrow AB^{-\cdot} \tag{2.14}$$

$$AB + e^-(0-15 \text{ eV}) \rightarrow A^- + B^{\cdot} \tag{2.15}$$

$$AB + e^-(> 10 \text{ eV}) \rightarrow A^- + B^+ + e^- \tag{2.16}$$

Of these three processes, the most useful approach is resonance electron capture (REC), which occurs with electrons of near thermal energies (Eq. 2.14). These electrons are generated in an EI source in the presence of a moderating gas such as H_2, CH_4, i-C_4H_{10}, NH_3, N_2, or Ar. With electrons of energies in excess of the thermal range, the dissociative electron capture (Eq. 2.15) and ion-pair formation (Eq. 2.16) are the dominating processes in the generation of negative ions; both, however, are of little structural significance.

An 100–1000-fold increase in the detection sensitivity is realized in the negative-ion REC-CI for compounds that have high electron affinities [21,22]. Compounds that contain a nitro group, a halogen atom, or a conjugated π-electron system all fall in this category. The electron affinity of other compounds can be augmented by derivatizing them with certain electrophilic groups, such as perfluoroacyl, perfluorobenzoyl, pentafluorobenzyl, or nitrobenzoyl [23,24].

If a compound does not have a strong affinity for the REC process, then its negative ions can be generated via ion–molecule reactions with specific reagent anions. This scheme is exemplified in the following reaction:

$$B^- + M \rightarrow [M - H]^- + BH \tag{2.17}$$

This reaction occurs because the B^- ion is a stronger Brønsted base than the sample molecules. Some of the reagent ions that produce abundant negative molecular ions, with decreasing PA, include NH_2^-, OH^-, $O^{-\cdot}$, CH_3O^-, F^-, $O_2^{-\cdot}$, and Cl^- ions. Of these, the OH^-, $O^{-\cdot}$, and CH_3O^- ions have been employed more frequently. EI of the respective reagent gas is used to generate these reagent ions:

N_2O/CH_4 (or i-C_4H_{10}) mixture:
$$N_2O + e^- \rightarrow O^{-\cdot} + N_2$$
$$O^{-\cdot} + CH_4 \rightarrow OH^- + CH_3^{\cdot} \tag{2.18}$$

N_2O/N_2 mixture:
$$N_2O + e^- \rightarrow O^{-\cdot} + N_2 \tag{2.19}$$

CH_3NO_2:
$$CH_3NO_2 + e^- \rightarrow CH_3O^- + NO \tag{2.20}$$

2.2.4. Limitations of Chemical Ionization

Like the EI technique, a major limitation of CI is its inability to handle nonvolatile compounds, and thus, it is restricted to compounds with molecular mass below 1000 Da. Many compounds of biomedical relevance that contain polar functional groups are excluded from the protocol of CI analysis. Another obvious limitation is the lack of structural information in the CI spectrum. An innovative approach to circumvent this difficulty is to acquire alternatively EI and CI spectra on the same injection of the sample. In order to acquire a pure EI spectrum, the reagent gas is evacuated from the chamber on every alternate scan.

2.3. DESORPTION IONIZATION

As discussed above, EI and CI are both restricted to volatile compounds. A large number of biological compounds are nonvolatile and thermally unstable. For such compounds, other methods must be used. As a result of sustained efforts in several laboratories, a range of desorption ionization techniques that can simultaneously volatilize and ionize condensed phase samples have emerged. These developments have brought a variety of biomolecules into the realm of mass spectrometry analysis, and have extended its applications beyond 100,000-Da mass range.

A unifying aspect of these ionization techniques (except field desorption) is that the sample is mixed with a suitable matrix, and bombarded with a high-energy particle beam to produce, just above the point of impact, a high concentration of the gas-phase neutral and ionic species. The underlying rationale in these sudden-energy methods is that, by depositing a large amount of energy into the analyte molecule on a timescale that is fast compared with the vibrational time period, vaporization may occur before thermal decomposition begins. Although all the desorption ionization methods described below embrace this concept, they differ in the way a primary beam is produced.

2.3.1. Field Desorption

Field desorption (FD) was one of the first serious attempts to ionize nonvolatile and thermally labile substances [25,26]. This technique differs from other desorption ionization methods in that no primary beam is used to bombard a sample. Instead, the sample ions are directly desorbed from the condensed phase sample under the influence of a strong electric field gradient of the order of 10^8 V/cm. A convenient way to create such intense electric fields is to apply a voltage of 10–20 kV to the tip of a sharp metal object similar to a razor blade or a thin wire. In practice, the sample is deposited on specially prepared numerous microneedle-like structures or whiskers. These very fine (approximately of 1 μm diameter) multiple ionization spots are grown on a thin wire filament by pyrolyzing the vapors of benzonitrile. The emitter electrode is placed approximately 1 mm away from the cathode. For some samples, it is necessary to heat the emitter to bring them to the molten state and to facilitate the movement of dissolved salts such as NaCl and KCl.

Field desorption of the analyte molecules produces ions mainly of the types $[M + H]^+$ and $[M + Na]^+$. The actual mechanism of FD ionization has been the subject of much discussion. The broad consensus is that the strong electric field distorts the electron cloud around the sample molecule, and lowers the barrier for the removal of an electron from the molecule by quantum-mechanical tunneling. As the electron from the molecule tunnels to the conduction bands of the emitter, the $M^{+\cdot}$ is left behind. The ion–molecule reactions of $M^{+\cdot}$ with the sample molecules produces $[M + H]^+$-type ions. The $[M + Cat]^+$-type ions are also produced because of the presence of cationic impurities. The influence of a strong electric field desorbs these ions from the emitter surface. The negative-ion formation involves the capture of an electron from the negatively charged emitter tip. When a sample is introduced in the ion source as a vapor, the ions formed are mainly of the $M^{+\cdot}$ type, and the ionization process is called *field ionization* (FI) [27].   *field ionization*

Because only a little excess internal energy is imparted to the analyte molecule during FD, fragmentation is usually absent; thus, FD and FI spectra contain mainly the molecular ion species and almost no fragment ions. A major contribution of FD has been to determine the molecular mass of nonvolatile compounds. The lack of fragmentation, however, is a concern for the structure determination applications. Attempts have been made to induce fragmentation of the FD-produced ions by heating the filament to higher temperatures, or alternatively by collision activation in the intermediate region of a tandem mass spectrometer.

Several applications of FD have been reported for the analysis of carbohydrates [28], peptides [29], organometallics [30], sugars [31], and industrial polymers [32]. Despite this success, FD has not enjoyed wide popularity, mainly owing to operational difficulties in making reproducible and effective emitters. In addition, FD ion currents are low and short-lived. This technique was practically abandoned soon after the discovery in 1981 of the more user-friendly FAB technique. One major advantage of FD over other desorption/ionization techniques, however, is the virtual absence of any background chemical noise.

2.3.2. Plasma Desorption Ionization ^{252}Cf

Californium-252 (^{252}Cf)–plasma desorption (PD) ionization, introduced in 1974 by Macfarlane and co-workers, is one of a group of desorption/ionization methods that uses a high-energy particle beam for desorption/ionization of nonvolatile, polar, and thermally labile molecules [33]. This technique is used in conjunction with time-of-flight (TOF) mass spectrometry. The incident particle beam in this technique is generated from the spontaneous fission of the radioactive nuclide ^{252}Cf.

The principle of PD is depicted in Figure 2.2 [34,35]. The sample is deposited onto a thin aluminum or aluminized Mylar foil, and is bombarded by a beam that consists of a plasma of high energy (≈ 100-MeV range) ions. The plasma is generated from a thin 10-μCi ^{252}Cf film. Each fission event in this nuclide simultaneously releases, in opposite directions, several pairs of fission particles with a range of masses. For example, one such pair is ^{106}Tc^{22+} and ^{142}Ba^{18+}, with kinetic energies of 100 and 80 MeV, respectively. One of the fission fragments bombards the sample target to emit secondary ions and neutrals, and the second fragment strikes the start detector. The secondary ions are accelerated into the flight tube of a TOF mass spectrometer by applying a potential of 10–20 kV between the sample target and the source grid. A time-to-digital converter (TDC) is used to

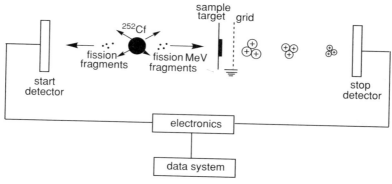

Figure 2.2. A schematic of the use of ^{252}Cf plasma desorption ionization with time-of-flight mass spectrometry.

measure the flight times of the ions. The sample ion flux is usually low; therefore, to obtain sufficiently high ion statistics, a large number of scans $(5 \times 10^5 - 10^6)$ are summed and averaged.

Sample Preparation. Sample preparation is very critical for the success of PDMS analysis. The presence of impurities has an adverse effect on the generation of a good signal; in particular, alkali metal salts may reduce or quench entirely the molecular ion signal. A recommended procedure to remove such salts is to adsorb the sample on a thin film of nitrocellulose, and to wash it with an ultapure solvent [36].

Applications. Plasma desorption has been used primarily for the determination of the molecular mass of biomolecules [36]. The observed peaks in a PD spectrum of proteins are the protonated molecule ion species with different charged states. Although a temperature of several thousand degrees is attained at the point of impact of the plasma beam, the molecular ions of proteins do not decompose because energy transfer is a very fast event, and is mediated by the absorption, excitation, and relaxation processes that take place in the matrix. The peaks are generally broad because they include contributions from metastable ions. Despite the broadness of peaks, PDMS has been used successfully to determine the molecular mass of proteins and peptides with a precision of ±0.03%. To date, the mass of the highest mass protein analyzed by this method was 45,000 Da [37]. For small peptides, fragmentation takes place at the peptide amide bonds to reveal the sequence of a peptide [38]. The glycosidic bonds are also cleaved in polypeptides. PD has also been used to determine posttranslational modifications in proteins and peptides [38], and to obtain information about tryptic maps [36,38,39] and disulfide bonds [40]. The introduction of a dedicated and relatively inexpensive commercial PDMS instrument in the late 1980s had prompted many laboratories to add a mass spectrometry facility to their arsenal of biochemical analysis. However, with the advent of ESI [2] and MALDI [3], the popularity of PDMS is waning.

2.3.3. Fast-Atom Bombardment

The discovery of fast-atom bombardment (FAB) in 1980 by Barber and colleagues was a major breakthrough in the analysis of condensed-phase samples [1]. The mass spectrometry community accepted FAB immediately with great enthusiasm for a wide range of applications. For the first time, the sequencing of peptides and analysis of other biomolecules became a routine task by mass spectrometry. FAB is well adapted to compounds that have some degree of polarity or that contain acidic or basic functional groups. Central to the success of FAB is the use of a liquid matrix which helps to dissolve the sample.

Fast-atom bombardment is an offshoot of secondary ionization mass spectrometry (SIMS) [41]. SIMS has enjoyed a wide popularity as a surface characterization technique. In SIMS, a beam of high-energy ions, usually of Ar^+ ions, directly bombards the surface of the sample. However, its major disadvantage with respect to analytical applications is the short-lived secondary ion current. Barber made two innovative changes to overcome this limitation. First, he dissolved the sample in a polar and relatively less volatile liquid matrix (e.g., glycerol), with the result that the sample current could be observed for longer periods. Second, he used a beam of high-energy (6–10 keV) atoms rather than ions (Figure 2.3) [1]. This change allowed FAB to be used with magnetic sector instruments, a popular form of mass spectrometry of that time. Soon, it became apparent that the nature of the secondary ions is not affected whether the sample is sputtered from a liquid matrix by bombardment with a beam of atoms or of ions [42]. It is the liquid matrix that made a dramatic difference in the yields of secondary ions. Because a matrix can absorb the impact of a high-flux atom (or ion) beam, damage to the sample layer is reduced commensurably. Furthermore, the uppermost layer of the matrix is replenished continuously with the sample molecules to prolong sample ion currents. A constant and steady emission of secondary ions is essential in many situations such as to acquire mass spectra, high-resolution mass measurements, quantitative applications, and tandem mass spectrometry experiments.

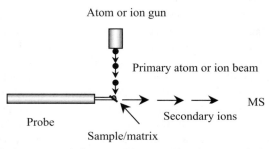

Figure 2.3. A conceptual diagram of fast-atom bombardment ionization.

A separate term, *liquid–secondary ionization mass spectrometry* (liquid–SIMS), has been coined for the desorption technique that employs high-energy ions as the primary beam in conjunction with a liquid matrix, even though it is mechanistically identical to FAB. The two techniques differ in the way a primary beam is generated. The current generations of mass spectrometers use Cs^+ ion guns for liquid–SIMS experiments. These guns contain a pellet of cesium aluminum silicate, which, when heated to an incandescent temperature, emits Cs^+ ions. The acceleration of these ions through an electric field that is applied between the emitter electrode and the grounded electrode provides a beam of high-energy (25–40-keV) Cs^+ ions. In the past, some liquid–SIMS guns also used liquid metals as the source of a primary ion beam [43].

In contrast, a FAB gun uses inert gas atoms (Ar or Xe) to produce a beam of fast atoms. Xe is preferred over Ar because the impact of massive atoms can enhance the yields of secondary ions. In an atom gun, the neutral gas atoms are first ionized by collisions with electrons that are moving in a saddle-field configuration. The ions formed are accelerated to the required potential (2–10 kV), and are neutralized in a dense cloud of the excess neutral gas atoms by a resonance electron-capture process; thus, fast ions are converted to fast atoms to generate a continuous beam of high-translational-energy atoms (Eq. 2.21). Any residual undischarged ions are deflected by a positive potential on a deflector plate:

$$\underset{\text{slow atoms}}{Xe} \xrightarrow{\text{ionization}} \underset{\text{slow ions}}{Xe^+} \xrightarrow{\text{acceleration}} \underset{\text{fast ions}}{Xe^+} \xrightarrow{\text{neutralization}} \underset{\text{fast atoms}}{Xe^0} \qquad (2.21)$$

Liquid–SIMS, however, offers certain advantages over FAB. Because of the high-translational-energy ion beams used in liquid–SIMS, high-mass ions are easily sputtered. Furthermore, it is easier to obtain a well-defined electrically focused primary ion beam (which is not possible with a neutral atom beam), resulting in enhanced secondary ion yields. Also, there is no gas load in a liquid–SIMS ion source, and thus, background noise is lower. These advantages lead to an improvement in the detection sensitivity and high-mass analysis capability. As a result of these advantages, more mass spectrometers are outfitted with liquid–SIMS sources than FAB sources.

Mechanism of Ion Formation. The elucidation of the mechanism of ion formation in FAB has been a painstaking process [44–47]. Several conflicting ideas have been put forward. The current thinking is that the impact of a high-energy atom or ion initiates a collision cascade to promote the formation of a high-temperature, high-density gas in the cavity that is formed at the point of impact. This intermediate pressure region between the liquid surface and mass spectrometry vacuum has been termed *selvedge*. The impact of the fast-moving atoms sputters from the solution neutrals and preformed ions. Additional ions, along with secondary electrons, radicals, and excited neutral species, are generated in the collision cascade. The protonated and deprotonated matrix ions are formed during these events. The sputtered ions are often associated with solvent aggregates. The desolvation of ion–solvent clusters begins in the selvedge region and continues in the gas phase to produce free sample ions. Additional ionization takes place in the selvedge region by gas-phase ion–molecule reactions. Thus, the observed mass spectrum is a reflection of preformed ions that are present in the matrix, the ions and electrons that are formed in the collision cascade, and the acid–base and electrochemical reactions that occur in the solution and gas phase.

The Nature of the FAB Mass Spectrum. Fast-atom bombardment analysis can be performed in positive- and negative-ion modes with equal ease. FAB is considered a soft ionization technique. As a consequence, the molecular ions are the predominant species in the FAB spectra, and are of the type $[M + H]^{+}$ or $[M - H]^{-}$, respectively. In the positive-ion mode, cationized (cat) clusters of the type $[M + cat]^{+}$ are also observed because the sample usually contains metallic salts as impurities. In special cases, radical cations have also been observed [48]. Unlike other desorption ionization techniques discussed in this chapter, the energy imparted to many samples during FAB ionization/desorption is sufficient to induce fragmentation. Thus, in favorable cases, a FAB mass spectrum may include fragments of structural significance. The matrix also has a control on the extent of fragmentation [49]. For example, the FAB spectrum of peptides, when acquired with glycerol as the matrix, contains a large number of sample-specific fragment ions [49]. In contrast, a very few sample-specific fragment ions are formed with α-thioglycerol as the matrix. In addition to the sample-related ions, a FAB spectrum also contains the matrix-related ions that are formed as a result of radiation damage and association/dissociation reactions [49–53]. The prominent ions from glycerol are m/z 15, 19, 27, 29, 31, 43–45, 56, 57, 61, 74, 75, 91, 93, 185, 277, 365, and higher [53,54]. This matrix-related chemical noise is a source of nuisance in many applications of FABMS.

The Role of the Liquid Matrix. A liquid matrix plays a significant role in the FAB analysis. It reduces damage to the sample by absorbing the impact of the primary beam, prolongs the sample ion current by constantly replenishing the upper layer with the fresh sample, and reduces the binding energy of the sample molecules. More importantly, it provides a medium in which the ionization of the sample can be facilitated. The proper choice of a liquid matrix is very crucial to obtain a good-quality FAB mass spectrum and to optimize the sample ion current. The criteria for the matrix selection have been discussed [55,56], and the use of several matrices has been reviewed [56–58]. A compilation of several important physicochemical properties of some typical matrices is also available [59]. Any liquid substance that is reasonably viscous, chemically inert, and nonvolatile; exhibits good solvent and electrolytic properties; and is mass-spectrally transparent can be used as a matrix. Glycerol is by far the most widely accepted matrix. Some other commonly used matrices are α-thioglycerol, a mixture of dithiothreitol (DTT) and dithioerythritol (DTE; 5:1), thiodiglycol, 3-nitrobenzyl alcohol, tetraglyme, sulfolane, diethanolamine, triethanolamine and crown ethers (e.g., 18-crown-G).

The acid–base and oxidizing–reducing nature of the matrix, the surfactancy of the solute relative to the matrix, and solute–solvent interactions all have a profound impact on the mass-spectral appearance [52,60,61]. The $[M + H]^{+}$ ion signal intensity is enhanced when the matrix is more acidic than the sample molecule. Likewise, $[M - H]^-$ ion production is facilitated in basic matrices. Glycerol, α-thioglycerol, and DTT/DTE are suitable acidic matrices. Also, α-thioglycerol, which is more acidic than glycerol, provides improved $[M + H]^+$ ion yields. Similarly for negative-ion analysis, basic matrices such as diethanolamine and triethanolamine are preferred. The matrix can also influence the extent of fragmentation of the analyte ions [49]. Matrices are also known to participate in chemical reactions with the sample. Reduction [62–65], transmethylation [66], halogen replacement [67,68], radiation-induced radical addition [51,53], hydride abstraction [69], and adduct formation [52] are some of the sample–matrix reactions that have been documented.

A FAB mass spectrum is also influenced by the surface composition of the matrix. Hydrophobic compounds tend to occupy the upper portion of a relatively hydrophilic matrix, and are ionized preferentially relative to the hydrophilic compounds, which prefer to remain embedded within the matrix [70]. Thus, certain compounds remain undetected by FAB when present along with more hydrophobic compounds. Also, the presence of alkali salts tends to suppress ionization. Similarly, compounds with negative surface activity relative to the matrix may not produce any FAB signal. The surfactancy of a solute, however, can be manipulated by chemical derivatization with a large alkyl group or by adding surface-active agents to the matrix–analyte mixture [71].

Sample Preparation. Good sample preparation is a key to the success of FABMS analysis. As discussed above, chemical properties of a matrix have a profound influence on the spectral appearance. The detection sensitivity in the FAB analysis is a function of the chemical composition of the solvent–solute mixture, and of the presence of other unwanted analytes. The alkali salts must be removed from the sample. Improvement in positive- and negative-ion sample signals can be realized by the addition of an acid or a base, respectively. The addition of trifluoroacetic acid (TFA) to the sample–matrix mixture also helps maintain the integrity of disulfide bonds in peptides.

Sample Introduction. In a routine operation, the matrix and sample are loaded onto a stainless steel probe tip, which usually is an integral part of a solid insertion probe. The probe is introduced into the ion source via a vacuum lock. Although this procedure is very simple, the detection sensitivity is limited because of an excessive matrix background that appears as a result of the high matrix : sample ratio used in the analysis [72]. In order to overcome this problem, a continuous-flow (CF)FAB sample introduction probe has been developed [73,74]. One of the benefits of this device is that the matrix : sample ratio is reduced significantly. Typically, a CFFAB probe consists of a hollow stainless steel shaft that contains a narrow fused-silica capillary (Figure 2.4) [75]. A stream of a very dilute aqueous solution of the matrix (usually 1–10% by volume of glycerol) flows continuously through the capillary over a FAB target. For stable operation, the rate of liquid that enters the probe must be balanced with the rate of evaporation at the tip of the probe. A wick made of porous cellulose fiber is added to absorb an excess liquid at the tip [73,74]. In

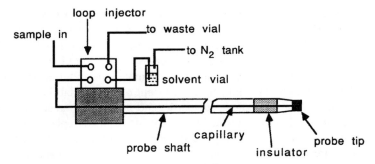

Figure 2.4. A schematic of continuous-flow FAB probe. [Reprinted from Ref. 75 by permission of Plenum Press, New York (copyright 1994).]

another design, the capillary terminates into a frit or a metal screen to disperse the liquid [76,77]. In both designs, the matrix forms a very thin layer at the probe tip. The sample is injected into this stream, and is bombarded by a high-energy beam in a conventional way. The matrix background is reduced significantly. As a consequence, a several-fold increase in the detection sensitivity is realized, as has been demonstrated by Caprioli and Moore for the analysis of the peptide angiotensin II [78].

The CFFAB probe is also useful for the on-line direct analysis of biochemical reactions [79]. In addition, this probe can be used as an interface for the on-line coupling of HPLC [80,81], capillary electrophoresis (CE) [82–84], and a micro-dialysis probe [85] with mass spectrometry. The last configuration especially enables the in situ analysis of dialyzable compounds (e.g., neurotransmitters) from body tissues. Additional information can be obtained from a monograph written on this subject [86].

Applications. One major area of applications of FABMS is to obtain molecular mass information for a variety of compounds. Although a FABMS signal in the mass range as high as 35,000 Da has been observed for CsI clusters, 4000 Da can be considered to be an upper mass limit for a routine operation of FAB. Above this mass, the molecular ion signal usually is very weak. Another major area of application is the characterization of peptide fragments of proteolytic digests of proteins [87,88]. FAB has also been used to obtain characteristic spectra of various classes of compound. The technique has been applied for the analysis of peptides [89–91], phosphopeptides [92,93], proteins [94,95], antibiotics [96], fatty acids [97,98], lipids [99,100], carbohydrates [101,102], nucleotides and nucleic acids [103,104], DNA adducts [105,106], organometallics [107], surfactants [108], and several other compound classes [109].

2.3.4. Matrix-Assisted Laser Desorption/Ionization

The irradiation by an intense laser beam is another suitable mode of depositing a large amount of energy into sample molecules for their desorption into the gas phase

[110–112]. In early mass spectrometry applications of lasers, the sample was irradiated directly by a laser beam. Infrared (IR) [e.g., the Nd/YAG (neodynium/yttrium aluminum garnet; 1.06 μm) laser, the pulsed CO_2 laser (10.6 μm)], and UV lasers (frequency-quadrupled Nd/YAG laser that emits light at 266 nm) were used. In this direct mode, termed *laser desorption/ionization* (LDI), the extent of energy transfer is difficult to control and often leads to excessive thermal degradation. Another pitfall of the direct mode is that not all compounds absorb radiation at the laser wavelength. As a consequence, the direct LDI mode is applicable to a limited number of compounds, usually those with molecular mass of < 1000 Da.

Currently, LDI is practiced in the MALDI mode [3,113,114]. Two research groups, Karas and Hillenkamp [3] and Tanaka and co-workers [115], nearly simultaneously reported this development in 1988. MALDI has significantly revolutionized the approaches to the study of large biopolymers. This development has provided a unique opportunity to apply mass spectrometry to the analysis of proteins and other biomolecules with masses in excess of 200 kDa. The key to the success of this landmark development is mixing of the sample with a matrix. In their pioneer work, Karas and Hillenkamp used nicotinic acid as the matrix [3], whereas Tanaka and co-workers embedded the sample in the slurry of finely divided platinum (10 nm size) in glycerol [115].

In practice, the sample (guest) is admixed with an excess of the host matrix material, and is irradiated with a laser beam of short pulses of 10–20 ns duration and $\sim 10^6$ W/cm^2 irradiance power. The main criterion of choosing a host matrix is that it absorbs energy at the wavelength of the laser radiation. This matrix–wavelength combination permits a large amount of energy to be absorbed efficiently by the matrix, and subsequently transferred to the sample in a controlled manner. Absorption of energy from the laser beam causes evaporation of the matrix. The analyte molecules are entrained in the resultant gas-phase plume and become ionized via gas-phase proton-transfer reactions. An astonishing feature of this process is that very large molecules are desorbed into gas phase without undergoing thermal degradation as is shown for a monoclonal antibody in Figure 2.5.

Positive- and negative-ion analyses can both be performed with MALDI. The MALDI mass spectra of proteins and peptides typically contain signals due to singly protonated target molecules and their oligomeric ions (e.g., $[M + H]^+$, $[2M + H]^+$). In some cases, doubly and triply charged protonated ions of low abundance are also formed. With increasing mass of the analytes, multiply charged ions increase in abundance. In addition, the Na^+ and K^+ adducts are also a common feature of MALDI spectra of biological extracts.

Mechanism of Desorption and Ion Formation. The process by which large molecules are desorbed and ionized by absorption of photons is not clearly understood [116–121]. Three different models have been proposed to explain desorption of the matrix–sample material from the crystal surface: (1) quasithermal evaporation as a result of increased molecular motion, (2) expulsion of upper lattice layers, and (3) an increase in the hydrodynamic pressure due to rapidly expanding molecules in

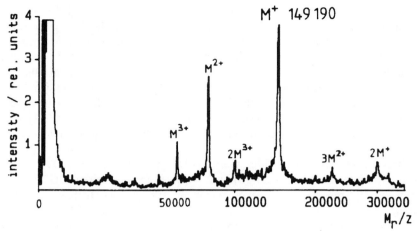

Figure 2.5. MALDI mass spectrum of a monoclonal antibody (IgG of mouse against a specific human lymphokine). [Reprinted from Ref. 113 by permission of Academic Press, New York (copyright 1990).]

the crystal lattice [116]. The analyte species are entrained in a dense plume of the desorbed matrix molecules. The initial velocity is considered an important value in characterizing the desorption process [117]. However, there is no consensus yet as to how the sample molecules are ionized. The widely accepted view is that, following their desorption as neutrals, the sample molecules are ionized by acid–base proton transfer reactions with the protonated matrix ions in a dense phase just above the surface of the matrix [118]. This view has been supported by the fact that the spectra of proteins obtained with UV- or IRMALDI are similar [122]. The protonated matrix molecules are generated by a series of photochemical reactions [119,120]. An alternate view is that the singly excited matrix molecules and not the frequently invoked photoionized matrix molecules are common precursors for all subsequent ionization events [123]. According to this model, two excited matrix molecules are required for the generation of free gaseous ions from the sample molecules. A 1998 study by Beavis and colleages has advocated that the ionization of the sample molecules occurs in a warm polar fluid that is formed from the matrix–sample crystals by absorption of laser energy [124]. The ionization of acidic matrices in this warm fluid creates a population of free protons. The acceptance or loss of a proton in this fluid results in the formation of the sample ions. This proposal, however, does not overrule the possibility of gas-phase ionization. The unified model presented by Karas et al. assumes the formation of initial clusters that comprise of matrix, protonated or deprotonated sample molecules, and counter ions and desolvation of the desorbed clusters to free ionic species via proton transfer and evaporation of neutrals [125]. Highly charged clusters are neutralized to produce singly charged ions by capture of the electrons that are generated during photoionization of the matrix. The electron capture process is also assumed to be the cause of fragmentation reactions.

Instrumentation. Current applications of MALDI overwhelmingly use TOF instruments because the pulsed nature of a laser beam matches well with the pulsed scanning mode of TOFMS [126]. In addition, the unlimited mass range, short duty cycle, high ion transmission, and multichannel detection features of TOFMS are also highly desirable for MALDIMS experiments. A schematic diagram of MALDI-TOFMS is presented in Figure 2.6. Linear and reflection TOF instruments have both been used for MALDIMS. The details of the design and performance characteristics of TOF mass spectrometers are discussed in Chapter 3 (Section 3.4). The combinations of MALDI with magnetic sector [127], quadrupole ion trap [128,129], and FTICR [130–132] instruments have also emerged. Especially FTICR has made a significant contribution to the accurate mass measurements of proteins.

Various laser systems have been used to rapidly deposit energy into the matrix–sample combination. Most applications have used UV lasers, such as the N_2 laser (337 nm), the frequency-tripled (355 nm) and frequency-quadrupled (266 nm) Nd : YAG laser, and the ArF excimer laser (193 nm) [3,113,114]. IR lasers have also been used to produce the MALDI effect [122,133,134]. The TEA CO_2 laser (10.6 µm), the Q-switched Er : YAG laser (2.94 µm), and the Cr : LiSAF or Nd : YAG pumped optical parametric oscillators (OPO) laser (3.28 µm) are the most common IR lasers. UV and IR lasers both yield similar spectra for proteins, although a better resolution has been obtained for some proteins with an IR laser [122,133].

The Role of a Matrix. The matrix performs two important functions: (1) it absorbs photon energy from the laser beam and transfers it into excitation energy of the solid system; and (2) it serves as a solvent for the analyte, so that the intermolecular forces are reduced and aggregation of the analyte molecules is held to a minimum. Some desirable characteristics of a typical MALDI matrix are

- A strong light absorption property at the wavelength of the laser flux.
- The ability to form microcrystals with the sample.

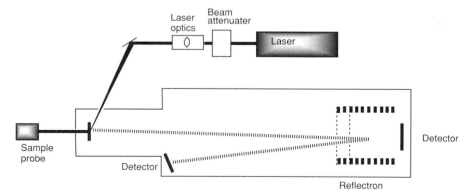

Figure 2.6. A schematic of MALDI-TOFMS.

- A low sublimation temperature, which facilitates the formation of an instantaneous high-pressure plume of matrix–sample material during the laser pulse duration.
- The participation in some kind of a photochemical reaction so that the sample molecules can be ionized with high yields.

Several matrix–laser combinations have been tested successfully [135]. Some commonly used matrices, along with the wavelengths at which they are used, the solvents in which they can be dissolved, and fields of their applications, are listed in Table 2.2. For peptides and small-molecular-mass proteins ($< 10,000$ Da), good results are obtained with α-cyano-4-hydroxycinnamic acid, whereas high-mass proteins are analyzed with sinapinic acid. The use of 3-amino-4-hydroxybenzoic acid and 2,5-dihydroxybenzoic acid has been recommended for the analysis of oligosaccharides. Ice has been used as a matrix for IRMALDI of proteins [136]. 3-Hydroxypicolinic acid (HPA) has gained a wide acceptance for the analysis of oligonucleotides [137].

Sample Preparation. As with the other desorption ionization methods, the preparation of the sample for MALDI analysis requires utmost care [114]. The homogeneity of the sample–matrix mixture is a critical factor to obtain good sample ion yields. Fortunately, MALDIMS is somewhat more tolerant of impurities, buffers, salts, and mixtures [138]. Several techniques have emerged for the sample preparation. These techniques include the

- Dried-droplet technique
- Fast evaporation technique
- Sandwich matrix technique
- Spin-dry technique
- Seed-layer technique

A saturated solution (or at least a mmol/mL concentration) of the matrix is first prepared in the deionized water. Other solvents, such as acetonitrile or methanol, may be added to increase its solubility. The solvent used to dissolve the matrix must also be compatible with the sample to be analyzed. For example, for protein analysis, a solvent mixture of water–acetonitrile that contains 0.1% TFA is recommended. A few microliters of a μmol/liter sample solution in 0.1% TFA is mixed with equal volume of the matrix solution to give a matrix : sample molar ratio of 1000–10,000 : 1. In the dried-droplet technique, a 1-μL portion of this mixture is applied to a stainless-steel or gold-coated sample well [139]. In order to obtain a fine-grained morphology of the crystal formation, the sample spot is evaporated slowly in the ambient air or by a gentle stream of cold air. Other researchers have advocated a fast evaporation of the matrix–sample mixture [140]. In this procedure, a 0.5-μL drop of the matrix solution in acetone that contains 1–2% water is rapidly deposited on the probe. The acetone rapidly evaporates to leave behind a homogeneous surface of

Table 2.2. The matrices used in MALDIMS

Matrix	Mass (Da)	Solvents	Laser λ (nm)	Applications
3-Amino-4-hydroxybenzoic acid	153	ACN,[a] water	337	Oligosaccharides
2,5-Dihydroxybenzoic acid (DHB)	154	ACN, water, methanol, acetone, chloroform	266, 337, 355	Oligosaccharides, peptides, nucleotides, oligonucleotides
5-Hydroxy-2-methoxybenzoic acid	168	ACN, water	337	Lipids
2[4-hydroxyphenylazo]benzoic acid (HABA)	242	ACN, water, methanol	266, 337	Proteins, lipids
Cinnamic acid	148	ACN, water	337	General
α-Cyano-4-hydroxycinnamic acid	189	ACN, water, ethanol, acetone	337, 355	Peptides, lipids, nucleotides
4-Methoxycinnamic acid	178	ACN, water	266, 337, 355	Proteins
Sinapinic acid (3,5-dimethoxy-4-hydroxy-cinnamic acid)	224	ACN, water, acetone, chloroform	266, 337, 355	Lipids, peptides, proteins
Ferlulic acid (4-hydroxy-3-methoxycinnamic acid)	194	ACN, water, propanol	266,337, 355	Proteins
6,7-Dihydroxycoumarin (esculetin)	178	Water	337	Lipids, peptides
3-Hydroxypicolinic acid (HPA)	139	Ethanol	337, 355	Oligonucleotides
Picolinic acid (PA)	123	Ethanol	266	Oligonucleotides
3-Aminopicolinic acid	138	Ethanol	266, 337, 355	Oligonucleotides
6-Aza-2-thiothymine (ATT)	143	ACN, water, methanol	266, 337, 355	Oligonucleotides, lipids
2,6-Dihydroxyacetophenone	152	ACN, water	337, 355	Proteins, oligonucleotides
2,4,6-Trihydroxyacetophenone	168	ACN, water	337, 355	Oligonucleotides
Nicotinic acid	123	Water	266, 337, 355	Proteins, oligonucleotides
1,5-Diaminonaphtalene	158	ACN, water, methanol	337	Lipids
Succinic acid	118	ACN, water	2940	Oligonucleotides
Urea	46	ACN, water	2940	Oligonucleotides

[a] ACN = acetonitrile. The laser light of 337 nm is produced with a nitrogen laser, and 266 and 355 nm, with frequency-quadrupled and -tripled Nd : YAG laser.

small crystals. This procedure yields enhanced resolution, sensitivity, and mass accuracy. In the sandwich matrix technique, a thin layer of the matrix is applied first, followed by 0.1% aqueous TFA, the sample solution, and an additional layer of the matrix [141]. This mixture is allowed to dry. In the spin-dry technique, a solution that contains equal volumes of a nitrocellulose (NC) membrane and a matrix is applied to the rotating target. The solution is immediately spin-dried and yields a uniform NC–matrix layer [141]. In the seed-layer technique, a seed layer of the matrix is prepared by depositing a droplet (0.5 μL) of the matrix solution on a sample target [142]. The drop is allowed to spread and dry in ambient air. A 0.5-μL drop of 1 : 1 (v,v) analyte–matrix mixture is deposited on the seed layer, and allowed to dry in ambient air. This technique allows rapid crystallization with a high degree of sample homogeneity.

Analysis of Solution-Phase Samples with MALDI. Several innovative approaches have been utilized to analyze biological compounds that are present in solutions [143–147]. Solutions can be introduced via a continuous-flow probe interface similar to that used for CFFAB analysis [143]. The matrix solution is composed of 3-nitrobenzoyl alcohol, 0.1% TFA, 1-propanol, and ethylene glycol (3 : 3 : 5 : 9 by volume). In another novel approach, the analyte stream that has been premixed with a suitable matrix is introduced into the ion source at liquid flow rates of 100–400 nL/min, and deposited on a rotating disk [144]. The solvent is rapidly evaporated in the vacuum region to leave a thin trace of sample–matrix mixture on the disk, which is transported to the repeller region, and bombarded with a laser beam. The detection sensitivity at the amol level has been reported. Chang and Yeung have also transported sample solution directly into the ion source via a narrow capillary column [145]. A UV laser bombards the tip of the capillary, and serves to vaporize and ionize the sample solution. Other approaches to the solution-phase sample analysis are the pulsed sample introduction (PSI) [146] and aerosol generator [147].

Applications. MALDIMS has been used primarily for the molecular mass determination of proteins (see Chapter 8). Two innovative approaches—delayed extraction of in-source fragmentation [148] and postsource decay (PSD) process [149]—have been developed to sequence peptides. The PSD data have been used in combination with a database search to characterize proteins [150]. Several other important classes of compounds, such as oligonucleotides (Chapter 15) [151], lipids (Chapter 13) [152], and oligosaccharides (Chapter 10) [153], are also accessible to MALDIMS. It is also an effective technique to characterize synthetic polymers [154,155]; the molecular mass and end-group information can both be obtained.

Desorption/Ionization on Silicon. Although MALDI has become one of the most successful ionization techniques, it suffers from the disadvantage of excessive matrix background, which creates problems in the analysis of small molecules. In order to solve this problem, Suizdak et al., have introduced a new technique, called *desorption/ionization on silicon* (DIOS) [156]. In this technique, instead of mixing with a matrix, the samples are directly placed on a modified porous silicon surface, and bombarded with a beam of laser (see Figure 2.7). The silicon surface is prepared

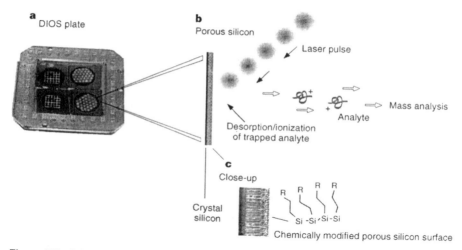

a DIOS plate

b Porous silicon

Laser pulse

Desorption/ionization of trapped analyte

Analyte

Mass analysis

c Close-up

Crystal silicon

R R R R

Si -Si ⁻Si-Si

Chemically modified porous silicon surface

Figure 2.7. Schemetic representation of desorption/ionization on silicon. [Reprinted from Ref. 156 by permission of *Nature*, Macmillan, New York (copyright 1999).]

from flat crystalline silicon by galvanostatic etching. Small pores and hydrophobic surfaces produce the more intense signals. The usefulness of this approach was demonstrated with the analysis of carbohydrates, peptides, glycolipids, natural products, and small drug molecules in the mass range of 150–12,000 Da. Peptides could be analyzed at levels as low as 700 amol. Also, excessive amounts of salts do not interfere in the analysis.

2.4. SPRAY TECHNIQUES

The direct sampling of solutions is often necessary in a variety of circumstances such as biological fluids and eluents from liquid chromatography and capillary electrophoresis separation devices. However, liquid solutions are difficult to handle by the mass spectrometry vacuum system. To overcome this difficulty, some novel ionization methods suitable for direct introduction of sample solutions have been devised. These techniques are also capable of direct analysis of LC effluents. In such devices, the solvent is selectively removed before the sample ions enter a mass spectrometer. These techniques differ from the aforementioned desorption methods in that the ions are generated in the presence of an ambient bath gas rather than in vacuum.

2.4.1. Thermospray Ionization

Thermospray has been one of the most popular techniques for handling samples in the liquid phase [157–159]. This device acts as an inlet system as well as an ionization source for samples from liquid solutions. Thermospray is best suited for the analysis of polar compounds, and has an upper molecular mass limit of 1000 Da.

It has proved to be a dependable interface for the coupling of HPLC with mass spectrometry [158–161]. Conventional columns with flow rates of $\leq 2\,mL/min$ can be combined with this interface. Currently, its use is being phased out in favor of more robust and universally accepted atmospheric pressure ionization (API)-based techniques (discussed below). Also, it is limited with respect to sensitivity (at least two orders of magnitude less) compared to API-based techniques.

As the term implies, *thermospray* uses thermal energy to produce a spray of fine liquid droplets. Figure 2.8 is a schematic representation of a typical thermospray source. The heart of this system is a probe, which contains an electrically heated capillary through which the sample solution is passed at a flow rate of 0.5–2 mL/min. The probe can be directly introduced into a specially designed heated ion source chamber. The resulting adiabatic expansion and cooling of the liquid generates a superheated mist that emerges from the capillary in the form of a supersonic jet of fine liquid particles admixed with the solvent vapors. Further vaporization occurs as the droplets strike the walls of the heated ion source. Although no external nebulizing gas is used, the solvent vapors provide the gas dynamic forces to assist further nebulization of the liquid droplets. The evaporated solvent molecules are pumped away, and the ejected ions are pushed through a skimmer into a conventional quadrupole or amagnetic sector mass analyzer for mass analysis. Because of the rapid heating and protective effect of the solvent, nonvolatile samples are converted to gas-phase ions without undergoing pyrolysis. The ion source is a modification of the conventional CI source. It is fitted with a high-throughput mechanical pump, which is placed directly opposite to the capillary. It also contains an EI gun and/or a low-current discharge electrode.

Ionization of the sample molecules takes place via one of the three processes [161,162]: direct ion evaporation, acid–base proton transfer CI, or solvent-initiated CI. The direct ion evaporation is favored for the solutions of ionic and polar sample molecules. Some of the sprayed droplets from such solutions become electrically charged because of statistical fluctuations in the distribution of positively and

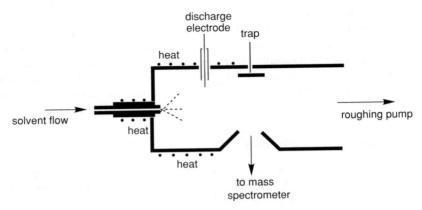

Figure 2.8. A block diagram of thermospray ionization source.

negatively charged preformed sample and solvent ions. The mechanism of the ion ejection from the droplets is similar to that explained for the ESI process (discussed in Section 2.4.3). Briefly, as the solvent is evaporated, the charge density on smaller droplets reaches a limit for free ions to be expelled from the droplets. The acid–base proton transfer CI mode of ion formation occurs if the solution contains volatile electrolytes such as ammonium acetate. The NH_4^+ and CH_3COO^- ions that are formed by the dissociation of ammonium acetate both participate in the proton transfer reactions to form positive and negative ions, respectively. The third process, solvent-initiated CI, is used for volatile samples and for mobile phases that contain a large fraction of an organic solvent such as that encountered in the normal-phase operation of HPLC. Either EI or a low-current discharge of the solvent vapors generates the plasma of solvent reagent ions, which are in turn used to ionize the sample molecules. Two special terms, *filament-on operation* and *plasmaspray ionization*, have been coined for this mode of ionization. In many situations, the filament-on operation is often combined with the normal thermospray process.

Polar (water, methanol, acetonitrile, isopropanol, etc.) and nonpolar (dichloromethane and hexane) liquids can both be effectively used as solvents. Nonvolatile additives should be avoided because they form cluster ions and, in addition, clog the capillary and skimmer pinholes. The recommended buffer additives include ammonium acetate, ammonium formate, acetic acid, triethylamine, and diethylamine.

Thermospray is a gentle mode of ionization, and produces mostly intact molecular ions and adduct ions. Fragmentation is relatively sparse, and thus little structural information is available. Increasing the repeller voltage, however, can induce fragmentation. Multiply charged ions of some analytes are also formed. Thermospray has been used effectively for the analysis of peptides [163,164], dinucleotides [162], prostaglandins [165], diquaternary ammonium salts [166], pesticides [167], drugs [168,169], dyes [170], and environmental pollutants [171].

2.4.2. Atmospheric Pressure Chemical Ionization

Atmospheric pressure ionization (API) is a means of analyzing nonvolatile samples without first converting them to the gaseous state [172]. A liquid solution of such samples can be sprayed directly into the ion source. This source was originally developed for the analysis of volatile compounds such as those that emerge from GC columns, present in ambient air, or introduced into the carrier-gas flow [173]. API is operated in two modes: (1) atmospheric pressure chemical ionization (APCI) and (2) electrospray ionization (described in Section 2.4.3). APCI is used for low-mass polar compounds, whereas ESI is suitable for low- and high-mass compounds. In addition, both these API-based techniques have emerged as robust and dependable interfaces to couple HPLC with mass spectrometry [174,175]. Successful combinations of API with CE [176] and supercritical fluid chromatography (SFC) [177] have also been demonstrated.

The principle of ionization in APCI is identical to that described for low-pressure CI, where ionization takes place via gas-phase ion–molecule reactions between reagent gas ions and the analyte molecule (Section 2.2). In APCI, a similar process occurs at atmospheric pressure; the difference is that at the prevailing pressure conditions, ions participate in several more ion–molecule collisions than at low-pressure CI conditions [172]. A greater sensitivity is realized in APCI because of the increased number of ion–molecule reactions and increased collisional deactivation of the initially formed energetically hot molecular ions.

The design of a typical APCI ion source is shown in Figure 2.9. It consists of a removable heated ($>500°C$) nebulizer probe, an ionization region, and an intermediate pressure ion transfer region [178]. The solvent flows through a fused-silica capillary tube, and is pneumatically nebulized into very fine droplets by a concentric flow of nitrogen. An additional flow of the nitrogen sheath gas sweeps the droplets through a heated tube. The transfer of heat from the heated tube converts the droplets and the analytes into a gas stream. This mixture of the warm gas and sample molecules flows toward the ionization region, where the charge transfer reaction from the reagent gas ions converts the sample molecules into ions. In early designs, reactant ions were produced by the interaction of electrons that are emitted from a ^{63}Ni β-particle emitter. In most current commercial APCI sources, the external source of electrons for the initial ionization of the nebulized aerosol particles is a corona discharge, which is formed by applying a potential of 2–3 kV between the corona pin (discharge electrode) and the exit aperture of the counter electrode. The ions are sampled through a small orifice skimmer, called the *nozzle* or *sampling cone*. One or two more additional reduced-pressure regions, each separated by skimmers, are added before the free ions enter the mass analyzer for subsequent mass analysis. In order to increase the transport efficiency, a lens assembly that consists of either an octopole or hexapole is added before the mass analyzer. Another common design for the APCI operation is the ion spray [179].

The reaction sequence responsible for the formation of reactant ions in ambient air has been investigated thoroughly [172,180]. The major primary ions formed by the interaction of the corona-created electrons are N^{2+}, N^{4+}, and H_2O^+. The charge transfer reactions of these primary ions with H_2O, present in the aerosol,

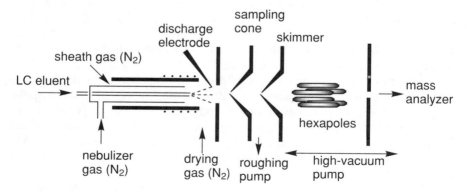

Figure 2.9. A schematic representation of APCI source of heated nebulization design.

produce a set of proton hydrates, $H_3O^+(H_2O)_n$, as the reactant ions. When methanol and acetonitrile are present in the aerosol, the $CH_3OH^+(CH_3OH)_n$ and $CH_3CN^+(CH_3CN)_n$ reactant ions, respectively, are also formed. In the positive-ion mode, the proton transfer from these solvated ions ionizes the sample molecules. In the negative-ion mode, the solvated oxygen anions serve as the reagent ions to ionize the sample molecules by the electron or proton transfer process. Thus, the APCI mass spectra are dominated by the intact $[M + H]^+$ or $[M - H]^-$ ions, the m/z value of which can provide the molecular mass information. During a free-jet expansion from the atmosphere to the vacuum region, ions undergo extensive clustering with water and other polar solvent molecules. Maintaining the source at a higher temperature, using a flow of a curtain gas, and inducing fragmentation at a higher sampling cone potential, either alone or in combination, can minimize this problem.

The technique of APCI is limited to the analysis of low molecular mass compounds (< 1200 Da). It is ideally suited for the molecular mass determination of compounds that display certain degree of polarity. For structural studies, fragmentation is deliberately induced in the ion transfer region by increasing the potential of the sampling cone. Typical applications of APCI include the analysis of steroids [181,182], pesticides [183], and pharmaceutical drugs [184].

2.4.3. Electrospray Ionization

Electrospray ionization (ESI) is an API technique applicable to a wide range of compounds that are present in liquid matrices. The emergence of ESI represents a significant advance in the capabilities of mass spectrometry for the characterization of large biomolecules and their noncovalent interactions. It has also become the most widely used interface to combine HPLC with mass spectrometry. The wide popularity of ESI in these fields is due to its continuous-flow operation, tolerance to different types of solvent, acceptance of wide solvent flow rates, and ability to generate intact multiply charged ions of fragile chemical and biochemical species. ESI uses a novel concept first introduced by Dole in 1968 for the generation of gas-phase ions from electrically charged liquid droplets, which are produced by electrospraying the solution of an analyte at atmospheric pressure [185]. Dole used ESI with an ion-drift spectrometer to produce molecular beams of very high-mass macroions of polystyrene [186,187]. The coupling of ESI to mass spectrometry was achieved in mid-1980s by Fenn and co-workers [187] and Aleksandrov et al. [188]. Several research groups, notably of Fenn [2,189], Henion [179], and Smith [190,191], later developed the applications of ESIMS for the analysis of biomolecules. With ESIMS, the molecular masses of these biopolymers in the mass range of over 100 kDa can be determined with an accuracy of > 0.01%.

The basic principle of electrospray ionization can be explained through a simple schematic of a typical ESI source as shown in Figure 2.10. The heart of the ESI source is a stainless-steel capillary tube through which a solvent flows continuously at the rate of 2–5 µL/min. The solvent consists of a mixture (usually 1 : 1) of water and an organic solvent (typically methanol, acetonitrile, or isopropanol), and

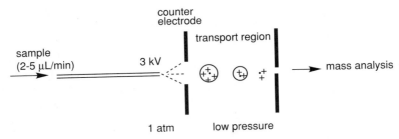

Figure 2.10. A pictorial depiction of the concepts of electrospray ionization.

typically contains <1% of acetic acid or another suitable acid. A solution of the sample is injected into this solvent stream. A potential difference of 3–4 kV between the tip of the capillary and walls of the surrounding atmospheric pressure region produces an electrostatic field sufficiently strong to disperse the emerging solution into a fine mist of charged droplets. A flow of hot-bath gas, usually nitrogen, is added to the interface to assist in the evaporation of the solvent from those charged droplets. The ions are transported from the atmospheric pressure region to the high-vacuum region of the mass analyzer via a low-pressure transport region, which consists of two or more successively pumped chambers. Before entry into a mass analyzer, all sample ions are stripped off the solvent molecules.

Two different designs of the transport region have become popular. In one design (shown in Figure 2.11), the transport region is equipped with a several-centimeters-long heated metal or glass capillary. This heated tube allows effective ion sampling, and provides declustering of ions. The other design, shown in Figure 2.12, is similar to that used in the APCI source (Figure 2.9). Here, sample is introduced through a fused-silica tube and nebulization is assisted by a sheath flow of nitrogen. Ions are sampled through a small-orifice skimmer (sampling cone). Positive and negative ions can both be generated. The polarity of the ions analyzed is selected by the capillary voltage bias.

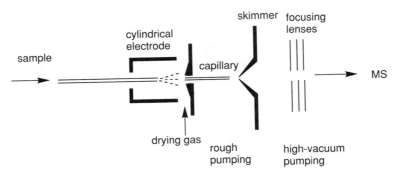

Figure 2.11. A block diagram of electrospray ionization source that uses heated capillary for droplet transfer.

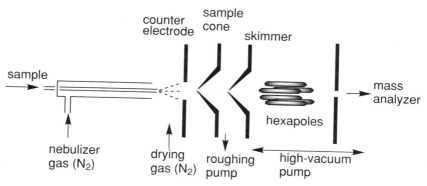

Figure 2.12. A block diagram of electrospray ionization source, in which ions are sampled through a small-orifice skimmer.

The sensitivity achievable in ESI operation is limited by the transmission efficiency of the transport region. The factors that limit transmission are gas-phase collisions and charge–charge repulsion, which lead to an expansion of the ion cloud. The effective sampling and focusing of this ion cloud can improve transmission efficiency. Earlier designs used ion lenses. A dramatic improvement in transmission efficiency can be achieved by use of radiofrequency (rf) multipoles, such as quadrupoles, hexapoles, and octapoles. The higher order multipoles allow a larger acceptance aperture. A further improvement in transmission efficiency has been realized by applying rf fields to a stack of ring electrodes [192,193]. Ion confinement within this "ion funnel" depends on the diameter and spacing between the rings and on the rf frequency and amplitude. The ion funnel developed by the Smith group (see Figure 2.13) consists of a series of ring electrodes with progressively decreasing diameter. The rf and direct-current (dc) potentials are both applied to the rings, and successive rings are out of phase by 180°.

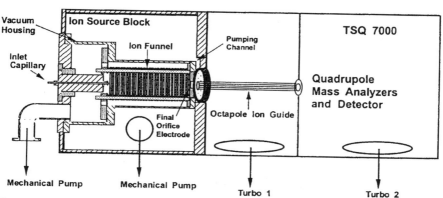

Figure 2.13. A block diagram of electrospray ionization source that uses ion funnels for transfer of ions. [Reprinted from Ref. 193 by permission of American Chemical Society, Washington, DC (copyright 1999).]

A design of an ESI source that uses a cylindrical capacitor has also emerged [194]. This source consists of a fused-silica tube that is enclosed in a grounded stainless-steel tube. A platinum wire is inserted in the fused-silica tube, and is maintained at a high voltage. Near the end of the metal tube, a second fused-silica tube of smaller i.d. is connected with the first one. Ions are produced as liquid flows through this concentric cylindrical capacitor. This arrangement eliminates the possibility of corona discharge to improve the ES operation in the negative-ion mode. Stable operation in the flow range of 50–500 nL/min can be realized. This design also provides stable operation in the flow range of 50–200 μL/min.

Mechanism of Ion Formation. Electrospray ionization encompasses three different processes: droplet formation, droplet shrinkage, and gaseous ion formation. At the onset of electrospray process, the electrostatic force on the liquid causes it to emerge from the tip of the capillary as a jet in the shape of a "Taylor cone" [195]. A thin liquid extends from this cone, which breaks into a mist of fine droplets. A number of factors such as the applied potential, the flow rate of the solvent, the diameter of the capillary, and solvent characteristics influence the diameter of the initially formed droplets.

The exact mechanism of the formation of gas-phase ions from the charged droplets is a widely debated topic [196–202]. Current thinking is based on two widely accepted mechanisms: the charge residue model (CRM) and the ion desorption model (IDM) [197–199]. According to the CRM proposal (depicted in Figure 2.14a), as the droplets shrink in size due to the solvent evaporation, the charge density on their surface increases until it reaches the Rayleigh instability limit. At this point, the repulsive Coulombic forces exceed the droplet surface tension to cause the droplets to break into smaller and highly charged offspring droplets. The sequence of the solvent evaporation and fission of the droplets is

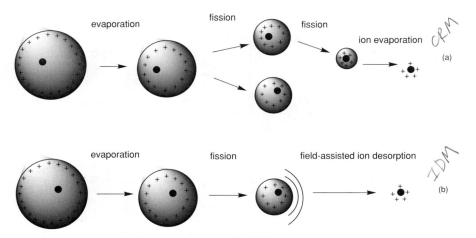

Figure 2.14. Conceptualization of the processes that are involved in the conversion of ions from droplets into the gas phase: (*a*) charge residue model; (*b*) ion desorption model.

repeated several times until the drop size becomes so small that they contain only one solute molecule. As the last of the solvent is evaporated, this molecule is dispersed into ambient gas, retaining some of the charge of the droplets. It has been shown that different predominant fission pathways may occur depending on the size of the initial droplets [202]. The IDM mechanism (illustrated in Figure 2.14b) is due to the proposal of Iribarne and Thomson [203]. In this proposal also, the sequence of solvent evaporation and droplet fission is repeated. However, instead of droplets becoming so small that they contain only one solute molecule, at some intermediate droplet size the electric field due to the surface charge density is sufficiently high to overcome the droplet cohesive forces leading to direct ion desorption. The IDM is much more widely accepted than the CRM proposal.

The Nature of the ESI Spectrum. Electrospray analysis can be performed in either positive or negative ionization mode. One unique feature of ESI is that the degree of fragmentation of the analyte molecules can be controlled at will. Under certain experimental conditions, only intact molecular ions are formed. Because the analyte ion current is centered only in the molecular ion(s), this feature helps increase the detection sensitivity of quantitative measurements and tandem mass spectrometry studies. In order to obtain the structural information, fragmentation can be induced deliberately in the intermediate region of a tandem mass spectrometer, or in the ion transport region of the ES source by increasing the sampling cone voltage. In this second mode, ions gain extra kinetic energy and as a result undergo dissociation on collision with the bath gas. A single-quadrupole mass analyzer can be used to perform the mass analysis of the products of the in-source collision-induced dissociation (CID). However, for this approach to be of some value in the structure determination studies, the analyte sample must be of high purity. A typical in-source CID mass spectrum of an anthracyline antibiotic is displayed in Figure 2.15. The inset shows the spectrum obtained at the low sampling cone voltage.

Another key feature of the ES process is the formation of a series of multiply charged ions for large biopolymers. The positive ions of the general nature $[M + nH]^{n+}$ are formed by the protonation of basic sites in biopolymers. A typical ESI mass spectrum of a protein is shown in Figure 2.16, which is the spectrum of horse heart myoglobin; each peak in this figure differs from its neighbor by one charge. In the negative-ion mode, the $[M - nH]^{n-}$-type ions are formed. Because a mass spectrometer analyzes ions on the basis of their m/z ratios rather than their masses, the effect of multiple charging is to reduce significantly the m/z of the intact macromolecule, a process that brings high-mass compounds within the usable mass range of an ordinary mass spectrometer. The charge state of a particular peak and hence the molecular mass of a biopolymer can be calculated by solving the following simultaneous equations:

$$n = \frac{m''H}{m' - m''} \tag{2.22}$$

$$M = n(m' - H) \tag{2.23}$$

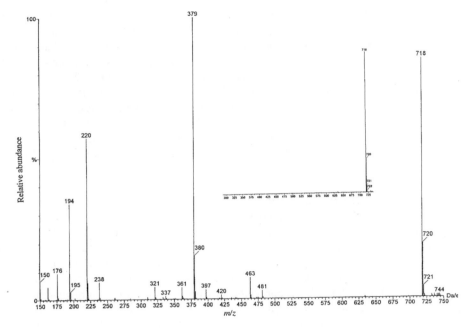

Figure 2.15. ESI mass spectrum of an anthracyline antibiotic.

Here, m' is the mass of the $[M + nH]^{n+}$ ion ($H = 1.007829$ is the mass of the proton) and m'' ($m' > m''$) of the adjacent $[M + (n + 1)H]^{n+1}$ ion (see Figure 2.16). Algorithms have been developed that can compute the value of M from successive pairs of adjacent ions, and provide the average value. In favorable cases, the molecular mass of macromolecules can be calculated with a precision of at least $\pm 0.005\%$.

Scalf et al. have introduced a new technique, termed as *charge reduction electrospray mass spectrometry* (CREMS), to reduce in a control fashion the multiple charge state into singly charge state [204]. The modified ESI ion source is shown in Figure 2.17, in which the electrospray-generated multiply charged ions undergo charge reduction through reactions with bipolar ions that are generated in a neutralization chamber by ionization of the components of the bath and sheath gases with a ^{210}Po α particle source. This technique can be beneficially used to analyze complex mixtures of biological compounds as is demonstrated in Figure 2.18 for the analysis of an equimolar, seven component protein mixture.

Electrospray at Unusual Flow Rates. Several parameters must be controlled to obtain a stable spray of the emerging solvent. The flow rate, conductivity, and surface tension of the solvent are the factors that profoundly influence the operation of the electrospray process. The optimum operation of a normal ESI source is achieved at solvent flow rates of 2–10 μL/min. As the flow is increased, large coarse

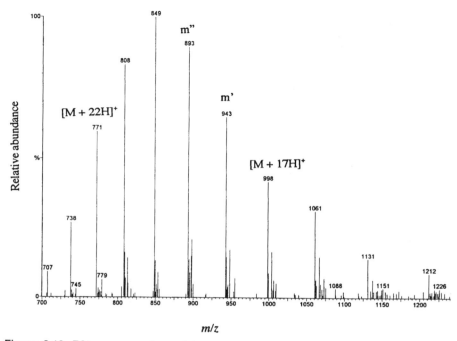

Figure 2.16. ESI mass spectrum of horse heart myoglobin; each peak in the spectrum represents a different charge state.

droplets are formed to result in electrical breakdown, and finally the disappearance of the spray. At lower flow rates, a constant and uniform stream of charged droplets cannot be maintained. Higher solution conductivities also can pose problems to maintain a stable spray. Such solutions can provide a stable electrospray signal only at lower liquid flow rates. Because of the high surface tension, highly aqueous solutions are also problematic in sustaining a stable signal.

Stable Operation at Higher Flow Rates. In order for ESI to be applicable for a wide range of solvent flows, several modifications to the conventional ESI source have been proposed. For lower flow rate operation ($< 20\,\mu$L/min), the addition of a sheath liquid improves the ES process [205]. The sheath liquid can be chosen to counteract the effects of the surface tension and conductivity of the sample-bearing solution. For stable operation at higher flow rates, some form of an additional source of energy is provided to disperse the liquid into fine droplets. As an example, in pneumatically assisted ESI source, a high-velocity annular flow of gas provides enhanced nebulization of the ES liquid [179,206]. This pneumatically assisted ES process has been alternatively termed *ionspray.* It has the advantage that a stable spray can be sustained at as high flow rates as $200\,\mu$L/min. As a consequence, ion spray can be coupled to microbore liquid chromatographic (LC) columns without the need for a flow splitter. *Ultrasonically assisted* ESI source also uses a mechanical

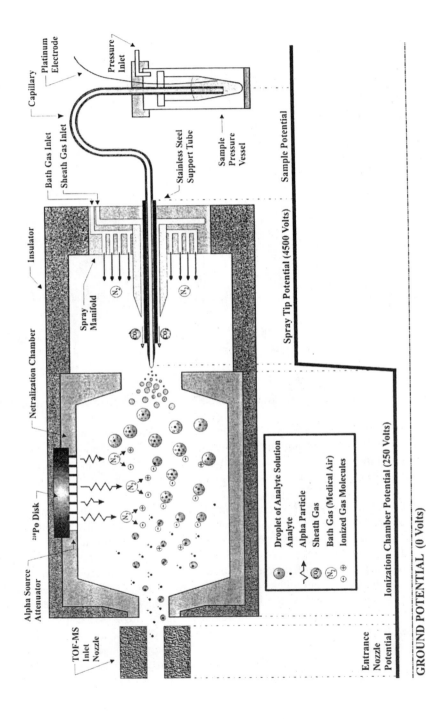

Figure 2.17. Schematic diagram of a charge reduction electrospray ion source. [Reprinted from Ref. 204 by permission of American Chemical Society, Washington DC (copyright 2000).]

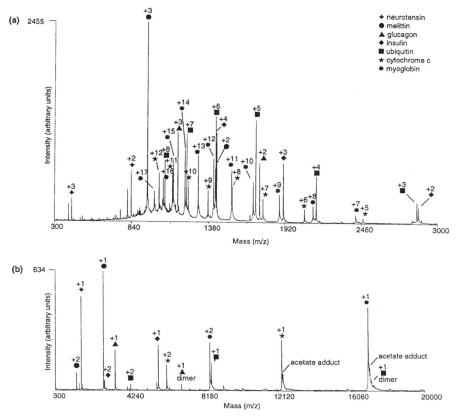

Figure 2.18. ESI mass spectra of a mixture of seven proteins: (*a*) conventional spectrum; (*b*) charge reduction spectrum. [Reprinted from Ref. 204 by permission of American Chemical Society, Washington, DC (copyright 2000).]

force to shear larger droplets [207]. Here, the ESI needle is held in an ultrasonic nebulizer. This source can produce a stable spray from any mobile-phase composition. At flow rates greater than 200 µL/min, larger droplets are formed, the rate of evaporation of which seriously affects the ionization efficiency. The rate of evaporation and hence the rate of ion ejection from the larger droplets can be enhanced by raising the temperature of the needle and its surroundings [208,209]. In order to accommodate mL/min flow rates of wide-bore columns directly into an ESI source, a grounded metal shield is added after the ESI needle and the ion sampling capillary is heated to higher temperatures [210].

Stable Operation at Submicro-liter Flow Rates. Since the mid-1990s, attention has focused on miniaturizing the ES ion source [195,211–217]. The motivation behind this work is to adopt it for small volumes and low amount of samples, as normally is the situation with most biological samples, or when there is a need to couple microseparation devices such as packed-capillary HPLC and CE to mass spectrometry. One miniature design commonly dubbed as nanoelectrospray (nanoES) is realized with a gold-coated fine-drawn capillary, the spraying tip aperture of which is adjusted to 1–3 μm [195,212,213]. Unlike in a conventional ES capillary, there is no need to force the sample solution through the sprayer by pumps. The solution emerges from the tip at the rate of 20–40 nL/min by the action of the electrical field on the tip. In order to obtain the maximum transfer efficiency of the sample, the ES needle is placed within 2 mm of the nozzle orifice of an ES source. Sample utilization is very efficient in the nanoES design. A total sample volume typically is as small as 1 μL, which can last up to 30 min. The extended lifetime of the analyte signal allows prolonged accumulation of signal and an opportunity for multiple tandem mass spectrometry experiments. An improvement in the ESI efficiency by a factor of ∼ 100 over conventional ESI is achieved [212]. McLafferty and colleagues have achieved subnanoliter (0–1.5-nL/min) flow rates with their miniaturized ES needle [214]. Impressive sensitivity has been realized with this system [215]. In another miniaturized design, a fused-silica capillary, one end of which is etched to a tapered tip by dipping it into HF, has been used as an ES needle [211]. The voltage is applied upstream of the sprayer tip at the metal junction that connects the sprayer with the transfer line. With this design, the pumping of the solution can be implemented at the flow rates of 300–800 nL/min [216]. Zeptomole–attomole sensitivity has been achieved for the analysis of peptides. A survey of the methods that are used to prepare metal-coated tips for low flow applications has been reported in an excellent article [218]. A new method, known as "fairy dust technique," to provide metal contact at the tip of the emitters, has been described [219]. The technique involves gluing 2-μm gold dust to freshly polyimide coated fused-silica capillaries. Stable electrospray at the solvent flow rates of as low as 30 nL/min can be achieved with this technique.

Microfabricated Glass Chips. Further efforts in reducing the sample size involve the use of microfabricated glass chips [220–226]. A further advantage of these devices is that the separation step can be combined on-chip with other sample manipulation steps such as dilution, derivatization, and enzyme digestion. These devices are constructed from glass, using standard photolithographic/wet chemical etching techniques, and usually contain the sample inlet ports, preconcentration sample loops, separation channels, and a port for ESI coupling. A chip developed by Karger's research group contains multiple parallel solvent delivery channels, each connected to its own solvent reservoirs [221]. High voltage is provided separately to each reservoir. The solvent is delivered at a flow of 100–200 nL/min by a syringe pump. The same group has developed and evaluated two different approaches to interface microfabricated devices with ESIMS; one design used a liquid junction and a fused-silica transfer line [225]. The second approach used a miniaturized on-chip pneumatic nebulizer, which results in a minimal dead volume. The development of a low-dead-volume microfabricated chip device has also been reported by one another group [226].

Mass Spectrometry Systems. To achieve full benefits of ESI and LC/ESIMS combination, a mass spectrometer should be inexpensive. In addition, it should have high scan speed, adequate mass range, reasonable mass resolution, and high sensitivity. Although several types of mass analyzers are available, a quadrupole mass filter offers most of these desirable features. In addition, the low-voltage environment of the quadrupole ion source is compatible with the high-voltage and high-pressure conditions of the ES process. Therefore, quadrupole-based ESIMS and LC/ESIMS instruments are considerably more popular.

Electrospray has also been implemented with other types of mass analyzer, including ion-trap-, magnetic-sector-, FT-ICR-, and TOF-based mass spectrometry systems. These analyzers possess certain unique features that are not available with quadrupoles. For example, ion-trap and FT-ICRMS can both be operated in the tandem mass spectrometry mode. Magnetic sector and FT-ICRMS have high-resolution accurate mass measurement capability; and TOFMS has unlimited mass range, extremely high scan speed, and multichannel detection capability. However, the coupling of an ES ion source with these mass spectrometry systems is problematic. High acceleration potentials that are required for ion extraction in magnetic sector and TOFMS pose risks of an electrical discharge. High pressure is also detrimental for the optimal performance of FT-ICRMS. Unlike scanning instruments, ions must be introduced in TOF mass analyzer in pulses, whereas the ion formation by ESI is a continuous process. The signal-averaging feature of FT-ICRMS and TOFMS is attractive for improving absolute detection limits.

The coupling of an ES ion source with an ion trap is very attractive because ion traps are more compact, cheaper, and sensitive than quadrupoles. An ESI interface was designed to combine with an ion trap [227], and the first application of LC/ESIMS with an ion trap was reported by Van Berkel et al. [228]. In this interface, an electrosprayed plume of HPLC effluent is drawn through an evacuated region (0.3 torr), which is enclosed by two aperture plates of 100- and 800-µm orifices, respectively. A three-element lens assembly guides the ions into the ion trap. Since then, a simple benchtop LC/ESI ion trap with MS/MS facility has been developed and marketed by commercial vendors.

The coupling of ESI with TOFMS has been accomplished by using the orthogonal ion extraction approach [229,230]. ESI-produced ions are stored between each duty cycle, and are pushed into the flight tube in the pulse mode. The duty cycle of this design is approximately 3% only. The use of an ion trap as an ion storage device has been shown to improve the duty cycle further [231,232].

The accurate mass measurement is a highly desirable feature for many applications. To take advantage of the high-resolution and accurate mass measurement capabilities of the double-focusing magnetic sector instruments, efforts have been made to combine ESI to these instruments [233,234]. In addition, these instruments can also be used for high-energy CID of the mass-selected ESI-produced ions. Some modifications, however, are made to the ESI interface. For example, additional differential pumping and vacuum baffles are added to the interface to prevent electrical discharge.

FT-ICRMS also holds immense potential for the analysis of ESI-produced ions [235,236]. This combination has the capability to yield high-mass resolution, high accuracy in mass measurement, and high-mass range. The sample is ionized external to the FT-ICR cell. However, the ion trapping efficiency is a limiting factor for the externally introduced ions. To achieve effective trapping of the ions in the ICR cell, a bath gas is momentarily introduced into the cell and pumped out immediately. During this step, the ions are translationally cooled by collisions with the bath gas. This arrangement is not ideal for LCMS applications because of the long duty cycle. Recently, Senko et al., have used an octopole to externally cool the ions before their ejection into the ICR cell [237,238]. Using this arrangement, Emmet et al., have combined microES/LC with FT-ICRMS [239].

Sample Consideration. Ultrahigh detection sensitivity (attomole to femtomole) is the hallmark of ESIMS. However, several factors must be controlled to achieve such high levels of sensitivity. The ease of charge transfer processes determines the sensitivity of an analysis. The sample concentration, its pK_a or pK_b value, the presence of electrolytes or other species, pH, and the shape and size of the sample molecule are some of the factors that generally influence the analyte ion current. The analyte detection sensitivity decreases as the total electrolyte concentration increases in the electrospray solvent. High concentrations of salts, ion-pairing reagents, plasticizers, or detergents, often encountered in protein chemistry, are harmful to the ESI process. The presence of dilute acids (acetic, formic) and bases (NH_4OH) substantially increases the sample ion current in the positive- and negative-ion modes, respectively.

Applications. Electrospray ionization has revolutionized the manner in which mass spectrometry is performed. A growing range of compounds are being analyzed by ESIMS. A list of applications is too long to be mentioned here. The readers are advised to consult timely volumes [240,241], a review by the author on this subject [242], and Chapters 8–16 in this volume. ESIMS has been applied to the analysis of proteins, peptides, combinatorial libraries, drug and drug metabolites, immunologic samples, nucleotides, DNA adducts, environmental pollutants, and many more. A typical example from our laboratory is the molecular mass determination of porcine kidney leucyl aminopeptidase (pkLAP) and its two large segments. LAP is an exopeptidase that catalyzes the hydrolysis of N-terminal amino residue of a polypeptide chain. The ESI mass spectrum of pkLAP contains several multiply charged ions, from which the molecular mass was estimated to be 52,700 Da. The two large segments were generated by limited tryptic cleavage of pkLAP. The cleavage occurs at the connecting loop –Lys–Arg–Lys–. The molecular mass of the N-terminal segment was found to be 14,608 Da and that of the C-terminal segment, 38,094 Da.

2.5. COMPARISON OF DIFFERENT IONIZATION METHODS

A concise comparison of different ionization method is presented in Table 2.3. For small and volatile molecules, EI and CI are the methods of choice. Both of these ionization methods are ideal for their coupling with GC for the analysis of complex mixtures. Thermospray can be used with small-size nonvolatile compounds. The current preference of an ionization technique for such compounds is APCI. Thermospray and APCI are both well adapted to the analysis of pharmaceutical drugs and their metabolites. In addition, they can be readily coupled to HPLC. ESI and MALDI can be used for a variety of medium and large polar molecules such as peptides, proteins, carbohydrates, lipids, and oligonucleotides. ESI has a clear edge over MALDI with respect to LCMS or CEMS applications. FAB is suitable for the analysis of small to medium polar molecules, but is less sensitive than ESI and MALDI. Also, its coupling with LC [79,80] and CE [81–83], although already demonstrated, is not straightforward.

Table 2.3. Comparison of various ionization methods

Ionization Method	Ion Type	Sample Type	Separation Technique
EI	$M^{+\cdot}$, fragments	Nonpolar and some polar organic compounds	GC
CI	$[M + H]^+$, $[M - H]^-$, $M^{-\cdot}$	Nonpolar and some polar organic compounds	GC
Thermospray	$[M + H]^+$, $[M - H]^-$, $[M + NH_4]^+$	Polar compounds	LC
FAB	$[M + H]^+$, $[M - H]^-$	Peptides, proteins, lipids, carbohydrates, oligosaccharides, nucleotides, oligonucleotides	LC, CE
APCI	$[M + H]^+$, $[M - H]^-$	Polar compounds, drugs	LC
ESI	$[M + nH]^{n-}$, $[M - nH]^{n-}$	Peptides, proteins, lipids, carbohydrates, oligosaccharides, oligonucleotides	LC, CE
MALDI	$[M + H]^+$, $[M - H]^-$	Peptides, proteins, lipids, carbohydrates, oligosaccharides, oligonucleotides	LC, CE

REFERENCES

1. M. Barber, R. S. Bordoli, R. D. Sedgwick, and A. N. Tyler, *J. Chem. Soc. Chem. Commun.* 325–327 (1981).

2. J. B. Fenn, M. Mann, C. K. Meng, S. F. Wong, and C. M. Whitehouse, *Science* **246**, 64–71 (1989).

3. M. Karas and F. Hillenkamp, *Anal. Chem.* **60**, 2299–2301 (1988).

4. A. E. Ashcroft, *Ionization Methods in Organic Mass Spectrometry*, The Royal Society of Chemistry, Cambridge, England, 1997.

5. F. W. McLafferty and F. Turecek, *Interpretation of Mass Spectra*, University Science Books, Mill Valley, CA, 1993.

6. M. Rosenstock, M. B., Wallenstein, A. L. Wahrhaftig, and H. Eyring, *Proc. Natl. Acad. Sci. (USA)* **38**, 667–678 (1952).

7. R. A. Marcus, *J. Chem. Phys.* **20**, 359–362 (1952).

8. M. S. B. Munson and F. H. Field, *J. Am. Chem. Soc.* **88**, 2621–2630 (1966).

9. A. G. Harrison, *Chemical Ionization Mass Spectrometry*, CRC Press, Boca Raton, FL, 1992.

10. F. W. McLafferty, *Org. Mass Spectrom.* **15**, 114–121 (1980).

11. T. Cairns, E. G. Seigmund, and J. J. Stamp, *Org. Mass Spectrom.* **21**, 161–162 (1986).

12. G. G. Lias, J. E. Bartmess, J. F. Liebman, J. L. Holmes, R. D. Levin, and W. G. Mallard, *J. Phys. Chem. Ref. Data* **17**, Suppl. 1 (1988).

13. M. Vairamani, U. A. Mirza, and R. Srinivas, *Mass Spectrom. Rev.* **9**, 235–258 (1990).

14. J.-P. Morizur, B. Desmazieres, J. Chamot-Rooke, V. Haldys, P. Fordam, and J. Tortajada, *J. Am. Soc. Mass Spectrom.* **9**, 731–734 (1998).

15. D. A. Peake and M. L. Gross, *Anal. Chem.* **57**, 115–120 (1985).

16. M. S. B. Munson, *Anal. Chem.* **59**, 466–471 (1987).

17. M. V. Buchanan, *Anal. Chem.* **54**, 570–574 (1982).

18. N. Einolf and B. Munson, *Int. J. Mass Spectrom. Ion. Phys.* **9**, 141–160 (1972).

19. C. Dass and M. L. Gross, *J. Am. Chem. Soc.* **105**, 5724–5729 (1983).

20. C. Dass and M. L. Gross, *Org. Mass Spectrom.* **20**, 34–40 (1985).

21. D. F. Hunt, G. C. Stafford, Jr., F. W. Crow, and J. W. Russell, *Anal. Chem.* **48**, 2098–2105 (1976).

22. R. C. Daugherty, J. Dalton, and F. J. Biros, *Org. Mass Spectrom.* **6**, 1171–1181 (1972).

23. S. Murray and D. Watson, *J. Steroid Biochem.* **25**, 255–264 (1989).

24. R. Strife and R. C. Murphy, *J. Chromatgr.* **305**, 3–12 (1984).

25. H. D. Beckey, *Principles of Field Ionization and Field Desorption Mass Spectrometry*, Pergamon, Oxford, England, 1977.

26. L. Prokai, *Field Desorption Mass Spectrometry*, Marcel Dekker, New York, 1990.

27. R. P. Lattimer and H. R. Schulten, *Anal. Chem.* **61**, 1201A–1215A (1989).

28. K. L. Olsen and K. L. Rinehart, *Methods Carbohdr. Chem.* **9**, 143–164 (1993).

29. D. M. Desiderio, J. Z. Sabbatini, and J. L. Stein, *Adv. Mass Spectrom.* **8**, 1198 (1980).

30. H. R. Schulten, *Int. J. Mass Spectrom. Ion. Phys.* **32**, 97–283 (1979).

31. J.-C. Prome and G. Puzo, *Org. Mass Spectrom.* **12**, 28–32 (1977).

32. K. Rollins, J. H. Scrivens, M. J. Taylor, and H. Major, *Rapid Commun. Mass Spectrom.* **4**, 355–359 (1990).

33. D. F. Torgerson, R. P. Skowronski, and R. D. Macfarlane, *Biochem. Biophys. Res. Commun.* **60**, 616 (1974).

34. B. Sundqvist and R. D. Macfarlane, *Mass Spectrom. Rev.* **2**, 421–460 (1985).

35. P. Roepstorff, in M. L. Gross, ed., *Mass Spectrometry in the Biological Sciences: A Tutorial*, Kluwer, Dordrecht, The Netherlands, 1992, pp. 213–227.

36. P. Roepstorff, in D. M. Desiderio, ed., *Mass Spectrometry of Peptides*, CRC Press, Boca Raton, FL, 1991, pp. 65–86.

37. G. Jonsson, A. Hedin, P. Hakansson, S. U. R. Sundqvist, H. Bennich, and P. Roepstorff, *Rapid Commun. Mass Spectrom.* **3**, 190–191 (1989).

38. R. J. Cotter, L. Chen, and R. Wang, in D. M. Desiderio, ed., *Mass Spectrometry of Peptides*, CRC Press, Boca Raton, FL, 1991, pp. 17–40.

39. P. F. Nielsen and P. Roepstorff, *Biomed. Environ. Mass Spectrom.* **18**, 131–137 (1989).

40. A. Tsarbopoulos, J. Varnerin, S. Cannon-Carlson, D. Wylie, B. Pramanik, J. Tang, and T. L. Nagabhushan, *J. Mass Spectrum.* **35**, 446–453 (2000).

41. A. Benninghoven, F. G. Rudenauer, and H. W. Werner, *Secondary Ion Mass Spectrometry—Basic Concepts, Instrumental Aspects, Applications and Trends*, Wiley, New York, 1987.

42. W. Aberth, K. Straub, and A. L. Burlingame, *Anal. Chem.* **54**, 2029–2034 (1982).

43. D. Barofsky, in P. A. Lyon, ed., *Desorption Mass Spectrometry—Are SIMS and FAB the Same?* ACS (American Chemical Society) Symposium Series 291, Washington, DC, 1985, pp. 113–124.

44. J. Sunner, *J. Mass Spectrom.* **28**, 805–823 (1993).

45. P. Kebarle and L. Tang, *Anal. Chem.* **65**, 972A–986A (1993).

46. J. Sunner, M. G. Ikonomou, and P. Kebarle, *Int. J. Mass Spectrom. Ion. Proc.* **82**, 221–237 (1988).

47. G. J. C. Paul, R. Theberge, M. J. Bertrand, R. Feng, and M. D. Bailey, *Org. Mass Spectrom.* **28**, 1329–1339 (1993).

48. C. Dass, *J. Am. Soc. Mass Spectrom.* **1**, 405–412 (1990).

49. C. Dass, *J. Mass Spectrom.* **31**, 77–82 (1996).

50. F. H. Field, *J. Phys. Chem.* **86**, 5115–5123 (1982).

51. T. Keough, F. S. Ezra, A. F. Russel, and J. D. Pryne, *Org. Mass Spectrom.* **22**, 241–247 (1987).

52. C. Dass and D. M. Desiderio, *Anal. Chem.* **60**, 2723–2729 (1988).

53. K. A. Kidwell and M. L. Gross, *J. Am. Soc. Mass Spectrom.* **5**, 72–91 (1994).

54. C. Dass, *Org. Mass Spectrom.* **29**, 475–482 (1994).

55. J. L. Gower, *Biomed. Mass Spectrom.* **12**, 191–196 (1985).

56. E. DePauw, in J. A. McCloskey, ed., *Methods in Enzymology*, Academic Press, New York, 1990, Vol. 193, pp. 201–204.

57. E. DePauw, *Mass Spectrom. Rev.* **5**, 191–212 (1991).

58. E. DePauw, *Mass Spectrom. Rev.* **10**, 283–301 (1986).

59. K. D. Cook, P. J. Todd, and D. H. Frier, *Biomed. Environ. Mass Spectrom.* **18**, 492–497 (1989).

60. K. L. Busch, in D. M. Desiderio, ed., *Mass Spectrometry of Peptides*, CRC Press, Boca Raton, FL, 1991, p. 173–200.

61. K. L. Busch, *J. Mass Spectrom.* **30**, 233–240 (1995).

62. A. M. Buko and B. A. Fraser, *Biomed. Environ. Mass Spectrom.* **12**, 577–585 (1985).

63. R. Yazdanparast, P. C. Andrews, D. L. Smith, and J. E. Dixon, *Anal. Biochem.* **153**, 348–353 (1986).

64. M. G. O. Santana-Marques, A. J. V. Ferrer-Correia, and M. L. Gross, *Anal. Chem.* **61**, 1442–1447 (1989).

65. J. N. Kyranos and P. Vouros, *Biomed. Environ. Mass Spectrom.* **19**, 628–634 (1990).

66. A. Liguori, G. Sindona, and N. Uccella, *J. Am. Chem. Soc.* **108**, 7488–7491 (1986).

67. J. A. McCloskey, S. A. Sethi, and C. C. Nelson, *Anal. Chem.* **56**, 1975–1977 (1984).

68. R. Theberge and M. J. Bertrand, *J. Mass Spectrom.* **30**, 163–171 (1995).

69. M. A. Baldwin, K. J. Welham, I. Toth, and W. A. Gibbons, *Org. Mass Spectrom.* **23**, 697–699 (1986).

70. S. Naylor, A. F. Findeis, B. W. Gibson, and D. H. Williams, *J. Am. Chem. Soc.* **108**, 6359–6363 (1986).

71. W. V. Ligon and S. B. Dorn, *Anal. Chem.* **58**, 1889–1892 (1986).

72. C. Heine, J. F. Holland, and J. T. Watson, *Anal. Chem.* **61**, 2674–2682 (1989).

73. R. M. Caprioli, T. Fan, and J. S. Cottrell, *Anal. Chem.* **58**, 2949–2954 (1986).

74. R. M. Caprioli, *Anal. Chem.* **62**, 477A–485A (1990).

75. C. Dass, in D. M. Desiderio, ed., *Mass Spectrometry: Clinical and Biomedical Applications*, Plenum Press, New York, 1994, Vol. 2, pp. 1–52.

76. Y. Ito, T. Takeuchi, D. Ishii, and M. Goto, *J. Chromatogr.* **346**, 161–166 (1985).

77. D. B. Kassel, B. D. Musselman, and J. A. Smith, *Anal. Chem.* **63**, 1091–1097 (1991).

78. R. M. Caprioli and W. T. Moore, *Anal. Chem.* **63**, 1978–1983 (1991).

79. R. M. Caprioli, *Biochemistry* **27**, 513–520 (1988).

80. R. M. Caprioli, W. T. Moore, B. DaGue, and M. Martin, *J. Chromatogr.* **443**, 355–362 (1988).

81. M. A. Moseley, L. J. Deterding, K. B. Tomer, and J. W. Jorgenson, *Anal. Chem.* **63**, 1467–1473 (1991).

82. M. A. Mosely, L. J. Deterding, K. B. Tomer, and J. W. Jorgenson, *J. Chromatogr.* **480**, 197–209 (1989).

83. R. M. Caprioli, W. T. Moore, M. Martin, B. DaGue, K. Wilson, and S. Moring, *J. Chromatogr.* **480**, 247–257 (1989).

84. M. A. Mosely, L. J. Deterding, K. B. Tomer, and J. W. Jorgenson, *Rapid Commun. Mass Spectrom.* **3**, 87–93 (1989).

85. R. M. Caprioli and S. N. Lin, *Proc. Natl. Acad. Sci. (USA)* **87**, 240–243 (1990).

86. R. M. Caprioli, *Continuous-Flow Fast Atom Bombardment Mass Spectrometry*, Wiley, New York, 1990.

87. H. R. Morris, M. Panico, and G. W. Taylor, *Biochem. Biophys. Res. Commun.* **117**, 299–305 (1983).

88. B. W. Gibson and K. Beimann, *Proc. Natl. Acad. Sci. (USA)* **81**, 1956–1960 (1984).

89. C. Dass and D. M. Desiderio, *Anal. Biochem.* **163**, 52–66 (1987).

90. K. Biemann and S. Martin, *Mass Spectrom. Rev.* **6**, 1–75 (1987).

91. C. Dass, in D. M. Desiderio, ed., *Mass Spectrometry of Peptides*, CRC Press, Boca Raton, FL, 1991, pp. 327–345.

92. B. W. Gibson, A. M. Falick, A. L. Burlingame, L. Nadasdi, A. C. Nguyen, and G. L. Kenyon, *J. Am. Chem. Soc.* **109**, 5343–5348 (1987).

93. C. Dass and P. Mahalakshmi, *Rapid Commun. Mass Spectrom.* **9**, 1148–1154 (1995).

94. K. Beimann and I. A. Papayannopoulos, *Acc. Chem. Res.* **27**, 370–378 (1994).

95. M. Barber and B. N. Green, *Rapid Commun. Mass Spectrom.* **1**, 80–83 (1987).

96. C. Dass, R. Seshadri, M. Israel, and D. M. Desiderio, *Biomed. Environ. Mass Spectrom.* **17**, 37–45 (1988).

97. E. Davoli and M. L. Gross, *J. Am. Soc. Mass Spectrom.* **1**, 320–324 (1990).

98. J. S. Crockett, M. L. Gross, W. W. Cristie, and R. T. Holman, *J. Am. Soc. Mass Spectrom.* **1**, 183–191 (1990).

99. R. C. Murphy and K. A. Harrison, *Mass Spectrom. Rev.* **13**, 57–75 (1991).

100. C. E. Costello, in T. Matsuo, R. M. Caprioli, M. L. Gross, and Y. Seyama, eds., *Biological Mass Spectrometry, Present and Future*, Wiley, New York, 1994, pp. 437–462.

101. J. Peter-Katalinic, *Mass Spectrom. Rev.* **13**, 77–98 (1994).

102. B. Domon and C. E. Costello, *Glycoconjugate* **5**, 397–409 (1988).

103. C. Nelson and J. A. McCloskey, *J. Am. Soc. Mass Spectrom.* **5**, 339–349 (1994).

104. R. L. Cerny, K. B. Tomer, M. L. Gross, and L. Grotjahn, *Anal. Biochem.* **165**, 175–182 (1987).

105. J. Wellemans, R. L. Cerny, and M. L. Gross, *Analyst* **119**, 497–503 (1994).

106. P. S. Branco, M. P. Chiarelli, J. O. Lay, Jr., and F. A. Beland, *J. Am. Soc. Mass Spectrom.* **6**, 248–256 (1995).

107. J. L. Gower, *Biomed. Mass Spectrom.* **12**, 191–196 (1985).

108. P. Lyon, K. B. Tomer, and M. L. Gross, *Anal. Chem.* **57**, 2984–2989 (1985).

109. C. Dass, K. N. Thimmaiah, B. S. Jayashree, and P. J. Houghton, *J. Mass Spectrom.* **32**, 1279–1289 (1997).

110. R. J. Cotter, *Anal. Chem.* **56**, 485A–504A (1984).

111. F. P. Novak, K. Balasanmugan, K. Viswanadham, C. D. Parker, Z. A. Wilk, D. Mattern, and D. M. Hercules, *Int. J. Mass Spectrom. Ion. Proc.* **53**, 135–149 (1983).

112. J. Grotemeyer and J. Schelag, *Biomed. Environ. Mass Spectrom.* **16**, 143–149 (1988).

113. F. Hillenkamp and M. Karas, in J. A. McCloskey, ed., *Methods in Enzymology*, Academic Press, New York, 1990, Vol. 193, pp. 280–295.

114. R. C. Beavis and B. T. Chait, *Rapid Commun. Mass Spectrom.* **3**, 233–237 (1989).

115. K. Tanaka, H. Waki, H. Ido, S. Akita, and T. Yoshida, *Rapid Commun. Mass Spectrom.* **2**, 151–153 (1988).

116. A. Vertes, in K. G. Standing and W. Ens, eds., *Methods and Mechanisms of Producing Ions from Large Molecules*, Plenum Press, New York, 1991, pp. 275–286.

117. M. Glückmann and M. Karas, *J. Mass Spectrom.* **34**, 467–477 (1999).

118. K. Dreisewerd, M. Schurenberg, M. Karas, and F. Hillenkamp, *Int. J. Mass Spectrom. Ion. Proc.* **141**, 127–2393 (1995).

119. T.-W. D. Chan, A. W. Colburn, P. J. Derrick, D. J. Gardiner, and M. Bowden, *Org. Mass Spectrom.* **27**, 188 (1992).

120. B. H. K. Wang, K. Dreisewerd, U. Bahr, M. Karas, and F. Hillenkamp, *J. Am. Soc. Mass Spectrom.* **4**, 393–398 (1993).

121. A. Vertes, G. Irinyi, and R. Gijbels, *Anal. Chem.* **65**, 2389–2393 (1993).

122. S. Niu, W. Zhang, and B. T. Chait, *J. Am. Soc. Mass Spectrom.* **9**, 1–7 (1998).

123. R. Knochenmuss, F. Dubois, M. J. Dale, and R. Zenobi, *Rapid Commun. Mass Spectrom.* **10**, 871–877 (1996).

124. X. Chen, J. A. Carroll, and R. C. Beavis, *J. Am. Soc. Mass Spectrom.* **9**, 885–891 (1998).

125. M. Karas, M. Glückmann, and J. Schäfer, *J. Mass. Spectrom.* **35**, 1–12 (2000).

126. R. J. Cotter, *Anal. Chem.* **64**, 1027A–1039A (1992).

127. V. S. K. Kolli and R. Orlando, *Anal. Chem.* **69**, 327–332 (1997).

128. J. Qin and B. T. Chait, *Anal. Chem.* **68**, 2102–2107 (1996).

129. V. M. Doroshenko and R. J. Cotter, *J. Mass Spectrom.* **31**, 602–615 (1997).

130. J. A. Castro and C. L. Wilkins, *Anal. Chem.* **65**, 2621–2627 (1993).

131. Y. Li, R. L. Hunter, and R. T. McIver, *Int. J. Mass Spectrom. Ion. Proc.* **157/158**, 175–188 (1996).

132. D. L. Vollmer, D. L. Rempel, and M. L. Gross, *Int. J. Mass Spectrom. Ion. Proc.* **157/158**, 189–198 (1996).

133. A. Overberg, M. Karas, and F. Hillenkamp, *Rapid Commun. Mass Spectrom.* **4**, 128 (1990).

134. C. W. Sutton, C. H. Wheeler, U. Sally, J. M. Corbett, J. S. Cotrell, and M. J. Dunn, *Electrophoresis* **18**, 424 (1996).

135. F. Hillenkamp, M. Karas, R. C. Beavis, and B. T. Chait, *Anal. Chem.* **63**, 1193A–1203A (1993).

136. S. Berkenkamp, M. Karas, and F. Hillenkamp, *Proc. Natl. Acad. Sci. (USA)* **93**, 7003–7004 (1996).

137. K. J. Wu, A. Steding, and C. H. Becker, *Rapid Commun. Mass Spectrom.* **7**, 142–146 (1993).

138. R. C. Beavis and B. T. Chait, *Anal. Chem.* **62**, 1836–1840 (1990).

139. S. J. Doktycz, P. J. Savickas, and D. A. Krueger, *Rapid Commun. Mass Spectrom.* **5**, 145–148 (1990).

140. O. Vorm, P. Roepstorff, and M. Mann, *Anal. Chem.* **66**, 3281–3287 (1994).

141. M. Kussmann, E. Nordhoff, H. Rahbek-Nielsen, S. Haebel, M. Rossel-Larsen, L. Jakobsen, J. Gobom, E. Mirgorodskaya, A. Kroll-Kristensen, L. Palm, and P. Roepstorff, *J. Mass Spectrom.* **32**, 593–601 (1997).

142. A. Westman, C. L. Nilsson, and R. Ekman, *Rapid Commun. Mass Spectrom.* **12**, 1092–1098 (1998).

143. L. Li, A. P. L. Wang, and L. D. Coulson, *Anal. Chem.* **65**, 493–495 (1993).

144. J. Preisler, F. Foret, and B. L. Karger, *Anal. Chem.* **70**, 5278–5287 (1998).

145. S. Y. Chang and E. S. Yeung, *Anal. Chem.* **69**, 2251–2257 (1997).

146. A. P. L. Wang and L. Li, *Anal. Chem.* **64**, 769–775 (1992).

147. K. K. Murray and D. H. Russell, *J. Am. Soc. Mass Spectrom.* **5**, 1–9 (1994).

148. R. S. Brown, B. L. Carr, and J. J. Lennon, *J. Am. Soc. Mass Spectrom.* **7**, 225–232 (1996).

149. J. C. Rouse, W. Yu, and S. A. Martin, *J. Am. Soc. Mass Spectrom.* **7**, 822–835 (1995).

150. P. R. Griffin, M. J. MacCoss, J. K. Eng, R. A. Blevines, J. S. Aaronson, and J. R. Yates, III, *Rapid Commun. Mass Spectrom.* **67**, 3202–3210 (1995).

151. H. Wu, R. L. Morgan, and H. Aboleneen, *J. Am. Soc. Mass Spectrom.* **9**, 660–667 (1998).

152. D. J. Harvey, *J. Mass Spectrom.* **30**, 1333–1346 (1995).

153. R. Orlando and Y. Yang, in B. S. Larsen and C. N. McEwen, eds., *Mass Spectrometry of Biological Materials*, Marcel Dekker, New York, pp. 215–245, 1998.

154. M. Okamoto, K.-I. Takahashi, T. Doe, and Y. Takimoto, *Anal. Chem.* **69**, 2919–2926 (1997).

155. A. Jacson, H. T. Yates, J. H. Scrivens, M. R. Green, and R. H. Bateman, *J. Am. Soc. Mass Spectrom.* **9**, 269–274 (1998).

156. G. Siuzdak, J. Wei, and J. M. Buriak, *Nature* **399**, 243–246 (1999).

157. C. R. Blakley and M. L. Vestal, *Anal. Chem.* **55**, 750–754 (1983).

158. P. Arpino, *Mass Spectrom. Rev.* **9**, 631–669 (1990).

159. M. L. Vestal, *Science* **226**, 275–281 (1984).

160. J. Serrano, D. W. Kuehl, and S. Naumann, *J. Chromatogr. B* **615**, 203–213 (1993).

161. A. L. Yargey, C. G. Edmonds, I. A. S. Lewis, and M. L. Vestal, *Liquid Chromatography/Mass Spectrometry*, Plenum Press, New York, 1990.

162. M. L. Vestal, *Int. J. Mass Spectrom. Ion. Proc.* **46**, 193–196 (1983).

163. H.-Y. Kim, D. Pilosof, D. F. Dyckes, and M. L. Vestal, *J. Am. Chem. Soc.* **106**, 7304–7309 (1984).

164. C. Fenselau, D. J. Librato, J. A. Yargey, R. J. Cotter, and A. L. Yargey, *Anal. Chem.* **56**, 2759–2762 (1984).

165. E. Gelpi and J. Abian, *J. Mass Spectrom.* **30**, 608–616 (1995).

166. G. Schmelzeisen-Redeker, U. Giessmann, and F. W. Rollgen, *Angew. Chem,, Int. Ed. Engl.* **23**, 892–893 (1984).

167. D. Volmer and K. Levsen, in H. J. Stan, ed., *Analysis of Pesticides in Ground and Surface Waters II*, Springer-Verlag, New York, 1995, pp. 133–179.

168. D. A. Catlow, *J. Chromatogr.* **523**, 163–170 (1985).

169. J. Abian, M. I. Churchwell, and W. A. Korfmacher, *J. Chromatogr.* **629**, 267–276 (1992).

170. L. D. Betowski and J. M. Ballard, *Anal. Chem.* **56**, 2604–2607 (1984).

171. R. D. Voyksner, J. T. Bursey, and E. D. Pellizzari, *Anal. Chem.* **56**, 1507–1533 (1984).

172. *The API Book*, PE SCIEX (1990).

173. D. I. Carroll, I. Dzdic, R. N. Stillwater, and E. C. Horning, *Anal. Chem.* **47**, 1956–1959 (1975).

174. T. Wachs, J. J. Conboy, F. Garcia, and J. D. Henion, *J. Chromatogr. Sci.* **29**, 357–366 (1991).

175. M. H. Allen and B. I. Shushan, *LC-GC* **11**, 112–126 (1993).

176. F. Garcia and J. D. Henion, *Anal. Chem.* **64**, 985–990 (1992).

177. E. Huang, J. D. Henion, and T. R. Covey, *J. Chromatogr.* **511**, 257–270 (1990).

178. T. R. Covey, E. D. Lee, and J. D. Henion, *Anal. Chem.* **58**, 1451A–1463A (1986).

179. A. Bruins, T. R. Covey, and J. D. Henion, *Anal. Chem.* **59**, 2642–2646 (1987).

180. D. M. Garcia, S. K. Huang, and W. F. Stansbury, *J. Am. Soc. Mass Spectrom.* **7**, 59–65 (1996).

181. P. O. Edlund, L. Bowers, and J. D. Henion, *J. Chromatogr.* **487**, 341–356 (1989).

182. P. O. Edlund, L. Bowers, J. D. Henion, and T. R. Covey, *J. Chromatogr.* **497**, 49–57 (1989).

183. S. Pleasance, J. F. Anacleto, M. R. Bailey, and D. H. North, *J. Am. Soc. Mass Spectrom.* **3**, 378–397 (1992).

184. D. R. Doerge, S. Bajic, and S. Lowes, *Rapid Commun. Mass Spectrom.* **7**, 1126–1130 (1993).

185. M. Dole, L. L. Mack, and R. L. Hines, *J. Chem. Phys.* **49**, 2240–2249 (1968).

186. L. L. Mack, P. Kralik, A. Rheude, and M. Dole, *J. Chem. Phys.* **52**, 4977–4986 (1970).

187. M. Yamashita and J. B. Fenn, *J. Phys. Chem.* **88**, 4671–4675 (1984).

188. M. L. Aleksandrov, L. N. Gall, V. N. Krasnov, V. I. Nikolaev, V. A. Pavlenko, and V. A. Shkurov, *Dokl. Akad. Nauk SSSR* **277**, 379–383 (1984).

189. J. B. Fenn, M. Mann, C. K. Meng, S. F. Wong, and C. M. Whitehouse, *Mass Spectrom. Rev.* **9**, 37–70 (1990).

190. R. D. Smith, J. A. Loo, R. R. Ogorzalek Loo, M. Busman, and H. R. Udseth, *Mass Spectrom. Rev.* **10**, 359–451 (1991).

191. R. D. Smith, J. A. Loo, and C. G. Edmonds, in D. M. Desiderio, ed., *Mass Spectrometry, Clinical and Biomedical Appliactions*, Plenum Press, New York, 1992, Vol. 1, pp. 37–97.

192. S. A. Shaffer, K. Tang, G. A. Anderson, D. C. Prior, H. R. Udseth, and R. D. Smith, *Rapid Commun. Mass Spectrom.* **11**, 1813–1817 (1997).

193. S. A. Shaffer, A. Tolmachev, D. C. Prior, G. A. Anderson, H. R. Udseth, and R. D. Smith, *Anal. Chem.* **71**, 2957–2964 (1999).

194. H. Wang and M. Hackett, *Anal. Chem.* **70**, 205–212 (1998).

195. M. S. Wilm and M. Mann, *Int. J. Mass Spectrom. Ion. Proc.* **136**, 167–180 (1994).

196. P. Kebarle and L. Tang, *Anal. Chem.* **65**, 972A–986A (1993).

197. J. B. Fenn and M. Mann, in D. M. Desiderio, ed., *Mass Spectrometry, Clinical and Biomedical Appliactions*, Plenum Press, New York, 1992, Vol. 1, pp. 1–35.

198. J. B. Fenn, J. Rosell, T. Nohmi, S. Shen, and F. J. Banks, Jr., in A. P. Snyder, ed., *Biochemical and Biotechnological Applications of Electrospray Ionization Mass Spectrometry*, ACS, Washington, DC, 1995, pp. 60–80.

199. R. D. Smith and K. J. Light-Wahl, *Biol. Mass Spectrom.* **22**, 493–501 (1993).

200. G. Schmelzeisen-Redeker, L. Butfering, and F. W. Rollgen, *Int. J. Mass Spectrom. Ion. Proc.* **90**, 139 (1989).

201. R. Guevremont, J. C. Y. Le Blanc, and K. M. W. Siu, *Org. Mass Spectrom.* **28**, 1345–1352 (1993).

202. R. Juraschek, T. Dulcks, and M. Karas, *J. Am. Soc. Mass Spectrom.* **10**, 300–308 (1999).

203. J. V. Irbine and B. Thomson, *J. Chem. Phys.* **64**, 2287–2294 (1976).

204. M. Scalf, M. S. Westphall, and L. M. Smith, *Anal. Chem.* **72**, 52–60 (2000).

205. R. D. Smith, J. A. Loo, C. G. Edmonds, and H. R. Udseth, *Anal. Chem.* **62**, 882–899 (1990).

206. T. R. Covey, R. F. Bonner, B. I. Shushan, and J. D. Henion, *Rapid Commun. Mass Spectrom.* **2**, 249 (1988).

207. F. J. Banks, Jr., S. Shen, C. M. Whitehouse, and J. B. Fenn, *Anal. Chem.* **66**, 406–414 (1994).

208. S. K. Chowdhury, V. Katta, and B. T. Chait, *Rapid Commun. Mass Spectrom.* **4**, 81 (1990).

209. M. G. Ikonomou and P. Kebarle, *J. Am. Soc. Mass Spectrom.* **5**, 791–799 (1994).

210. G. Hofgartner, T. Wachs, K. Beans, and J. D. Henion, *Anal. Chem.* **65**, 439–446 (1993).

211. M. R. Emmet and R. M. Caprioli, *J. Am. Soc. Mass Spectrom.* **5**, 605–613 (1994).

212. M. Wilm and M. Mann, *Anal. Chem.* **68**, 1–8 (1996).

213. M. Wilm, A. Shevchenko, T. Houthaene, S. Breit, L. Schweigerer, T. Fotsis, and M. Mann, *Nature* **379**, 466–469 (1996).

214. G. A. Valaskovic, N. L. Kelleher, D. P. Little, D. J. Aaserud, and F. W. McLafferty, *Anal. Chem.* **67**, 3802–3805 (1995).

215. N. L. Kelleher, M. W. Senko, D. P. Little, P. B. O'Conner, and F. W. McLafferty, *J. Am. Soc. Mass Spectrom.* **6**, 220–221 (1995).

216. P. E. Andren, M. R. Emmet, and R. M. Caprioli, *J. Am. Soc. Mass Spectrom.* **5**, 867–869 (1994).

217. D. C. Gale and R. D. Smith, *Rapid Commun. Mass Spectrom.* **7**, 1017–1021 (1993).

218. M. S. Kriger, K. D. Cook, and R. S. Ramsey, *Anal. Chem.* **67**, 385–389 (1995).

219. D. R. Barnidge, S. Nilsson, and K. E. Markides, *Anal. Chem.* **71**, 4115–4118 (1999).

220. C. Henry, **69**, 359A–361A (1997).

221. Q. Xue, F. Foret, Y. M. Dunayevskiy, P. M. Zavracky, N. E. McGruer, and B. L. Karger, *Anal. Chem.* **69**, 426–430 (1997).

222. B. Zhang, F. Foret, and B. L. Karger, *Anal. Chem.* **72**, 1015–1022 (2000).

223. R. S. Ramsey and J. M. Ramsey, *Anal. Chem.* **69**, 1174–1178 (1997).

224. D. Figeys, Y. Ning, and R. Aebersold, *Anal. Chem.* **69**, 3153–3160 (1997).

225. B. Zhang, B. L. Karger, and F. Foret, *Anal. Chem.* **71**, 3258–3264 (1999).

226. J. Li, P. Thibault, N. H. Bings, C. D. Skinner, C. Wang, C. Colyer, and J. Harrison, *Anal. Chem.* **71**, 3036–3045 (1999).

227. G. J. Van Berkel, S. A. McLuckey, and G. L. Glish, *Anal. Chem.* **62**, 1284–1295 (1990).

228. S. A. McLuckey, G. J. Van Berkel, G. L. Glish, E. Huang, and J. D. Henion, *Anal. Chem.* **63**, 375–383 (1991).

229. J. G. Boyl and and C. M. Whitehouse, *Anal. Chem.* **64**, 2084–2089 (1992).

230. A. I. Verentchikov, W. Ens, and K. G. Standing, *Anal. Chem.* **66**, 126–133 (1994).

231. J.-T. Wu, L. He, M. X. Li, S. Parus, and D. M. Lubman, *J. Am. Soc. Mass Spectrom.* **8**, 1237–1246 (1997).

232. R. W. Purves and L. Li, *J. Am. Soc. Mass Spectrom.* **8**, 1085–1093 (1997).

233. B. S. Larsen and C. N. McEwen, *J. Am. Soc. Mass Spectrom.* **2**, 205–211 (1991).

234. J. A. Loo, R. R. Ogorzalek Loo, and P. C. Andrews, *Org. Mass Spectrom.* **28**, 1640–1649 (1993).

235. K. D. Henry, J. P. Quinn, and F. W. McLafferty, *J. Am. Chem. Soc.* **113**, 5447–5450 (1991).

236. J. E. Bruce, S. A. Hofstadler, B. E. Winger, and R. D. Smith, *Int. J. Mass Spectrom. Ion Proc.* **132**, 97 (1994).

237. M. W. Senko, C. L. Hendrickson, L. Pasa-Tolic, J. A. Marto, F. M. White, S. Guan, and A. G. Marshall, *Rapid Commun. Mass Spectrom.* **10**, 1824–1828 (1996).

238. M. W. Senko, C. L. Hendrickson, M. R. Emmett, S. D.-H. Shi, and A. G. Marshall, *J. Am. Soc. Mass Spectrom.* **8**, 970–976 (1997).

239. M. R. Emmett, F. M. White, C. L. Hendrickson, S. D.-H. Shi, and A. G. Marshall, *J. Am. Soc. Mass Spectrom.* **9**, 333–340 (1998).

240. A. P. Snyder, ed., *Biochemical and Biotechnological Appliactions of Electrospray Ionization Mass Spectrometry*, ACS, Washington, DC, 1995.

241. R. B. Cole, ed., *Electrospray Ionization Mass Spectrometry: Fundamentals, Instrumentation, and Appliactions of*, Wiley-Interscience, New York, 1997.

242. C. Dass, *Current Org. Chem.* **3**, 193–209 (1999).

3

MASS ANALYSIS AND ION DETECTION

The primary function of a mass spectrometer is to measure the mass of the ions that are generated in the ion source. A mass analyzer, which is the heart of a mass spectrometer, accomplishes this task. The performance of a mass spectrometer depends to a great extent on the design of the mass analyzer and the associated ion optics. On entering a mass analyzer, a collimated beam of ions is separated according to the mass-to-charge (m/z) ratios of individual ions. Another function of a mass analyzer is to maximize the transmission of all ions that enter it from the ion source. In addition, it acts as a focusing device to collect all mass-resolved ions at a single focal point. A moving charged particle can be distinguished on the basis of its momentum, kinetic energy, and velocity. A mass analyzer makes use of one or more of these properties to mass-resolve and to focus various ions.

The performance of a mass analyzer is gauged on the basis of the following criteria:

- *Mass range*—the maximum allowable mass that can be analyzed.
- *Resolution*—the ability to separate two neighboring mass ions.
- *Scan speed*—a fast scan speed is required for rapidly changing events, such as to monitor eluants from a chromatography step, or record a pulsed event. A slow speed is obligatory for exact mass measurements.
- *Detection sensitivity*—the smallest amount of an analyte that can be detected at a certain confidence level.

An additional criterion is the ease with which certain ionization and other ancillary devices, such as multichannel array detectors and chromatographic devices, can be outfitted.

Several different designs of mass analyzers have been developed; of these, magnetic sector, quadrupole, quadrupole ion-trap (QIT), time-of-flight (TOF), and ion cyclotron resonance (ICR) instruments have become popular [1]. This chapter provides a detailed account of the basic concepts and latest developments related to these mass analyzers.

3.1. MAGNETIC SECTOR MASS SPECTROMETERS

Magnetic sector instruments are the oldest types of mass spectrometer. Some magnetic sector instruments are constructed with only a magnetic analyzer, for example, those used in isotope ratio measurements. A vast majority of applications, however, require magnetic sector instruments in which an electrostatic analyzer is also incorporated either before or after the magnetic field.

3.1.1. Kinetic Energy of Ions

Before enumerating the basic principle of a magnetic sector mass spectrometer, it is pertinent to discuss the concept of kinetic energy as applicable to mass spectrometry. When a charged particle traverses in the direction of an electric field, it experiences acceleration. The kinetic energy gained in this process is given by Eq. 3.1. Here m is the mass of the ion in kilograms (1 kg = 6.023×10^{26} Da), v is the velocity of the ion in meters per second (m/s), z is the charge on the ion ($z = ne$), n is the integral number of units of charge, and e is the fundamental unit of charge = 1.602×10^{-19} coulombs), and V, in volts, is the ion acceleration voltage. In practice, all ions are made to exit the ion source at nearly the same specified kinetic energy (KE) irrespective of their mass by subjecting them to a large acceleration voltage:

$$\text{KE} = zV = \tfrac{1}{2}mv^2 \tag{3.1}$$

3.1.2. Magnetic Analyzer

The separation of different mass ions by a magnet takes place as a result of their deflection by the magnetic field. As shown in Figure 3.1, when a charged particle is injected into a magnetic field, it is forced to travel in a curved path of a certain radius (r) in a plane perpendicular to the direction of the field. Equation 3.2 describes the motion of an ion in the magnetic field; here B, in tesla, is the magnetic field strength [2,3]. Thus, the magnetic sector acts as a momentum analyzer, and disperses the ions according to their momentum : charge ratios.

$$\frac{mv^2}{r} = zvB \quad \text{or} \quad \frac{mv}{z} = Br \tag{3.2}$$

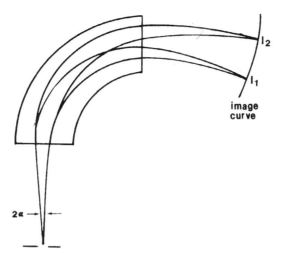

Figure 3.1. Schematic depiction of mass separation by a magnetic analyzer. Two ions of different m/z values that emanate from a point source with an angular divergence of 2α are brought to focus on the image curve at points I_1 and I_2. [Reproduced from Ref. 1 by permission of Plenum Press, New York (copyright 1994).]

When the value of v from Eq. 3.1 is substituted into this equation, it becomes clear that a magnetic sector separates ions according to their m/z ratios:

$$\frac{m}{z} = \frac{B^2 r^2}{2V} \tag{3.3}$$

This equation implies that all ions of the same m/z value follow the same trajectory and are collected at a single focal point. It also implies that a mass spectrum of the ions that enter the magnetic sector can be obtained by scanning the magnetic field while keeping V constant. Alternatively, by keeping B constant, all ions that follow different radii can be collected simultaneously in the same plane, as is the usual practice in the Mattauch–Herzog geometry double-focusing mass spectrometers. Rearranging Eq. 3.3 in terms of radius r gives

$$r = \left(\frac{1}{B}\right)\left(\frac{2mV}{z}\right)^{1/2} \tag{3.4}$$

The logarithmic differentiation of Eq. 3.4 leads to Eq. 3.5. This equation defines the requirements for the optimum separation of different mass ions; specifically, all ions must be homogeneous in energy (i.e., ΔV must be as small as possible) and the value of V should be as high as possible. For this reason, the source in the magnetic sector instruments is held at a high voltage of 8000 V.

$$\frac{\Delta r}{r} = \frac{1}{2}\left(\frac{\Delta m}{m}\right) + \frac{1}{2}\left(\frac{\Delta V}{V}\right) \tag{3.5}$$

3.1.3. Electrostatic Analyzer

An electrostatic analyzer (ESA) has become an integral part of a double-focusing mass spectrometer. Its function is to produce ions homogeneous in kinetic energy. An ESA consists of two parallel plates (Figure 3.2), one of which is held at a positive potential and the other at a negative potential of equal magnitude [2,3]. Ions that enter this radial electric field are forced to follow a circular trajectory, and are dispersed according to their kinetic energy. Thus, an ESA can be used as an energy filter to produce an ion beam of nearly homogeneous energy. This analyzer is also used for direction focusing. All ions of an identical energy content that emanate from the same point with small angular divergence are brought to focus at a single point (e.g., at points F_1 and F_2 in Figure 3.2). The motion of an ion in this analyzer is governed by Eq. 3.6, where r is the radius of curvature in meters, E is the electric field strength in volts per meter (V/m), and m, v, z, and V are as defined above.

$$\frac{mv^2}{r} = zE \qquad \text{and} \qquad r = \frac{2V}{E} \qquad (3.6)$$

3.1.4. Double-Focusing Magnetic Sector Mass Spectrometers

In the preceding discussion, it was assumed that all ions that enter a magnetic field have the same kinetic energy. In practice, ions that are produced in a typical ion source have a spread of kinetic energies, resulting in wider peak widths and lower resolution. However, when a proper combination of an ESA and a magnetic analyzer is used, it is possible to correct simultaneously for direction and velocity inhomogeneities [1–3]. These so-called double-focusing mass spectrometers are capable of achieving very high resolution, provided the aberrations from the second-order terms

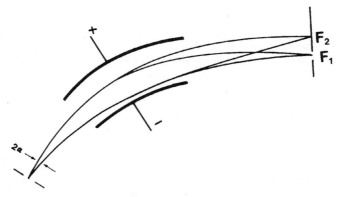

Figure 3.2. Resolution of ions by an electrostatic analyzer on the basis of their kinetic energy. All ions of the same kinetic energy that emanate from a point source with an angular divergence of 2α are brought to focus at a single point. For example, F_1 is the focal point of all ions of kinetic energy E_1, and F_2 of energy E_2. [Reproduced from Ref. 1 by permission of Plenum Press, New York (copyright 1994).]

caused by field inhomogeneities and fringing fields are held to a minimum. This condition is generally achieved by the use of additional electrostatic lenses.

Different arrangements of electrostatic and magnetic analyzers have become popular, each designed for a specific purpose. When an ESA (E) is placed before a magnetic analyzer (B), the combination results in a forward-geometry (EB) double-focusing mass spectrometer (Figure 3.3). In reverse-geometry instruments, a magnetic analyzer precedes an ESA (i.e., the BE geometry). In Figure 3.3, the two fields are arranged in the configuration of Nier–Johnson geometry, in which the two fields deflect the ions in the same direction (i.e., have a C-shape trajectory), and the point of double focusing is achieved when the velocity- and direction-focusing curves are made to intersect. A detector is placed at this point, and a mass spectrum is acquired by changing the magnetic field strength. On the other hand, when the two fields are arranged in the Mattauch–Herzog geometry, the ions are deflected in the opposite directions (i.e., have an S-shape geometry), and the double focusing of all masses results in the same plane (Figure 3.4). As a consequence, all mass-resolved ions are detected simultaneously by placing a focal plane detector (e.g., a photographic plate or a multichannel array detector) at the exit boundary of the magnet. In contrast to Nier–Johnson instruments, in which only one ion is detected at any one time, these instruments are uniquely qualified to provide increased sensitivity and accuracy in mass measurement.

Double-focusing mass spectrometers have the advantages of high-mass range, high resolution, accuracy of mass determination, reasonable scan speed, and high dynamic range. An additional unique quality of magnetic sector instruments is that they provide an opportunity to study reactions in field-free regions (FFRs). These include high-energy collision-induced dissociation (CID) reactions, charge permutation reactions, reactions of metastable ions, and kinetic energy release (KER) experiments. The disadvantages of these instruments are their exorbitant cost and a low ion transmission efficiency. In addition, the high voltage environment in the ion source creates arcing problems with atmospheric pressure ionization techniques (e.g., thermospray and electrospray) or when effluents from liquid chromatography columns are analyzed.

Despite their high cost, these state-of-the-art magnetic sector mass spectrometers have dominated the field for a long time primarily on the grounds of their high-

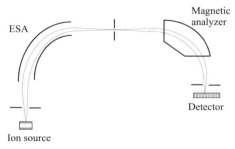

Figure 3.3. A schematic diagram of a double-focusing forward geometry Nier–Johnson-type magnetic sector mass spectrometer. The ions are detected at a single focal point.

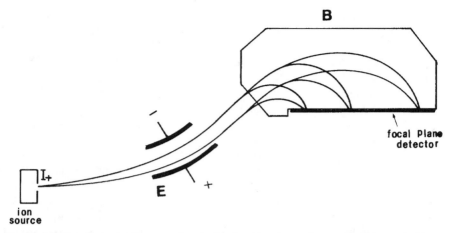

Figure 3.4. A schematic diagram of a double-focusing forward geometry Mattauch–Herzog design magnetic sector mass spectrometer. All ions are simultaneously detected on a focal plane. [Reproduced from Ref. 1 by permission of Plenum Press, New York (copyright 1994).]

resolution and exact mass measurement capabilities. With improved magnets and ion optics, they are capable of achieving a mass resolution of over 100,000 and a mass range in excess of 15,000 Da at the highest sensitivity (at 8–10 kV). The upper mass range can be extended further by reducing the accelerating potential (i.e., up to 120,000 Da at 1 kV) albeit at the cost of reduced sensitivity.

3.2. QUADRUPOLE MASS SPECTROMETERS

The quadrupole mass spectrometer is the most widely used type of mass spectrometer. The mass separation in this instrument is accomplished solely by using electric fields [4–8]. Quadrupoles are dynamic mass analyzers, which means that ion trajectories are controlled by a set of time-dependent forces that are generated by applying direct current (dc) and radiofrequency (rf) potentials to a set of electrodes. In contrast, ions experience constant force during mass analysis in static devices such as magnetic sector instruments discussed in Section 3.1.

3.2.1. Principle of Operation

A quadrupole mass analyzer is a two-dimensional quadrupole field device. As shown in Figure 3.5, it consists of four accurately aligned parallel rods that are arranged symmetrically in a square array. The field within the square array is created by electrically connecting opposite pairs together. Ions are injected at one end of the quadrupole structure in the direction of the quadrupole rods (z direction). The separation of different m/z ions is accomplished through the criterion of path stability within the quadrupole field. In other words, at a given set of operating parameters, ions of very narrow m/z range have stable trajectories (i.e., their motion

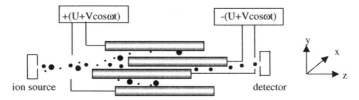

Figure 3.5. A quadrupole mass analyzer. At a certain values of dc potential U and rf potential V, ions of a specific m/z are made to have stable trajectories in the x and y directions, and travel in the z direction within the quadrupole field; all other ions are lost because they follow unstable trajectories.

is confined within the field-defining electrodes), whereas the remainder of the ions will have unstable trajectories (i.e., the amplitude of their motion exceeds the boundaries of the electrodes). To obtain a mass spectrum, the quadrupole field is varied to force other ions to follow the stable path. Thus, its function is analogous to a variable narrowband filter.

The quadrupole field is created by supplying a positive direct current (dc) potential U and a superimposed radiofrequency (rf) potential $V \cos \omega t$ (i.e., $U - V \cos \omega t$) to one pair of rods (where ω is the angular frequency related to the frequency f, in hertz, by $\omega = 2\pi f$, V the amplitude of rf voltage, and t the time). The other opposing pair of rods receives a dc potential of $-U$ and an rf potential of magnitude $V \cos \omega t$, but out of phase by $180°$ [i.e., $-(U - V \cos \omega t)$]. This arrangement creates an oscillating field such that the potential Φ at any point within the rods is given by

$$\Phi_{(x,y)} = \Phi_0 \frac{x^2 - y^2}{r_0^2} \tag{3.7}$$

where Φ_0, the applied potential, is equal to $U - V \cos \omega t$, r_0 the inscribed radius between the rods (i.e., one-half the distance between the opposite electrodes), and x and y the distances from the center of the field. It is obvious from this equation that the potential is zero at the center of the square array along the z axis (i.e., where x and y are equal to zero). The potential is also zero when the values of x and y are equal.

The motion of an ion in the x and y directions is described using a quadratic equation of the form

$$\frac{d^2u}{d^2\xi} + (a_u - 2q_u \cos 2\xi)u = 0 \tag{3.8}$$

which is commonly known to mathematicians as the *Mathieu equation*, where u represents the transverse displacement in the x and y directions from the center of the field, ξ is equal to $\omega t/2$, and the dimensionless variables a and q are given by

$$a_u = a_x = -a_y = \frac{8eU}{m\omega^2 r_0^2} \tag{3.9}$$

$$q_u = q_x = -q_y = \frac{4eU}{m\omega^2 r_0^2} \tag{3.10}$$

The solution to the Mathieu equation shows that trajectories for some ions are stable and for others unstable in the quadrupole field. For certain values of a and q, a region of simultaneously stable oscillations in the x and y directions is obtained for ions within a chosen m/z range. Several regions of simultaneous x and y stability are possible. However, the region close to the origin, dubbed as the *first stability region* (shown in Figure 3.6), is of practical interest for mass analysis. In this region, the maximum value of q_u is 0.908. The ions within this boundary can travel the entire length of the quadrupole assembly without interruption. A line with the slope $a/q = 2U/V$ that passes through the origin of the a, q diagram is the operating line or mass-scan line. The width of the stable region is a measure of mass resolution (R). Therefore, as shown in Figure 3.6 for the $R = 100$ mass-scan line, the optimum mass separation is obtained when the mass-scan line is made to intersect the tip of the stability diagram; this condition is attained by increasing the U/V ratio to a value of 0.167. At that point, only ions of one mass pass through the quadrupole assembly. Thus, by operating at the tip of the a, q diagram, a quadrupole behaves as a narrow-bandpass mass filter.

In order to obtain a mass spectrum, the quadrupole field is changed by simultaneously scanning U and V, while keeping their ratio and f constant. Less commonly, the applied frequency is changed at fixed values of U and V. This way, ions of different m/z can be brought into the stability region, and transmitted along the length of the quadrupole field.

3.2.2. Mass Range

The mass range and the mass resolution of a quadrupole mass spectrometer are both dependent on the length (L) and diameter (r_0) of the quadrupole rods, the supply

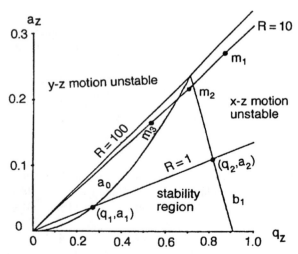

Figure 3.6. A diagram of the first stability region. Three scan lines are shown; the line that crosses the tip of the stability region provides maximum resolution. [Reproduced from Ref. 7 by permission of Elsevier Scientific Publishing Corporation (copyright 1980).]

voltage to the rods (V), the rf frequency (f), and the initial kinetic energy of ions. With the value of U/V equal to 0.167, the upper mass limit is given by

$$m = \frac{0.136\,V}{f^2 r_0^2} \tag{3.11}$$

where m is the mass in daltons, V is in volts, r_0 is in centimeters, and f is in megahertz. In principle, the upper mass limit can be increased by increasing the amplitude of the rf signal, decreasing its frequency, and using short-diameter rods. However, there are practical limitations in optimizing these parameters. As a consequence, an upper mass limit of only 4000 Da is accessible currently.

3.2.3. Mass Resolution

In a quadrupole field, resolution is proportional to n^2 given by Eq. 3.12, where n is the number of rf cycles that an ion spends within the boundaries of the quadrupole rods and zV_z is energy of ions that are injected into the quadrupole structure. From this relation and Eq. 3.11, it is obvious that the upper mass limit and resolution of a quadrupole mass filter are interdependent (i.e., when f is increased to improve mass resolution, the upper mass limit decreases). However, by scanning the rf frequency while maintaining the dc and rf amplitudes constant, the resolution can be made independent of m/z values [9]. In theory, it is possible to increase the resolution of a quadrupole device by decreasing the initial velocity of ions (i.e., by reducing V_z), increasing the frequency of the rf signal, and using longer rods. In practice, the field imperfections of the rods limit this value to only unit mass:

$$R \propto n^2 \quad \text{or} \quad \frac{mf^2 L^2}{zV_z} \tag{3.12}$$

The operation of quadrupoles in higher-stability regions (second and third) has also been exploited with the objective to provide high-resolution measurements. This approach has been successful in low-mass-range measurements [10–12]. A typical example is the separation of isobaric $^{56}Fe^+$ and $^{40}Ar^{16}O^+$ ions at a resolution of 9000 [full width at half maximum (FWHM)] [12].

Resolution also depends on the number of oscillations that an ion makes within the quadrupole field. Therefore, to increase the duration of ions in the quadrupole field, the velocity with which they enter the quadrupole structure must be kept low. Unlike magnetic sector instruments (in which ions are ejected out of the source at 6–10 kV), the source of a quadrupole mass filter is operated at very low accelerating potentials (usually 10–20 V). Reflecting the ions back and forth within the quadrupoles can also increase the number of oscillations. A resolution of up to 22,000 (FWHM) has been achieved using this provision [13].

Several useful attributes of a quadrupole mass filter are low cost, mechanical simplicity, high scan speeds, high transmission, increased sensitivity, independence from the energy distribution of ions, and linear mass range. Quadrupoles can tolerate relatively higher pressures in the ion source and mass analyzer regions. This feature has led to their widespread use as detectors for inductively coupled plasma (ICP), gas chromatography (GC), liquid chromatography (LC), and capillary electrophoresis (CE) techniques. In addition, quadrupole mass spectrometers are compact and lightweight. Therefore, they are ideally suited for field applications, study of upper atmosphere, and onboard space exploration research.

However, in contrast to magnetic sector instruments, quadrupole mass filters are inferior with respect to upper mass range, mass resolution, high-mass discriminating effect, and accuracy of mass measurement. Also, they produce high background noise because the ion source and detector are situated in the same plane (the line-of-sight geometry). In addition, ion trajectories within the quadrupole field are complex and difficult to understand.

3.2.4. RF-Only Quadrupole

In some situations, it is advantageous to operate the quadrupole in the rf- only mode. When the dc component of Φ_0 is made zero (i.e., $U = 0$), then under all operating conditions the Mathieu coordinate a_u also becomes zero, which implies that a large portion of the mass-scan line falls within the stability region along the q_u axis (see Figure 3.6). Now, a quadrupole field device behaves as a wideband mass filter. As a consequence, ions of a wide mass range for which q_u is < 0.908 can be contained within the rf-only quadrupole field. The rf-only quadrupole has been used as an intermediate reaction region in modern triple quadrupole tandem mass spectrometers, as an ion containment region in TOF and ICR-FTMS instruments, and as pre- and postfilters for high-performance quadrupole mass analyzers.

3.3. QUADRUPOLE ION-TRAP MASS SPECTROMETER

Ion-trap mass spectrometers work on a principle different from that of the beam-type instruments. Unlike the latter, they store and manipulate ions in time rather than in space. Two versions of an ion trap are currently very popular. One version, called the *quadrupole ion trap* (QIT), uses an oscillating electric field for the storage and mass analysis of ions. Another version is based on the ion cyclotron resonance principle, and uses a magnetic field to trap ions. A more popular version of this type of ion trap is a Fourier transform mass spectrometer, which is described in Section 3.5. QIT is discussed here. Paul and Steinwedel first introduced QIT in 1953 [14], but its use as a mass spectrometer gained popularity only after Stafford and co-workers developed a new method of mass analysis, known as the *mass-selective instability mode* [15]. The need for a sensitive, inexpensive detector for GC and HPLC, and an inexpensive tandem mass spectrometry (MS/MS) device has provided impetus for the development of QIT mass spectrometers. The publication of several reviews and books attests to the popularity of this device [16–28]. In the near future, QITMS technology is expected to play a significant role in the high-resolution separation and characterization of biological compounds, especially in the proteome field.

3.3.1. Principle of Operation

A QIT, also referred to as a *quadrupole ion store* (QUISTOR), is a three-dimensional analog of a quadrupole mass filter. The trap, as shown in Figure 3.7, consists of three electrodes—a doughnut-shaped central electrode and two end-cap electrodes, each with a hyperbolic cross section. One of the end-cap electrodes has a small opening through which a gated electron beam can enter the trap for in situ ionization of the sample molecules. The other end-cap electrode usually has several perforations through which ions escape for external detection. By applying the appropriate dc (U) and rf (V) voltages to the electrodes, ions of a broad m/z range can be trapped within the boundaries of the electrodes. This process is further assisted by introducing helium gas at a pressure of 10^{-3} torr into the trap. The trapped ions precess in the trapping field with a frequency that is dependent on their m/z ratio. The mass spectrum of the trapped ions is obtained by increasing the magnitude of the dc and rf voltages and the frequency of the rf signal (ω), either singly or in combination, such that ions of higher m/z values become sequentially unstable in the axial direction. Those ions exit the trap through perforations in the end-cap electrodes, and are detected by a detector placed just outside this electrode. This form of mass analysis is called the *mass-selective instability* or *mass-selective axial ejection* mode.

In practice, the dimension of electrodes used in ion traps is chosen in such a manner that the internal radius, r_0, of the central ring electrode is related to the closest distance, z_0, from the center to one of the end-cap electrodes by the expression $r_0^2 = 2z_0^2$. The three-dimensional quadrupole field is created by applying a potential $\Phi_0 = U - V \cos \omega t$ to the central electrode and maintaining the end-cap electrodes at ground potential. The potential at any point in the Cartesian coordinates (x, y, z) within the space bounded by the three electrodes is given by

$$\Phi_{r,z} = \Phi_0 \frac{r^2 - 2z^2}{r_0^2} + \frac{\Phi_0}{2} \tag{3.13}$$

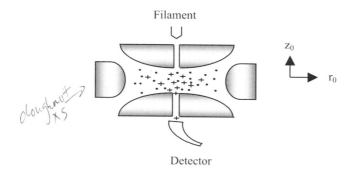

Filament

Detector

Figure 3.7. A sketch of a quadrupole ion-trap mass analyzer.

where $r^2 = x^2 - y^2$ and z are the radial and axial dimensions, respectively. This equation shows that an ion at the center (i.e., $r_0 = 0$) senses a potential of $\Phi_0/2$, and experiences a relative potential of $-\Phi_0/2$ on the end-cap electrodes and a potential of $\Phi_0/2$ on the ring electrode.

Under the influence of this potential, ions move independently in the radial and axial directions. At a certain rf potential, an ion experiences acceleration in the radial direction, while it is confined to the central axis in the z direction. During the next phase of the rf signal, this situation is reversed. Therefore, to store ions within the field-defining electrodes, radial and z-directional stabilities must both be simultaneously maintained by changing the potential within the electrodes at very high frequencies. The Mathieu equation of the form discussed above (Eq. 3.8) can also be written to describe the motion of ions in the three-dimensional quadrupole field.

The solution of that Mathieu equation gives a stability diagram (shown in Figure 3.8), the boundaries of which are defined by the Mathieu coordinates a_z and q_z. Under a given set of values of the dc voltage, the amplitude and angular frequency of the rf signal, and the electrode dimensions (r_0, z_0), ions of a certain range of m/z are confined within the boundaries of the stability diagram. By changing the dc and rf voltages, the ions can be moved from one point to another in this zone and yet remain confined in it as shown in Figure 3.8 for an ion of m/z 400.

In a simple and more commonly used variation (mass-selective instability mode) of this procedure, only rf voltage is applied to the ring electrode and the end caps are held at ground potentail (i.e., U and a_u are both equal to 0), so that the containing field is purely oscillatory. In this *rf-only mode* of operation, the locus of all trapped

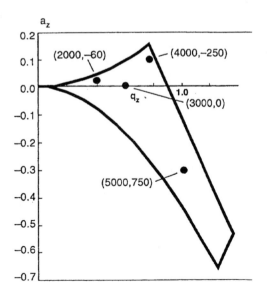

Figure 3.8. Depiction of the stability region closest to the scan axis. [Reproduced from Ref. 18 by permission of American Chemical Society, Washington, DC (copyright 1991).]

ions is on the q_z axis. Consequently, the trap behaves as a total ion storage device. The basis of mass analysis is the relation given by

$$\frac{m}{z} = \frac{4V}{q_{max}\omega^2 r_0^2} \tag{3.14}$$

where q_{max} (equal to 0.908) is the maximum value of q_z at which ions exit the stability diagram. At a given value of V, all ions for which q_z lies between 0 and 0.908 are trapped within the quadrupole field. By increasing V, ions of successively increasing m/z are forced to adopt unstable trajectories, and to exit the ion trap along the z direction (Figure 3.7).

Another useful operational mode of an ion trap is the *resonance ion ejection*, which is highly effective for the selective accumulation and isolation of species of interest [19,22]. In this mode, the trapped ions are excited and eventually ejected out of the trap by applying an auxiliary rf signal of relatively small voltage (<50-V peak) to the end-cap electrodes. All ions within the boundaries of the trap have a natural tendency to undergo oscillation with a characteristic frequency (known as the *secular frequency*) in the radial and axial directions. This natural motion of a trapped ion is strongly affected when it is brought in resonance with an auxiliary rf signal. At that point, energy is absorbed by the ion from the resonant signal to promote it to higher orbits until it is ejected from the trap along the z direction. In this way, ions of interest are allowed to accumulate while others are selectively removed. Several approaches to resonance ion ejection have been used, of which the single-frequency, broadband, and stored waveform inverse Fourier transform (SWIFT) excitation is a more common approach [29,30].

The application of a fast dc pulse also forces ions into a coherent motion with characteristic secular frequencies [31,32]. This approach has led to the development of nondestructive and mass-selective mode of ion detection, in which ions are detected by monitoring the image current as is normally done in FT-ICRMS.

For some applications, the ion trap is operated in the *mass-selective stability mode*, an operation analogous to that used in a quadrupole mass filter [16]. In this mode, by applying a potential Φ_0 to the ring electrode and $-\Phi_0$ to the end-cap electrodes, the quadrupole field is generated. The applied potential is chosen so as to operate with the a, q coordinates close to the apex of the stability diagram. Under these conditions, only one species is stored at a time. Sweeping the dc and rf voltages, while keeping their ratio constant, provides the mass spectrum of the ions.

One major advantage of a QIT is its high efficiency in conventional and tandem mass spectrometry operations. It is perhaps the most sensitive of all types of mass spectrometers. In favorable cases, the detection sensitivity in the attomole range can be achieved. Like quadrupole mass filters, ion-trap mass spectrometers are compact, simple to operate, and less costly, and have high scan speed. Because of their small size and low pumping requirements, these instruments have gained wide acceptance for field applications. An ion trap is also a useful device to conduct gas-phase ion–molecule reactions and to couple with the condensed-phase separation techniques (such as HPLC or CE). The use of a membrane introduction system for the analysis of volatile pollutants present in aqueous solutions has also been demonstrated with the trap [22]. The major drawback of this device, however, is its lower dynamic range compared to scanning instruments. This limitation is the consequence of the space-charge effect that occurs as a result of increased density of the trapped ions.

3.3.2. Extension of Mass Range and Resolution of an Ion Trap

The upper mass limit of most commercial ion traps is approximately 650 Da. Many applications require mass spectrometers to have a much higher mass range, as, for example, in characterization of biopolymers. In a manner similar to quadrupole mass filters, the upper mass limit of ion traps can be increased by the manipulation of certain operational parameters [22,25]. Referring to Eq. 3.14, the upper mass limit of an ion trap can be extended by operating the trap at higher rf voltages, reducing the rf frequency, reducing the dimensions (i.e., r_0), and forcing instability at lower values of q_z than the usual stability limit of 0.908. Of these proposals, the first two are not very practical because it is difficult to generate high voltages and to operate the trap at low rf signal frequencies. Although a decrease in the radius of electrodes to half the magnitude that is used in common commercial ion traps has helped increase the upper mass limit fourfold, further reduction in the physical dimensions is not practical. The last approach is potentially more promising. Using the ion ejection approach, the ions are forced to adopt unstable trajectories at the q_z values below 0.908. In this method, ion motion is modulated via the application of a bipolar supplementary rf signal of a fixed amplitude and frequency across the end-cap electrodes. During the voltage scan of the main rf signal, the ions that are in resonance with the auxiliary signal gain additional translational energy, and as a consequence, they are ejected from the trap at the q_z values far below the normal ejection point. By using this approach, it has been possible to extend the upper mass limit close to 100 kDa [33].

Ion traps are usually limited to unit mass resolution. This limit clearly is a disadvantage compared to other types of mass spectrometers. However, substantial improvements over unit mass resolution can be realized by manipulation of the scanning speed, and of the frequency and amplitude of the resonance ejection signal [34,35]. Several groups have obtained impressive results with these approaches [34–37]. A resolution in excess of 30,000 (FWHM) at m/z 502 has been demonstrated by decreasing the scan speed and using the resonance ejection at the appropriate frequency and amplitude [34]. A resolution of 1.13×10^6 was achieved for the cluster ions of CsI at 3510 Da with a 2000-fold decrease in the scan speed [35]. The reduction of the scan rate to 0.1 m/z units per second has led to a further improvement in resolution to 1.2×10^7 for m/z 614 [36].

3.3.3. Coupling with ESI and MALDI Techniques

In order to make use of the high resolution, high-mass, and MS/MS capabilities of QITMS for the analysis of biological macromolecules, electrospray ionization (ESI) and matrix-assisted laser desorption/ionization (MALDI) techniques have been coupled with QIT [38–41]. This coupling is usually achieved by attaching ion guides to QIT instruments to allow entry of the externally formed ions into the trap.

3.4. TIME-OF-FLIGHT MASS SPECTROMETER

A time-of-flight (TOF) mass spectrometer is one of the simplest mass-analyzing devices. Since the 1990s, it has reestablished itself as a mainstream technique and is becoming increasingly useful in meeting the demands of contemporary research in biomedical sciences. The recent successes of TOFMS can be attributed to the development of MALDI, high-speed data processing devices, and focal plane detectors. A few reviews and books can be referred to as additional useful reading material on this subject [42–48].

3.4.1. Principle of Operation

A TOF mass spectrometer behaves as a velocity spectrometer, in which ions are separated on the basis of their velocity differences. A short pulse of ions, after exiting the source, is dispersed in time by allowing it to drift in an FFR of a long flight tube. The principle behind the mass analysis is that after acceleration to a constant kinetic energy (equal to zV, where z is the charge on the ion and V the accelerating potential), ions travel at velocities, v, that are an inverse function of the square root of their m/z values:

$$v = \left(\frac{2zV}{m}\right)^{1/2} \tag{3.15}$$

The lighter ions travel faster and reach the detector placed at the end of the flight tube (of length L) earlier than do the heavier ones. Thus, a short pulse of ions is dispersed into packets of isomass ions (Figure 3.9). Therefore, mass analysis of ions that enter the flight tube can be accomplished by determining their time of arrival given by

$$t = \frac{L}{v} = L\left(\frac{m}{2zV}\right)^{1/2} \tag{3.16}$$

In order to convert the time spectrum into a mass spectrum, the instrument is mass-calibrated by measuring the flight times of two different known mass ions.

A primary requirement in the operation of a TOFMS is that all ions enter the flight tube precisely at the same time. Generating ions in short bursts fulfills this condition. In this respect, TOF instruments are well matched to ^{252}Cf-plasma desorption (PD) and MALDI ion sources. The continuous ion beam sources (e.g., electron ionization and ESI), however, can be coupled with a TOF mass spectrometer, but only after conversion of the generated ions into discrete packets. Pulsing

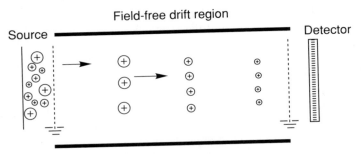

Figure 3.9. Principle of the mass separation by a time-of-flight mass analyzer. Ions are separated on the basis of their size; high-mass ions (big circles) travel more slowly than the lighter ions.

of the accelerating potential converts a continuous ion beam into discrete ion packets.

3.4.2. Mass Resolution

In the past, poor mass resolution was one of the major limitations of TOFMS. The mass resolution of simple linear TOF mass spectrometers is usually less than 500. In TOFMS, this term is given by Eq. 3.17, where Δt is commonly measured in terms of FWHM. The spatial, temporal, and velocity dispersions of ions are the limiting factors in achieving higher resolution in TOF instruments [42,44,48]. A higher resolution is obtained only when all ions are formed in the source at the same time and the same location (i.e., their temporal and spatial distributions are minimum), and all have the same kinetic energy. The temporal distribution, which is a combined effect of uncertainties in the time of ion formation as well as limitations of ion detection and time-recording devices, can be minimized with the use of very short ionization pulses and also by increasing the difference in arrival times of two adjacent ions. Increasing the flight path of ions or reducing their velocities (i.e., with lower accelerating potentials) has the effect of increasing the arrival times of different mass ions. The use of lower accelerating potentials is counterproductive because of the concomitant loss of transmission, and of the greater impact of the energy spread on resolution at lower accelerating potentials. The pulsed ion extraction with very fast rise time can also be used to correct the temporal distribution. The spatial distribution degrades resolution because ions formed in different regions of the ion source are accelerated to different kinetic energies. Ions that are formed to the left of the central line are accelerated to a higher velocity than are ions formed to the right of the central line, resulting in loss of mass resolution. The use of ionization methods that produce ions from a surface, such as ^{252}Cf-PD and MALDI, eliminates the spatial width to some extent because in these ionization techniques the plane of ion formation is well defined. Another way to correct for the spatial distribution is to use dual-stage ion extraction optics:

$$R = \frac{m}{\Delta m} = \frac{t}{\Delta t} \tag{3.17}$$

The dominating factor that restricts the resolution in TOF instruments, especially with desorption ionization methods, is the kinetic energy inhomogeneity within the ion beam. The higher-initial-energy ions arrive at the detector sooner than do the same-mass lower initial energy ions. The acceleration region energy spread can be eliminated by the prompt ejection of ions from the source (i.e., by using high accelerating potentials, usually $> 10 \, kV$). A further reduction in the energy spread is achieved through an energy-correcting device, known as the *reflectron* (described below). Another factor that limits resolution is the *turnaround time* taken by the ions that are traveling initially away from the exit slit. These ions take extra time in exiting the source than ions that have the same initial velocity, but are facing the exit slit. Longer flight tubes and longer flight times can reduce the effect of turnaround time.

In TOF instruments, the time difference in the arrival of various ions at the detector is very short. As an example, m/z 2500 after acceleration through a potential of 6000 V will reach the end of the one-meter-long flight path 5.01 µs after the arrival of m/z 2000. The difference in the arrival times of ions differing by one dalton (say, 2000 and 2001) is even shorter (in nanoseconds). Therefore, the mass selectivity of TOF instruments is also limited by the accuracy with which short intervals of time can be measured.

Time-of-flight mass spectrometers have a number of attractive features, such as theoretically unlimited mass range, high ion transmission, very high spectrum acquisition rate, multiplex detection capability, simplicity in instrument design and operation, reasonable mass resolution, and low cost. The detection sensitivity of TOF instruments is much higher than in scanning instruments because they can record all the ions that reach the detector after each ionization event, and because they have a high ion transmission efficiency. A major asset of TOF mass spectrometers is their ability to record a complete mass spectrum in time intervals as short as 25 µs. These attributes make TOFMS an attractive research instrument as well as a valuable analytical tool.

3.4.3. Reflectron

A reflectron is a new development that corrects for any initial position and energy dispersions in the accelerator region of a TOF instrument [49,50]. This elegant device is in fact an electrostatic mirror that consists of a series of electrical lenses, each with progressively increasing repelling potential (Figure 3.10). The initial spatial spread is translated into a velocity spread, which can be readily corrected by the mirror. A reflectron works on a principle that the ions that enter this device after traversing the first FFR (L_1) are slowed down until they come to rest, and then their direction of motion is reversed, and finally, they are reaccelerated into a second FFR (L_2). Qualitatively, the faster-moving ions (i.e., ions with excess energy $zV + U_0$) spend less time in the drift regions, but penetrate to a greater depth (d) into the reflecting field and consequently, spend more time there. This extra time in the mirror compensates for the shorter flight times of faster ions in the drift regions, with

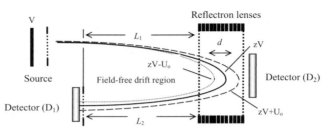

Figure 3.10. A sketch of a reflectron time-of-flight mass analyzer. All ions of the same mass, but that differ in kinetic energy, are made to arrive at the same time at a detector (D_1) that is located at the end of second field-free region (L_2).

the result that all isomass ions arrive simultaneously at the detector that is located at the end of the flight path L_2. Mass resolution is, thus, improved. An additional contributing factor in improving the mass resolution is the longer pathlength of the reflectron (i.e., $L_1 + L_2$). A mass resolution of $>20,000$ (FWHM) has been achieved with a grid-free reflectron [51]. A considerable improvement in performance is realized over linear TOFMS instruments for <10-kDa molecular mass compounds. The total flight time of an ion in the reflectron is given by

$$t = \left(\frac{m}{2zV}\right)^{1/2}(L_1 + L_2 + 4d) \tag{3.18}$$

The single-stage and dual-stage reflectrons are both in common use. A *single-stage reflectron* is a simple ion mirror in which a single retarding–reflecting field is used. It consists of an entrance grid electrode and a series of ring electrodes as shown in Figure 3.10. The single-stage reflectron provides a first-order correction for the kinetic energy distribution. The *dual-stage reflection* contains two linear retarding voltage regions that are separated by an additional grid. The first stage is smaller than the second stage, but with considerably higher field strength capable of reducing ion kinetic energies to about one-third of their initial values. The dual-stage reflectrons provide higher mass resolution than do the single-stage devices. The gain in resolution, however, is at the expense of sensitivity because transmission losses occur when ions pass through the additional grid lenses. The reflecting TOF instruments are usually outfitted with an additional detector behind the reflectron. A conventional linear TOF mass spectrum is recorded with this detector when the reflectron voltage is turned off.

Usually, the detection region of a reflectron is separated from the ionization region by reflecting the ions at a small angle (Figure 3.10) with respect to the incoming ions. In an alternative design, called *axial symmetry reflectron*, the reflected ions travel back along the flight axis of the incoming ions, but with a slight angular divergence [46]. The multichannel plate (MCP) detector used in this design is of annular shape, and is located just outside the ion source.

3.4.4. Orthogonal Acceleration TOF Mass Spectrometer

In order to realize the potential of TOFMS for a wide range of applications, the orthogonal ion extraction approach is used to combine the continuous-mode ion formation techniques with these instruments [52–56]. The ion source of a TOF instrument is modified to contain an additional field-free ion-sampling region. Ions that are produced between each duty cycle of mass analysis are stored in this region (see Figure 3.11). A short pulse of an orthogonal accelerating field is applied to efficiently eject those ions out of the source. A high efficiency (duty cycle) in gating ions from an external source is achieved with this device. The orthogonal gating of ions from a continuous ion beam has an additional advantage in that it reduces the spatial and energy spreads of the source-formed ions. Also, by very quickly switching on the pushout pulse, a minimal temporal dispersion is achieved. Mass resolution of 4000 (FWHM) has been reported for orthogonal acceleration (oa)-TOFMS [53].

A QIT has also been used as an interface between a continuous ion source and a TOF mass spectrometer [57]. In this combination, the trap serves as a front-end storage device, and converts a continuous ion beam into pulsed ion packets for mass analysis by TOFMS. Also, it provides an opportunity to reduce the kinetic energy spread of the ions via collisional cooling with the background trapping gas. A similar system that uses an rf-only quadrupole in place of QIT has also emerged [55,58]. ESI [59,60], MALDI [61], and LC [62] have been successfully coupled to TOF via these interfaces. Another use of orthogonal TOF mass spectrometers is in hybrid tandem instruments. Several combinations have emerged where an orthogonal TOF mass spectrometer acts as an MS2 device (see Chapter 4) [63–65].

3.4.5. Time-Delay Ion Extraction for Structure Elucidation

Knowledge of the structure-specific fragment ions is essential for elucidation of structures of analytes. That information is missing in a conventional linear TOF mass

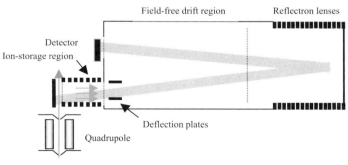

Figure 3.11. Principle of orthogonal time-of-flight mass analyzer. The ions from a continuous ionization source, such as ESI, are stored in the ion-storage region, and pushed into the field-free region in the pulse mode.

spectrum because the molecular ions are usually promptly extracted from the ion source before they have a chance to fragment. A modification in the procedure of ion extraction, in which ions are first stored for some time in a FFR region within the ion source and then ejected out of it by applying a delayed ion extraction pulse, affords detection of the source-formed fragment ions [66,67]. Delayed extraction (DE) has also been effective in improving the resolution and mass measurement accuracy of TOFMS analysis [68–71].

Applications. Because of the potential of unlimited mass range, TOFMS has become a workhorse for the analysis of macromolecules such as proteins, carbohydrates, and oligonucleotides. In combination with MALDI, TOFMS is capable of determining the molecular mass of proteins of > 300 kDa mass with an accuracy of $> 0.1\%$, and with a detection sensitivity in the attomole–femtomole range. TOFMS is routinely used for peptide mapping, verifying the homogeneity of synthetic peptides, assessing glycosyl, phosphoryl, and other posttranslational heterogeneity, and the determination of cleavage sites in protein processing. The instrument can also be used for amino acid sequence analysis of peptides via the postsource decay [72,73] and delayed extraction [66,67] processes. Its potential role in mapping the human genome is also anticipated.

3.5. FOURIER TRANSFORM ION CYCLOTRON RESONANCE MASS SPECTROMETER

Because of its unique high-mass, high-resolution, and multiplex detection capabilities, Fourier transform mass spectrometry (FTMS) is gaining wide acceptance as an analytical tool for the analysis of biomolecules. Several reviews and books have been published on this novel technique [74–82]. The advantages of FTMS include ultra-high-mass resolution, high-mass accuracy, multistage tandem mass spectrometry, and ability to trap ions for extended periods of time. In addition, FTMS is a useful tool for conducting ion–molecule reactions and for structure elucidation studies.

3.5.1. Principle of Operation

The technique is based on the principle of ion cyclotron resonance (ICR), in which ionization, mass analysis, and detection take place in the same region, generally in a 1-in. cubic cell that is placed in a strong magnetic field. Ions are constrained spatially by a combination of electric and magnetic fields. The basis of mass analysis in ICR is that an ion when placed in a magnetic field will precess at a frequency (ω_c) that is characteristic of its m/z value. The ions move in circular orbits, as described by Eq. 3.2. The time to complete a single revolution is given by Eqs. 3.19 and 3.20 (r, v, z, m, and B have been defined earlier), from which the cyclotron frequency, which is the number of revolutions per second (in cycles per sec or Hertz), is calculated according to Eq. 3.21.

$$t = \frac{2\pi r}{v} \tag{3.19}$$

$$t = 2\pi \frac{m}{zB} \tag{3.20}$$

$$\omega_c = \frac{zB}{2\pi m} \tag{3.21}$$

Three steps are performed to detect ions in the ICR cell: (1) ions are formed in the pulsed mode, and trapped in the cell by applying a few volts of electric potential to the front and rear plates of the cell (Figure 3.12a); (2) an excitation pulse is applied whereby ions whose precessional frequency matches with the excitation pulse absorb energy from the external pulse (Figure 3.12b); and (3) finally, the image current that is induced when ions are in close proximity of the receiving plates is detected (Figure 3.12c). In the FTMS version of the ICRMS, a broadband excitation pulse (also known as a "chirp") is applied to the transmitter plates of the cell. Ions of many different m/z values, whose cyclotron frequency falls within the applied frequency range, simultaneously absorb energy from that pulsed signal, and are forced to move in phase-coherent packets of larger orbits. When these coherently moving ions are in the proximity of the receiving plates, they transmit a complex rf signal (i.e., the image current) that contains frequency components characteristic of each ion. This induced signal, after passing through impedance and amplifying circuit, is converted to a time-domain free-ion decay signal. By applying a Fourier transform, this time-domain signal is converted to a frequency-domain signal to provide the simultaneous detection of all ions across a wide mass range. This multiplex (Fellgett) advantage is the driving force to combine Fourier transform with ICRMS, and is responsible for the speed and improved signal-to-noise ratio (S/N) of the analysis.

The full potential of FT-ICRMS can be realized only when the pressure inside the cell is very low ($< 10^{-7}$ torr). An elegant way to reduce pressure during analysis is to produce ions external to the ICR cell and then guide them into the cell [83]. The two regions are differentially pumped. Biological compounds can be ionized either within the cell by bombardment with a laser beam or outside the magnetic field (e.g., via ESI) and then transported into the cell.

The current trend in FT-ICRMS is to use high-field magnets. Marshall and Guan have discussed the advantages of such instruments [84]. Aside from the well-known advantage of increased resolving power, the data acquisition rate, upper mass limit, maximum kinetic energy of ions, maximum number of trapped ions, maximum ion-trapping duration, quadrupole axialization efficiency, and two-dimensional FT-ICRMS resolving power also increase with the magnetic field. As a consequence of these attributes, improvements in S/N ratio, dynamic range, mass accuracy, instrumental efficiency, and mass selectivity for tandem mass spectrometry operation are also realized.

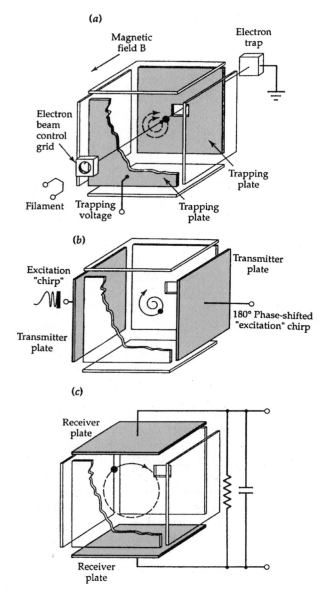

Figure 3.12. Operation of FT-ICRMS. The mass analysis involves three steps: (*a*) ion formation and storage; (*b*) excitation of the trapped ions by an external, broad frequency range pulse; and (*c*) the detection of the ions by measuring their image current. [Reproduced from Ref. 74 by permission of American Chemical Society, Washington, DC (copyright 1983).]

3.5.2. High-Mass Analysis Capabilities

FT-ICRMS is particularly very promising for high-mass analysis [75,81]. In the FT-ICR cell, ions of lower than a certain critical mass (m_c) have the possibility of being trapped and analyzed. For a cubic cell, this value is given by

$$m_c = \frac{zB^2a^2}{8V_{eff}\alpha} \tag{3.22}$$

where a is the length of the cell, V_{eff} is the trapping voltage that has been corrected for space charge effects, and a is the geometry constant. Although in theory an upper mass limit of over several hundred thousand daltons may be possible with current high-field magnets, the practical limit may be much lower than this value.

3.5.3. High-Mass Resolution Capability

FT-ICRMS is capable of providing ultrahigh resolving power because the basis for mass measurement lies in measuring frequency, which can be measured with great precision. From Eqs. 3.23 and 3.24, it is evident that R is limited by the time of observation (T) and the pressure in the ion trap (P). At higher pressures the collisions of ions with the background molecules can cause dephasing and scattering of ion pockets, and lead to a short transient signal. A resolution of 100,100,100 at m/z 18 has been demonstrated by FT-ICRMS. Although resolution degrades rapidly as the mass of the analyte increases, values well above that required to resolve carbon isotopes is routinely obtained with FT-ICRMS.

Using an external accumulation of ions from an ESI source, Marshall and co-workers have achieved 8,000,000 resolving power at 9.4 T, high enough to observe the isotopic fine structure of proteins as large as 15.8 kDa [85].

$$R = \frac{m}{\Delta m} \leq \frac{kzBT}{m} \tag{3.23}$$

$$R = \frac{m}{\Delta m} \leq \frac{kzB}{(mP)(\xi/n)} \tag{3.24}$$

3.6. ION MOBILITY MASS SPECTROMETRY

Ion mobility spectrometry (IMS) instrument uses mobility rather than m/z ratio as a criterion to separate ions [86–90]. In this technique, ions are allowed to drift under the influence of an electric field in a drift tube that contains a buffer gas, and ions are separated according to their size : charge ratio. The drift tube usually consists of a series of uniformly spaced electrodes that provide uniform electric field. The mobility of ions is a combined effect of ion acceleration by the electric field and retardation by collisions with the buffer gas. Several variations of IMS have emerged. A typical instrument is shown in Figure 3.13. It consists of an ESI source, a desolvation region, a drift tube, and a quadrupole mass analyzer. The

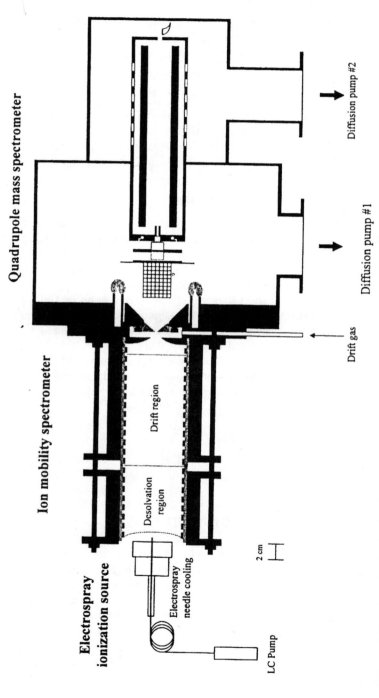

Figure 3.13. Schematic of an ion mobility mass spectrometer outfitted with an electrospray ionization source. [Reproduced from Ref. 86 by permission of American Chemical Society, Washington, DC (copyright 1998).]

desolvation region and the drift tube are both constructed of a set of electrically conducting stainless-steel ring electrodes, each electrode is separated by an insulating alumina ring. The conducting rings are connected via a series of high resistors. The instrument also contains two ion gates, one at the entrance and another at the exit of the drift region. In order to improve the duty cycle, and thus the sensitivity, a quadrupole ion-trap interface is connected between a continuous ion source and the drift tube [91]. The trap allows accumulation of ions between each pulse of the ion mobility experiment. A quadrupole has also been interfaced between a continuous ion source and the drift tube [88]. This arrangement allows mass selection of a particular ion. The resolution of an IMS is usually very low (10–12). The resolution can be increased to a 200–400 range by increasing the pressure of the buffer gas, connecting the ion source directly with the drift tube, increasing the length of the drift tube, and increasing the electric field gradient of the drift tube. An ion chromatogram can be obtained within less than a second.

The drift times obtained from IMS measurements can be used to calculate average collision cross section by using Eq. 3.25, which in turn can be used to estimate the molecular size (Eq. 3.26):

$$\Omega = \frac{\sqrt{18\pi}}{16} \cdot \frac{ze}{\sqrt{k_B T}} \left(\frac{1}{m} + \frac{1}{m_b} \right) \cdot \frac{t_D E}{L} \cdot \frac{760}{P} \cdot \frac{T}{273.2} \cdot \frac{1}{N} \qquad (3.25)$$

$$\Omega = \pi (r_{ion} + r_{neutral})^2 \qquad (3.26)$$

where Ω is the average collision cross section; ze is the charge on the ion; k_B is Boltzmann's constant; T is the temperature of the buffer gas; m and m_b are the masses of the ion and buffer gas, respectively; L is the length of the drift tube; t_D is the drift time of the ion; E is the electric field; P is the buffer gas pressure; N is its number density; and r_{ion} and $r_{neutral}$ are radius of the ion and its neutral, respectively.

Several applications of this technique have been reported and are enumerated in a tutorial on this subject [87]; these include detection of drugs, chemical warfare agents and environmental pollutants, size distribution of aerosol particles, structure information of gas-phase clusters, and conformational studies of proteins and oligonucleotides.

A variation of IMS, known as *field ion spectrometry* (FIS), has also emerged [92]. This new technique functions as an ion filter, which allows one type of ions to be transmitted continuously. In FIS, the electric field is applied as a high-frequency asymmetric waveform rather than a dc voltage. Ions travel in the axial direction of the drift tube in a flowing stream of a buffer gas, and the electric field is applied perpendicular to the direction of the gas flow.

3.7. A COMPARISON OF IMPORTANT CHARACTERISTICS OF MASS ANALYZERS

Obviously, TOFMS and FT-ICRMS are the instruments of choice for the analysis of high-mass biological compounds. These compounds can also be analyzed by other

Table 3.1. A comparison of different types of mass analyzer

Characteristic	Magnetic Sector	Quadrupole	QIT	TOF	FT-ICR
Mass range (Da)	15,000	4,000	100,000	Unlimited	$>10^6$
Resolution	200,000	Unit	30,000	15,000	$>10^6$
Dynamic range	++++	+++	++	+++	++
MS/MS	++++	+++	+++	+++	++
LC (or CE)MS	+	++++	+++	++	++
Cost	$$$$	$	$	$$	$$$$

mass spectrometers if coupled with ES ionization technique. Exact mass measurement is the strength of magnetic sectors and FT-ICR instruments. Reflectron TOF can also provide adequate mass measurement accuracy for small molecules. LCMS and CEMS applications currently dictate the use of quadrupoles and QIT instruments, although strides have been made to couple these separation devices with TOF and FT-ICR instruments. Magnetic instruments have limitations in this respect; but these instruments clearly stands out in the MS/MS applications, although quadrupoles, QIT, and TOF instruments are not far behind in their performance. If the cost of a mass spectrometer is the concern, then quadrupoles and QIT instruments have a clear edge over other types of instruments. Table 3.1 compares several important features of the various types of mass spectrometers.

3.8. ION DETECTION

Ions after their separation by a mass analyzer arrive at a detector for the detection of their mass and abundance. A detector measures the electric current in proportion with the number of ions striking it. Sensitivity, accuracy, resolution, and response time are the most important characteristics of any ion detector. Two basic designs of detectors are in common use [93,94]. The focal point detectors use the principle of one-at-a-time detection of all ions, whereas focal plane detectors (FPDs) detect the arrival of all ions simultaneously along a plane. The focal plane detectors provide an increase in sensitivity and mass accuracy. When the ion flux is low, either it is measured in the form of ion counts or the signal is integrated, as is done in focal plane detectors. High ion fluxes are conveniently measured in the form of direct current.

3.8.1. Faraday Cup Detector

The Faraday cup detector is not a common type of detector, and is used only in isotope ratio mass spectrometers. It is a very simple and robust device that consists of a cone-shaped metal cup that is connected to a high-impedance amplifier. Another design uses an inclined collector electrode that is enclosed in a metal cage. The charge transferred to the cup (or the collector electrode) by the incoming ion beam develops a voltage drop across a large feedback resistance ($10^{11} \, \Omega$). The high-impedance amplifier amplifies the resulting voltage drop. The response of a Faraday cup detector is slow to render it unsuitable for scanning mass spectrometers.

3.8.2. Secondary Electron Multipliers

Several mass spectrometry detectors make use of the phenomenon of secondary electron emission. The basic concept is that when a fast-moving ion strikes a specially coated metal surface, it leads to the emission of secondary electrons. The most common design is the electron multiplier (EM). Discrete- and continuous dynode versions are both in common use. The former consists of a series of dynodes (generally 16 or 20), usually made of copper/beryllium, that are connected together via a chain of resistors of equal value (Figure 3.14). A high voltage (up to -3000 V), applied across the first dynode (called the *conversion dynode*) and the final dynode (anode), is divided into equal steps between the dynodes. The impact of a beam of ions on the conversion dynode results in the emission of secondary electrons in direct proportion to the number of incident ions. The emitted electrons are accelerated toward, and made to strike, the second dynode. More secondary electrons are emitted from this dynode. This process is repeated at the next dynode to cause an amplification of secondary electrons at each successive stage. The current that results is amplified to provide a gain in excess of 10^7. Fast response time, high sensitivity, and high gain are the characteristics of these detectors.

Another popular version of an electron multiplier is the *channel electron multiplier* (CEM). This detector employs a continuous arrangement of electrodes. It is constructed from a special type of glass that has been either heavily doped with lead, or its inner surface has been coated with beryllium. The glass is drawn as a horn-shaped tube. A high voltage (1.8–2 kV) applied across the two ends of the tube creates a uniform field throughout the length of the tube. The beam of ions that emerges from the mass analyzer is made to strike the surface near the entrance of the detector, resulting in the ejection of several electrons, which are attracted along the surface farther on to the other end of the detector. With each impact on the inner surface of the tube, amplification of the current takes place as a result of the emission of increased number of electrons. A CEM is more compact and less expensive than a discrete EM, and provides gains as high as 10^8.

Figure 3.14. Principle of ion detection by an electron multiplier. [Reprinted from Ref. 1 by permission of Plenum Press, New York (copyright 1994).]

A *microchannel plates* (MCPs) assembly is a multichannel version of CEM [95]. It consists of millions of individual channels, each made of very small diameter ($\approx 10\,\mu m$) fiberoptic cables of metal-doped glass. The gain of each individual channel is much higher than the conventional CEM because of the significant increase in the length : diameter ratio of the tube. Usually, two MCPs are arranged in a chevron configuration. The ion beam strikes the front of the MCPs, and secondary electrons, after severalfold multiplication, emerge from the rear face of the MCPs. This device provides a two-dimensional image of the impinging signal. MCPs are used in TOF instruments [96] and as the front end of multichannel array detectors (see Section 3.8.5).

Some new generations of mass spectrometers use *photomultipliers* in place of an EM. The use of these detectors requires that the incoming ion beam is first converted to a photon beam (ion-to-photon conversion). This conversion is accomplished when ions first strike a scintillation material. The emitted photons are amplified and detected by a conventional photomultiplier.

3.8.3. Postacceleration Detector

The principle of secondary electron emission eventually breaks down as the mass of the ion increases. The yield of secondary electrons falls off exponentially as the velocity of the striking ions decreases [97]. Because the velocity of an ion is an inverse function of its mass (Eq. 3.1), the detection efficiency of high-mass ions decreases commensurately. This problem is very critical when high-mass biomolecules are detected with MALDI and ESI techniques.

One way to augment the velocity of the incident ions is to use a postacceleration device [98]. For positive-ion detection, the incoming beam is first deflected toward the cathode of an off-axis postacceleration detector (PAD) held at a potential of -5 to $-30\,kV$. The secondary electrons emitted from this detector are amplified, and detected in the usual manner by an EM.

Because negative ions are slowed down as they approach the first dynode (held at a large negative potential), postacceleration of negative ions becomes an essential step for their detection by an EM. For the detection of negative ions, the PAD target is held at a positive potential to accelerate them to kinetic energies sufficient to emit secondary positive ions after impinging the target. An EM then detects these emitted positive ions in the usual way.

A PAD-based detector used in Micromass (Beverly, MA) instruments contains two conversion dynodes, one for positive ions and one for negative ions, a phosphor, and a photomultiplier (Figure 3.15). For positive-ion detection, the conversion dynode is maintained at -10 to $-20\,kV$, and the emitted secondary electrons are accelerated toward the phosphor-coated electrode, the voltage of which is varied between 10 and 20 kV. When operated for the detection of negative ions, the incoming beam is first deflected toward a cylinder-shaped conversion dynode held at half the phosphor voltage. The positive charge on the phosphor attracts the emitted electrons toward it.

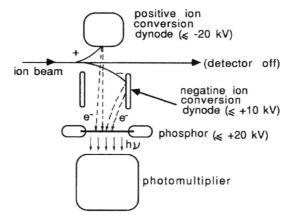

Figure 3.15. A combined positive- and negative-ion postacceleration detector. [Reprinted from Ref. 1 by permission of Plenum Press, New York (copyright 1994).]

Another combination of PAD and a photomultiplier is shown in Figure 3.16. This *ion-to-photon detector* is specially designed for the detection of high-mass ions [99]. It consists of several rings placed in front of a photomultiplier. The scintillator is mounted on ring 5. Rings 3, 4, and 5 are held at the same electrical potential, whereas ring 1 is grounded and isolated from the high voltage by a plastic ring (ring 2). A venetian blind copper/beryllium conversion dynode is attached to ring 1 to convert the incoming primary ion beam into electrons and secondary ions, which are accelerated to the scintillator surface by the post acceleration device that consists of rings 1–5. The detector has been successfully used to detect ions as large as 70 kDa mass.

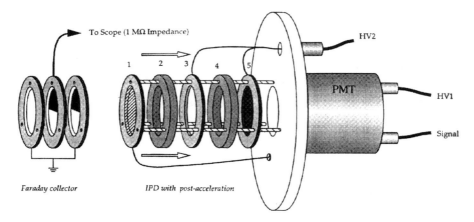

Figure 3.16. Schematic of the ion–to–photon detector (IPD). The Faraday collector is placed in front of the IPD to normalize the incoming ion current. [Reprinted from Ref. 99 by permission of John Wiley & Sons (copyright 1999).]

A copper/beryllium dynode has been effectively used as a postacceleration device in combination with a triple MCP detector for the detection of large DNA-mers by TOFMS [100]. In this assembly, one grounded grid is placed in front of and one behind the copper/beryllium dynode, which is floated to $-15\,$kV for the positive-ion detection and to $+15\,$kV for the negative-ion detection. The ions that impinge on the dynode sputter Cu^+ ions from its surface. The MCP detector then detects the sputtered species.

3.8.4. Low-Temperature Calorimetric Detectors for High-Mass Ions

This type of detector is still in experimental stage, but has shown promise for sensitive detection of high-mass ions [101]. The response of calorimetry detectors is independent of the mass of the ions and is related to their kinetic energy. It is based on the principle that the kinetic energy E of the impinging ions is transformed into heat, which results in the rise of temperature given by

$$\Delta = \frac{E}{C_T} \tag{3.27}$$

where C_T is the heat capacity, which at low cryogenic temperatures has a $1/T^3$ dependence for dielectrics and superconductors. A typical design uses a super-conductor–insulator–superconductor (SIS) junction [101]. It consists of two super-conducting films that are separated by a thin insulator, which is made of the oxide of the base metal (e.g., $Nb-Al_2O_3-Nb$ and $Sn-SnO_x-Sn$). The detector is operated at cryogenic temperatures. When a pulse of energetic ions impinges on a super-conducting film, nonthermal phonons are created in the film, which in turn breaks apart the weakly bound electrons (also called *Cooper pairs*). These electrons are excited to Fermi levels, where they quantum-mechanically tunnel through the insulator to produce a large tunnel current. A kinetic energy of only a few millielectron volts is required to break a Cooper pair, which means that an incident ion of 25 keV energy will produce a large number (in millions) of excited electrons. The usefulness of this detector has been demonstrated by TOFMS analysis of lysozyme (mass 14,300 Da) [102], human serum albumin (mass 66,000 Da) [103,104], and immunoglobulin (mass 150,000 Da) [104]. The MALDI-TOF mass spectrum of a mixture of human serum albumin and immunoglobulin is shown in Figure 3.17.

3.8.5. Focal Plane Detectors

A serious problem with focal point detectors is their low detection efficiency. In these detectors, at any one time only one ion is detected and the ion current due to all other ions is lost. On the other hand, the concurrent detection of all mass-resolved ions (*multiplex detection*) improves the detection efficiency, as a result of which several-order-of-sensitivity enhancement is realized. The simultaneous detection of all spatially dispersed ions is achieved by use of FPDs [105–107]. In the past, a

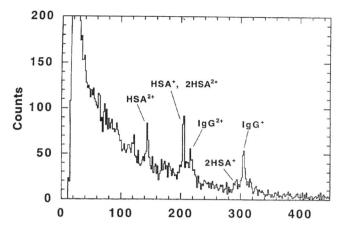

Figure 3.17. The MALDI-TOF mass spectrum of a mixture of human serum albumin and immunoglobulin. The spectrum was obtained with low-temperature calorimetric detector, and is the sum of 500 laser shots. [Reproduced from Ref. 104 by permission of Elsevier Scientific Publishing Corporation (copyright 1997).]

photoplate detector placed at the focal plane of Mattauch–Herzog double-focusing mass spectrometers served this purpose [108]. The ions that strike a photoplate produce a darkening in proportion to the number of incident ions. A microdensitometer measures the relative position and optical density of the darkened lines to provide mass and intensity of the ions, respectively. The idea of a photoplate has been discarded because the whole arrangement and operation is very tedious.

Development of the *multichannel array detector* has made multiplex detection in mass spectrometry a much simpler task. A typical multichannel array–based FPD consists of an MCP assembly and a one-dimensional array of detector electrodes. Each electrode acts as an independent detector, and contains its own charge sensor, an 8-bit counter, control logic, and bus interface. The ion beam that emerges from a mass analyzer first strikes the entrance face of the MCP assembly to emit secondary electrons. The MCP output falls on the electrodes, and is independently detected. Cottrell and Evans have developed an integrating electrooptical multichannel array detector that consists of 1024 pixels for use with magnetic sector tandem mass spectrometers [109]. It employs a CEM made of a pair of MCPs arranged in a chevron configuration, a phosphor screen, a fiberoptic coupling, and a 1024-channel photodiode array (Figure 3.18). The phosphor screen transforms secondary electron output from MCPs into photons, which travel through the fiberoptic cables into a charge- or plasma-coupled device (CCD or PCD), where they are converted to electric charge. This 25-mm array detector simultaneously covers 4% of the mass range. The FPD developed by Hill et al. uses a variable-dispersion 2048-pixel photodiode array [110].

The principle of position- and time-resolved ion counting (PATRIC) has been used in Finnigan MAT instruments. The ion beam is converted in the usual manner into electrons by using a pair of MCPs, but a multianode assembly, instead of a

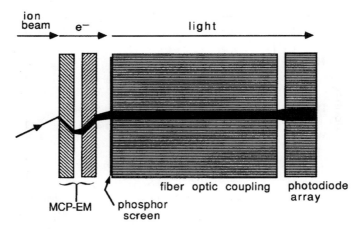

Figure 3.18. Principle of operation of multichannel array detector. [Reprinted from Ref. 1 by permission of Plenum Press, New York (copyright 1994).]

photodiode array, detects the produced electron cloud. The anode assembly consists of a large number of conductive strips interconnected by identical capacitors. Charge-sensitive amplifiers located at each end of the anode assembly provide position and time information for every ion. The detector covers a mass range of 8%. Because ions are detected individually, the array can also be used to obtain a mass spectrum by scanning over the complete mass range. This detector has been used along with ESIMS for sensitive and specific detection of proteins [111]. Interfering background ions can be attenuated relative to multiply charged sample ions by reducing the voltage applied to the MCP assembly.

A scanning-array detection system has also been used in mass spectrometry [112,113]. This system consists of an 1152-channel MCP assembly and a CCD device. In this mode, the ion beam that emerges from a mass spectrometer sweeps across the MCP by rapid scanning of the magnetic field. A particular channel integrates the ion current due to each particular m/z value, and transfers to the next channel, and ultimately the last channel provides an output. This arrangement makes feasible high-speed recording of CID spectra from chromatography columns.

REFERENCES

1. C. Dass, in D. M. Desiderio, ed., *Mass Spectrometry: Clinical and Biomedical Applications*, Plenum Press, New York, 1994, Vol. 2, pp. 1–52.
2. J. Roboz, *Introduction to Mass Spectrometry*, Interscience, New York, 1968.
3. C. A. McDowell, *Mass Spectrometry*, McGraw-Hill, New York, 1963.
4. P. H. Dawson, *Quadrupole Mass Spectrometry and Its Applications*, Elsevier, New York, 1976.

5. P. H. Dawson, *Quadrupole Mass Spectrometry and Its Applications*, American Institute of Physics, AIP-AVS Classics Series, AIP Press, Woodbury, NY, 1994.

6. P. H. Dawson, *Mass Spectrom. Rev.* **5**, 1–37 (1986).

7. J. E. Campana, *Int. J. Mass Spectrom. Ion Proc.* **33**, 101–117 (1980).

8. P. E. Miller and M. Bonner Denton, *J. Chem. Educ.* **63**, 617–622 (1986).

9. J. D. Williams, K. A. Cox, R. G. Cooks, R. E. Kaiser, and J. C. Schwartz, *Rapid Commun. Mass Spectrom.* **5**, 327–329 (1991).

10. S. Hiroki, K. Sakata, N. Sugiyama, S. Muramoto, T. Abe, and Y. Murakami, *Vacuum* **46**, 681–683 (1995).

11. Z. Duo, T. N. Olney, and D. J. Douglas, *J. Am. Soc. Mass Spectrom.* **8**, 1230–1236 (1997).

12. Z. Duo and D. J. Douglas, *Rapid Commun. Mass Spectrom.* **10**, 649–652 (1996).

13. M. H. Amad and R. S. Houk, *J. Am. Soc. Mass Spectrom.* **11**, 407–415 (2000).

14. W. Paul, and H. Steinwedel, *Z. Natureforsch., Teil A* **8**, 448 (1953).

15. G. C. Stafford, P. E. Kelly, J. E. P. Syka, W. E. Reynolds, and J. F. J. Todd, *Int. J. Mass Spectrom. Ion Proc.* **60**, 85–98 (1984).

16. R. E. March and R. J. Hughes, *Quadrupole Storage Mass Spectrometry*, Wiley-Interscience, New York, 1989.

17. P. K. Ghosh, *Ion Traps*, Clarendon Press, Oxford, 1995.

18. R. G. Cooks, G. L. Glish, S. A. McLuckey, and R. E. Kaiser, *Chem. Eng. News* 26–41 (1991).

19. R. G. Cooks and R. E. Kaiser, *Acc. Chem. Res.* **23**, 213–218 (1992).

20. J. F. J. Todd, *Mass Spectrom. Rev.* **10**, 3–52 (1991).

21. R. C. Dorey, in M. L. Gross, ed., *Mass Spectrometry in the Biological Sciences: A Tutorial*, Kluwer, Dordrecht, The Netherlands, 1992, pp. 79–92.

22. R. G. Cooks and K. A. Cox, in T. Matsuo, R. M. Caprioli, M. L. Gross, and Y. Seyama, eds., *Biological Mass Spectrometry*, Wiley, New York, 1994, pp. 179–197.

23. S. A. McLuckey, G. J. Van Berkel, D. E. Goeringer, and G. L. Glish, *Anal. Chem.* **66**, 689A–696A (1994).

24. R. E. March and J. F. J. Todd, *Practical Aspects of Ion Trap Mass Spectrometry: Chemical, Environmental, and Biomedical Applications*, CRC Press, Boca Raton, FL, 1996, p. 544.

25. R. E. March, *J. Mass Spectrom.* **32**, 351–369 (1997).

26. R. E. March, *Int. J. Mass Spectrom. Ion Proc.* **118/119**, 71–135 (1992).

27. J. C. Schwartz and I. Jardine, in B. L. Karger and W. S. Hancock, eds., *Methods in Enzymology*, Academic Press, San Diego, 1996, pp. 552–586.

28. S. A. McLuckey, G. J. Van Berkel, D. E. Goeringer, and G. L. Glish, *Anal. Chem.* **66**, 737A–743A (1994).

29. M. H. Soni and R. G. Cooks, *Anal. Chem.* **66**, 2488–2496 (1994).

30. M. H. Soni, P. S. H. Wong, and R. G. Cooks, *Anal. Chem. Acta.* **303**, 149–162 (1995).

31. R. K. Julian, M. Nappi, C. Weil, and R. G. Cooks, *J. Am. Soc. Mass Spectrom.* **6**, 57–70 (1995).

32. R. G. Cooks, C. D. Cleven, L. A. Horn, M. Nappi, C. Weil, M. H. Soni, and R. K. Julian, *Int. J. Mass Spectrom. Ion Proc.* **146/147**, 147–163 (1995).

33. R. E. Kaiser, Jr., R. G. Cooks, G. C. Stafford, J. E. P. Syka, and P. H. Hemberger, *Int. J. Mass Spectrom. Ion Proc.* **106**, 79–115 (1991).

34. J. C. Schwartz, J. E. P. Syka, and I. Jardine, *J. Am. Soc. Mass Spectrom.* **2**, 198–204 (1991).

35. J. D. Williams, K. A. Cox, R. G. Cooks, and J. C. Schwartz, *Rapid Commun. Mass Spectrom.* **5**, 327–329 (1991).

36. F. A. Londry, G. J. Wells, and R. E. March, *Rapid Commun. Mass Spectrom.* **7**, 43–45 (1993).

37. D. E. Goeringer, W. B. Whitten, J. M. Ramsey, and S. A. McLuckey, *Anal. Chem.* **64**, 1434–1439 (1992).

38. G. J. Van Berkel, S. A. McLuckey, and G. L. Glish, *Anal. Chem.* **62**, 1284–1295 (1990).

39. J. C. Schwartz and I. Jardine, *Rapid Commun. Mass Spectrom.* **6**, 313–317 (1992).

40. K. Jonscher, G. Currie, A. L. McCormack, and J. R. Yates III, *Rapid Commun. Mass Spectrom.* **7**, 20–32 (1993).

41. J. C. Schwartz and M. E. Bier, *Rapid Commun. Mass Spectrom.* **7**, 27–32 (1993).

42. R. J. Cotter, *Anal. Chem.* **64**, 1027A–1039A (1992).

43. H. Wollnik, *Mass Spectrom. Rev.* **12**, 89–114 (1993).

44. M. Guilhaus, *J. Mass Spectrom.* **30**, 1519–1532 (1995).

45. M. Guilhaus, V. Mlynski, and D. Selby, *Rapid Commun. Mass Spectrom.* **11**, 951–962 (1997).

46. R. J. Cotter, *Time-of-Flight Mass Spectrometry: Instrumentation and Applications in Biological Research*, ACS, Washington, DC, 1997.

47. C. Weickhardt, F. Moritz, and J. Grotemeyer, *Mass Spectrom. Rev.* **15**, 139–162 (1996).

48. R. J. Cotter, *Anal. Chem.* **71**, 445A–451A (1999).

49. B. A. Mamyrin, V. I. Karataev, D. V. Shmikk, and V. A. Zagulin, *Sov. Phys. JETP* **37**, 45–48 (1973).

50. B. A. Mamyrin, *Int. J. Mass Spectrom. Ion Proc.* **131**, 1–19 (1994).

51. R. Grix, R. Kutscher, G. Li, U. Gruner, and H. Wollnik, *Rapid Commun. Mass Spectrom.* **2**, 83–85 (1988).

52. J. H. J. Dawson and M. Guilhaus, *Rapid Commun. Mass Spectrom.* **3**, 155–159 (1989).

53. A. N. Verentchikov, W. Ens, and K. G. Standing, *Anal. Chem.* **66**, 126–133 (1994).

54. A. N. Krutchinsky, A. V. Loboda, V. L. Spicer, R. Dworschak, W. Ens, and K. G. Standing, *Rapid Commun. Mass Spectrom.* **12**, 508–518 (1998).

55. I. V. Chernushevich, W. Ens, and K. G. Standing, *Anal. Chem.* **71**, 452A–461A (1999).

56. M. Guilhaus, D. Selby, and V. Mlynski, *Mass Spectrom. Rev.* **19**, 65–107 (2000).

57. B. M. Chien, S. M. Michael, and D. M. Lubman, *Rapid Commun. Mass Spectrom.* **7**, 837–843 (1993).

58. A. N. Krutchinsky, I. V. Chernushevich, V. L. Spicer, W. Ens, and K. G. Standing, *J. Am. Soc. Mass Spectrom.* **9**, 569–579 (1998).

59. B. M. Chien and D. M. Lubman, *Anal. Chem.* **66**, 1630–1636 (1994).

60. R. W. Purves and L. Li, *J. Am. Soc. Mass Spectrom.* **8**, 1085–1093 (1997).

61. P. Kofel, M. Stockli, J. Krause, and U. P. Schlunegger, *Rapid Commun. Mass Spectrom.* **10**, 658–662 (1996).

62. J.-T. Wu, L. He, M. X. Li, S. Parus, and D. M. Lubman, *J. Am. Soc. Mass Spectrom.* **8**, 1237–1246 (1997).

63. H. R. Morris, T. Paxton, A. Dell, J. Langhorne, M. Berg, R. S. Bordoli, J. Hoyes, and R. H. Bateman, *Rapid Commun. Mass Spectrom.* **10**, 889–896 (1996).

64. A. Schevchenko, I. Chernuschevich, W. Ens, and K. G. Standing, *Rapid Commun. Mass Spectrom.* **11**, 1015–1024 (1997).

65. R. H. Bateman, M. R. Green, G. Scott, and E. Claton, *Rapid Commun. Mass Spectrom.* **9**, 1227–1233 (1995).

66. J. K. Olthof, I. A. Lys, and R. J. Cotter, *Rapid Commun. Mass Spectrom.* **2**, 171–175 (1988).

67. R. S. Brown and J. J. Lenon, *Anal. Chem.* **67**, 1998–2003 (1995).

68. S. M. Colby, T. B. King, and J. P. Reilly, *Rapid Commun. Mass Spectrom.* **8**, 865–868 (1994).

69. R. M. Whittal and L. Li, *Anal. Chem.* **67**, 1950–1954 (1995).

70. R. S. Brown and J. J. Lennon, *Anal. Chem.* **67**, 1998–2003 (1995).

71. M. L. Vestal, P. Juhasz, and S. A. Martin, *Rapid Commun. Mass Spectrom.* **9**, 1044–1050 (1995).

72. R. Kaufmann, D. Kirsch, and B. Spengler, *Int. J. Mass Spectrom. Ion Proc.* **131**, 355–385 (1994).

73. J. C. Rouse, W. Yu, and S. A. Martin, *J. Am. Soc. Mass Spectrom.* **6**, 822–835 (1995).

74. J. H. Holland, C. G. Enke, J. Allison, J. T. Stults, J. D. Pinkston, B. Newcombe, and J. T. Watson, *Anal. Chem.* **55**, 997A–1012A (1983).

75. M. L. Gross and D. L. Rempel, *Science* **226**, 261–268 (1984).

76. M. V. Buchanan, ed., *Fourier Transform Mass Spectrometry: Evolution, Innovations, and Applications*, ACS, Washington, DC, 1987.

77. C. B. Jacoby, C. L. Holliman, and M. L. Gross, in M. L. Gross, ed., *Mass Spectrometry in the Biological Sciences: A Tutorial*, Kluwer, Dordrecht, The Netherlands, 1992, pp. 93–116.

78. A. G. Marshall and P. B. Grosshans, *Anal. Chem.* **63**, 215A–229A (1991).

79. I. J. Amster, *J. Mass Spectrom.* **31**, 1325–1337 (1996).

80. C. Köster, M. S. Kahr, J. A. Castoro, and C. A. Wilkins, *Mass Spectrom. Rev.* **11**, 459–512 (1992).

81. C. L. Holliman, D. L. Rempel, and M. L. Gross, *Mass Spectrom. Rev.* **13**, 105–132 (1994).

82. T. Dienes, S. J. Paster, S. Schürch, J. R. Scott, J. Yao, S. Cui, and C. A. Wilkins, *Mass Spectrom. Rev.* **15**, 163–211 (1996).

83. P. A. Limbach, A. G. Marshall, and M. Wang, *Int. J. Mass Spectrom. Ion Proc.* **125**, 135–143 (1993).

84. A. G. Marshall and S. Guan, *Rapid Commun. Mass Spectrom.* **10**, 1819–1823 (1996).

85. S. D.-H. Shi, C. L. Hendrickson, and A. G. Marshall, *Proc. Natl. Acad. Sci. (USA)* **95**, 11532–11537 (1998).

86. C. Wu, W. F. Siems, G. R. Asbury, and H. H. Hill, Jr., *Anal. Chem.* **70**, 4929–4938 (1998).

87. D. E. Clemmer and M. F. Jarrold, *J. Mass Spectrom.* **32**, 577–592 (1997).

88. K. Shelimov, D. E. Clemmer, R. R. Hudgins, and M. F. Jarrold, *J. Am. Chem. Soc.* **119**, 2240–2248 (1997).

89. K. Shelimov and M. F. Jarrold, *J. Am. Chem. Soc.* **119**, 2987–2994 (1997).

90. S. J. Valentine, A. E. Counterman, and D. E. Clemmer, *J. Am. Soc. Mass Spectrom.* **8**, 954–961 (1997).

91. C. Hoaglund, S. J. Valentine, and D. E. Clemmer, *Anal. Chem.* **69**, 4156–4161 (1997).

92. R. Guevremont and R. W. Purves, *J. Am. Soc. Mass Spectrom.* **10**, 492–501 (1999).

93. S. Evans, in J. A. McCloskey, ed., *Methods in Enzymology*, Academic Press, San Diego, 1990, Vol. 193, pp. 61–68.

94. P. W. Geno, in M. L. Gross, ed., *Mass Spectrometry in the Biological Sciences: A Tutorial*, Kluwer, Dordrecht, The Netherlands, 1992, pp. 133–144.

95. W. Aberth, *Int. J. Mass Spectrom. Ion Phys.* **37**, 379–382 (1981).

96. D. Price and G. J. Milnes, *Int. J. Mass Spectrom. Ion Proc.* **60**, 61–84 (1984).

97. R. J. Beuhler and L. Friedman, *Int. J. Mass Spectrom. Ion Phys.* **23**, 81–97 (1977).

98. A. Hedin, P. Hakansson, and B. Sundqvist, *Int. J. Mass Spectrom. Ion Proc.* **75**, 275–289 (1987).

99. F. Dubois, R. Knochenmus, and R. Zenobi, *Rapid Commun. Mass Spectrom.* **13**, 1958–1967 (1999).

100. D. M. Lubman, J. Bai, Y.-H. Liu, J. R. Srinivasan, Y. Zhu, D. Siemieniak, and P. J. Venta, in B. S. Larsen and C. N. McEwen, eds., *Mass Spectrometry of Biological Materials*, Marcel Dekker, New York, 1998, pp. 405–634.

101. N. E. Booth, *Rapid Commun. Mass Spectrom.* **11**, 944–947 (1997).

102. D. Twerenbold, J.-L. Vuilleumier, D. Gerber, A. Tadsen, B. van den Brandt, and P. M. Gillevet, *Appl. Phys. Lett.* **68**, 3503 (1996).

103. M. Frank, C. A. Mears, S. E. Labov, W. H. Bener, D. Horn, J. M. Jaklevic, and A. T. Barfknecht, *Rapid Commun. Mass Spectrom.* **10**, 1946–1950 (1996).

104. W. H. Bener, D. Horn, J. M. Jaklevic, M. Frank, C. A. Mears, S. E. Labov, and A. T. Barfknecht, *J. Am. Soc. Mass Spectrom.* **8**, 1094–1102 (1997).

105. D. P. Langstaff and K. Birkinshaw, *Rapid Commun. Mass Spectrom* **9**, 703–706 (1995).

106. K. Birkinshaw and D. P. Langstaff, *Rapid Commun. Mass Spectrom.* **10**, 1675–1676 (1996).

107. K. Birkinshaw, *J. Mass Spectrom.* **32**, 795–806 (1997).

108. K. Biemann, P. Bommer, and D. M. Desiderio, *Tetrahedron Lett.* **38**, 1725–1731 (1964).

109. J. S. Cottrell and S. Evans, *Anal. Chem.* **59**, 1990–1995 (1987).

110. J. A. Hill, J. E. Biller, and K. Biemann, *Int. J. Mass Spectrom. Ion Proc.* **111**, 1–25 (1991).

111. J. A. Loo and R. Pesch, *Anal. Chem.* **66**, 3659–3663 (1994).

112. A. L. Burlingame, in T. Matsuo, R. M. Caprioli, M. L. Gross, and Y. Seyama, eds., *Biological Mass Spectrometry, Present and Future*, Wiley, New York, 1994, pp. 147–164.

113. S. Evans, R. Buchanan, A. Hoffman, F. A. Mellon, K. R. Price, S. Hall, F. C. Walls, A. L. Burlingame, S. Chen, and P. J. Derrick, *Org. Mass Spectrom.* **28**, 289–290 (1993).

4

TANDEM MASS SPECTROMETRY

Tandem mass spectrometry (MS/MS) refers to the coupling of two mass spectrometers in time and space with the objective to obtain further information of a more specific nature about the sample in question. Tandem mass spectrometry was first used in the late 1960s [1]. Since that time, the applications and popularity of this technique continue to grow. Its usefulness is unsurpassed in the structure elucidation of unknown compounds, identification of compounds in complex mixtures, elucidation of fragmentation pathways, and quantification of compounds in real-world samples. Several new generations of instruments have been designed for tandem mass spectrometry applications. A few reviews [2–5] and books [6,7] are recommended for additional reading.

4.1. BASIC CONCEPTS

Tandem mass spectrometry takes advantage of the fragmentation reactions that occur in field-free regions (FFRs) of multisector instruments. The concept of tandem mass spectrometry is illustrated in Figure 4.1, and involves mass-selection, fragmentation, and mass analysis. These three steps are performed using two stages of mass analysis. The first stage (MS1) performs the mass selection of a specified ion from a mixture of ions that are produced in the ion source. This mass-selected ion undergoes fragmentation in the intermediate region, usually via collisions with neutral gas atoms [8]. By definition, the mass-selected ion is called the *precursor*

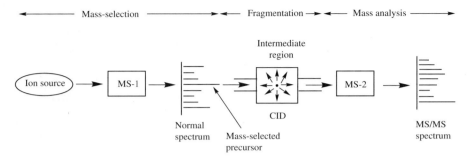

Figure 4.1. Basic concepts of tandem mass spectrometry.

ion, and the ionic fragments are called *product ions*. The second stage (MS2) of MS/MS is used to mass-analyze the product ions that are formed in the intermediate step. This operation is akin to the combination of a chromatography technique with mass spectrometry (e.g., gas or liquid chromatography). The first stage of MS/MS separates a mixture of ions into individual components as a chromatography technique resolves a mixture of compounds, and the second stage obtains mass spectra of each mass-resolved ion.

A unique attribute of tandem mass spectrometry is its molecular specificity, which is the result of an incontrovertible link between a precursor ion and all its product ions. Those product ions must derive exclusively from the preselected precursor. Since the introduction of various desorption ionization techniques (e.g., field desorption, fast-atom bombardment, matrix-assisted laser desorption/ionization, electrospray ionization) for the analysis of nonvolatile and polar compounds such as those encountered in biomedical fields, MS/MS has assumed even more critical significance. On one hand, these techniques do not provide required fragment ion information for structure elucidation. On the other hand, the mass spectra are complicated because of the presence of the chemical background noise that is derived from the matrix and solvent ions. In addition, many isolates of biological samples are mixtures, even after the best possible chromatographic separation. MS/MS overcomes most of those difficulties.

4.2. TYPES OF SCAN

Tandem mass spectrometry data are acquired in the following four scan modes:

- Product ion scan
- Precursor ion scan (the old terms for these two scans are "daughter ion" and "parent ion" scans, respectively)
- Constant neutral loss scan
- Selected-reaction monitoring

Figure 4.2 is a pictorial representation of the first three scan modes. Magnetic sectors, triple sector quadrupole, or hybrid-type tandem instruments all can be employed to perform these scans. The most common mode of MS/MS operation is the *product ion scan*, which provides the spectrum of the product ions that are formed by the dissociation of a mass-selected precursor. It involves setting up the first mass analyzer to transmit the chosen precursor ion, and scanning the second mass spectrometer over a certain mass range to obtain the product ion spectrum. This spectrum is useful in the structural analysis of the preselected ions. As an example, the identification of peptide antigens that are associated with class I or II major histocompatibility complex (MHC) from cell cultures is a daunting task. In combination with microcapillary liquid chromatography, tandem mass spectrometry has been successful to structurally characterize peptide antigens presented by several class I alleles [9].

Another popular MS/MS scan is the *precursor ion scan*, which provides the identity of all precursor ions that fragment to a preselected product ion. The precursor ion spectrum is obtained by adjusting the second mass spectrometer to transmit a certain specified product ion, whereas the first mass analyzer is scanned over a certain mass range to transmit only those precursor ions that generate the specified chosen product ion in the collision cell. This mode of operation is especially useful for the selective identification of a closely related class of compounds in a mixture. A typical example is the detection of phosphopeptides from biological samples. All phosphopeptides fragment to produce the PO_3^- ion at

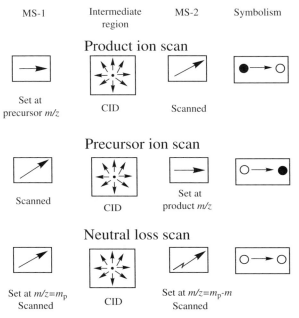

Figure 4.2. A pictorial representation of various scans used in tandem mass spectrometry.

m/z 79. The precursor ion scan monitoring of this m/z value will selectively detect the presence of phosphopeptides in a mixture of peptides from biological samples [10].

The third important scan is the constant neutral loss scan, in which, through monitoring of a specific neutral loss, the information about all precursors that undergo the loss of that specified common neutral is obtained. This scan mode is accomplished when both mass analyzers are scanned simultaneously, but with a specified mass offset. Similar to the precursor ion scan, this technique is also useful in the selective identification of closely related classes of compound in a mixture. The utility of this scan can be explained by using the example of opioid peptides. The sequence of the first four amino acids in all endogenous opioid peptides is TyrGlyGlyPhe [11]. These peptides undergo the cleavage of the Tyr side chain to exhibit the loss of a neutral moiety of 107 Da mass. The neutral loss scan set to monitor the loss of 107 Da will detect the presence of all members of the opioid peptide family in a mixture of peptides isolated from a body tissue.

Tandem mass spectrometers are also operated in the selected-reaction monitoring (SRM) mode. In this mode, a complete product ion spectrum is not acquired; instead, the two mass analyzers are adjusted to monitor only one or two chosen precursor/product pairs of the analyte. This operation is identical to the selected-ion monitoring (SIM) mode of data acquisition. The technique is useful in quantitative measurements of analytes present in complex mixtures (see Chapter 6). The monitoring of more than one reaction is termed *multiple-reaction monitoring* (MRM).

4.3. INSTRUMENTATION FOR TANDEM MASS SPECTROMETRY

A variety of instrument designs have emerged for tandem mass spectrometry research, and each fulfills special needs. Some instruments perform MS/MS in space, meaning that the mass selection, fragmentation, and mass analysis steps are carried out in different regions of the tandem instrument. In this category, two mass spectrometers are sequentially arranged as is done in magnetic sector and quadru-pole instruments. In contrast, some instruments perform these steps in the same region, but using a temporal sequence. Quadrupole ion trap (QIT) and Fourier transform (FT)–ion cyclotron resonance (ICR) mass spectrometers are the examples of this latter category of tandem instruments.

4.3.1. Magnetic Sector Tandem Mass Spectrometers

Magnetic sector instruments have enjoyed the status of mainstream tandem mass spectrometers for a long time. Their contribution to the chemistry of gas-phase ions is unsurpassed. The field of peptides and other biomolecules has also benefited extensively from these instruments. One of the unique attributes of magnetic sector tandem instruments is higher ion energies, typically of the order of 3–10 keV. Initially, double-focusing instruments were used for MS/MS studies [12]. Later, multisector instruments were developed for high-performance tandem MS studies.

As discussed in Chapter 3 (Section 3.3), electrostatic (E) and magnetic analyzers (B) are the main components of magnetic sector instruments. These components can be arranged in several ways to provide a viable tandem instrument. Figure 4.3 shows the schematics of various configurations of magnetic sector tandem mass spectrometers. Before giving an account of these instruments, it is pertinent to understand the fate of an ion of mass m_p that fragments in front of electrostatic and magnetic analyzers (Eq. 4.1). First, consider an ion that fragments in an FFR that is located in front of a magnetic analyzer. The fragmentation of a precursor ion of mass m_p produces an ion of mass m_1, which will not be detected at its true mass value, but at a value given by Eq. 4.2:

$$m_p^+ \rightarrow m_1^+ + m_2 \tag{4.1}$$

$$m^* = \frac{m_1^2}{m_p} \tag{4.2}$$

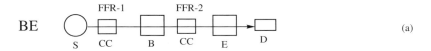

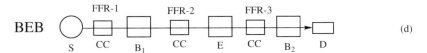

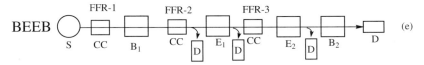

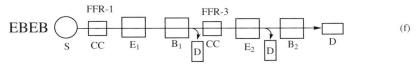

Figure 4.3. Various configurations of magnetic sector tandem mass spectrometers. Here, S represents ion source, CC collision cell, E electric field, B magnetic field, and D detector.

By scanning the magnetic field strength, one can obtain the mass values of all product ions of m_p. The product ion peaks usually are broad and of low intensities.

When a precursor ion fragments in front of an electric sector, the kinetic energy of the product ions will be different from their precursor, because the velocity of the product ions is conserved, but their mass is reduced. The electric sector will allow the newly formed ion m_1 to pass through it only when the electric field E_p is changed to a new value E_1 given by

$$E_1 = \frac{m_1}{m_p} E_p \qquad (4.3)$$

Thus, by scanning the electric sector, an *ion kinetic energy* (IKE) spectrum is obtained. The information about all product ions generated from m_p can be obtained from this spectrum. The resolution of the IKE spectrum is poor because the kinetic energy released during fragmentation causes broadening of peaks. Also, the mass values of the product ions cannot be read directly from this spectrum.

Mass-Analyzed Ion Kinetic Energy Spectroscopy. The design of a reverse geometry (a magnetic analyzer precedes an electrostatic analyzer; i.e., the BE geometry) magnetic sector instrument was the first serious attempt to develop a true tandem mass spectrometer [13]. With this type of instrument, the precursor ion is exclusively mass-selected by adjusting the magnetic field, and allowed to fragment in a region between the two analyzers (Figure 4.3a). The spectrum of fragment ions is obtained by scanning the electric sector (Eq. 4.3). This technique is known as *mass-analyzed ion kinetic energy spectroscopy* (MIKES) [13,14]. The precursor ion selection is unambiguous in this technique, but the poor mass resolution (<300) of the product ions is a major issue. The mass selection of a precursor by the magnetic sector has the benefit that it allows the direct assignment of the mass values to its product ions. An example of the MIKE spectrum is shown in Figure 4.4, which is the spectrum of 10-[3'-(N-bishydroxyethyl)amino]propyl-2-trifluoromethylphenoxazine, a putative chemosensitizer [15].

Linked-Field Scans. The forward geometry design (EB geometry) is the most popular type of double-focusing mass spectrometer, but a serious problem with this instrument is that it cannot be used as a true tandem instrument because the electrostatic analyzer is unable to provide the first stage of mass selection. However, a forward geometry instrument can be made to act as a tandem instrument by the implementation of several ingeniously designed linked-field scans [16]. All three tandem mass spectrum types—product ion, precursor ion, and neutral loss—can be obtained by simultaneously scanning two of the three fields, accelerating V, electric E, and magnetic B, while maintaining their ratio constant. More commonly used linked-field scans are listed in Table 4.1. The mass selection and fragmentation in the EB instruments occur in front (the first FFR) of the electrostatic analyzer (see Figure 4.3b). The most popular scan is the *linked-field scan at constant B/E*, in which B and E fields are both scanned while their ratio is held constant. It is designed to acquire a product ion spectrum. The double-focusing principle still applies to the product ion

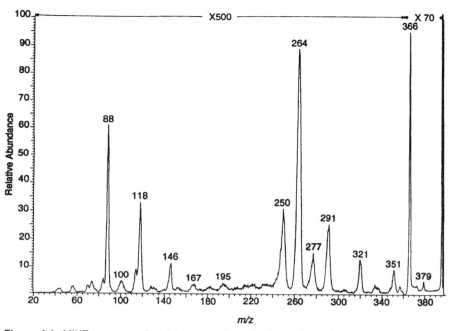

Figure 4.4. MIKE spectrum of 10-[3'-(N-bishydroxyethyl)amino]propyl-2-trifluoromethylphenox-azine. The [M + H]⁺ ion produced by liquid-SIMS was mass-selected by the EB portion of an EBE instrument. [Reproduced from Ref. 15 by permission of John Wiley & Sons (copyright 1994).]

Table 4.1. Linked-field scans for tandem mass spectrometry

Scan Type	Reaction Region	Type of Spectrum	Collision Energy	Instrument
V	FFR1	Precursor ion spectrum	High	EB
E	FFR2	Product ion (MIKES) spectrum	High	BE
	FFR3	MIKES	High	EBE
B/E	FFR1	Product ion spectrum	High	EB, BE, EBE, BEB, EBQQ, EB-TOF
B/E	FFR2	Product ion spectrum	High	EBE, BEB
B^2/E	FFR1	Precursor ion spectrum	High	EB, BE, EBE, BEB, EBQQ, EB-TOF
B^2/E	FFR2	Precursor ion spectrum	High	EBE, BEB
B^2E	FFR2	Precursor ion spectrum	High	BE, BEQQ
	FFR3	Precursor ion spectrum	High	EBE
$(B/E')^2(1 - E')$	FFR1	Neutral loss spectrum	High	EB, BE, EBE, BEB, EBQQ, EB-TOF
Q_3	Rf-only Q	Product ion spectrum	Low	TSQ
Q_1	Rf-only Q	Precursor ion spectrum	Low	TSQ
B	Q_1	Precursor ion spectrum	Low	EBQQ
Q_1–Q_3	Rf-only Q	Neutral loss spectrum	Low	TSQ
B_2–Q_2	Q_1	Neutral loss spectrum	Low	EBQQ

analysis. Therefore, the mass resolution is much higher compared to the MIKE spectrum. A major drawback of this type of scan, however, is that the mass selection of the precursor is at a lower resolution (typically at 300–400). Several applications of this scan mode have been reported for qualitative and quantitative analysis of biologically active neuropeptides [17–19].

A precursor ion spectrum can be obtained by a different linked-field scan, in which B and E fields are also scanned simultaneously, but now the B^2/E ratio is kept constant. The neutral loss spectrum is obtained when the B and E fields are scanned simultaneously, and the expression $B^2 (1 - E')/E'^2$ (where B is the magnetic field required to transmit m_2 and $E' = E_2/E_1$) is held constant throughout the scan.

These linked-field scans can also be acquired with a reverse-geometry instrument. The product ions and neutral loss spectra are obtained with the same scan laws as those used in a forward geometry instrument, and the fragmentation of the precursor also occurs in the first FFR. However, for a precursor ion spectrum, the mass selection and fragmentation are accomplished in the second FFR, and the spectrum is acquired by scanning the B and E fields, but holding the product B^2E constant [20].

Three- and Four-Sector Tandem Mass Spectrometers. The next step in extending tandem mass spectrometry applications was to develop three- and four-sector instruments (Figure 4.3c–f) [21–23]. The motivation behind this activity was to improve the mass resolution in precursor ion selection and product ion mass analysis. The most common designs are of the EBE, BEB, EBEB, BEEB, and BEBE geometries.

A unique feature of three-sector instruments (EBE and BEB) is their ability to mass-select precursor ions at high resolution (>100,000). The product ion analysis is still limited to low to modest resolution. However, with the BEB design, the product ion mass resolution can be improved significantly when the first magnet is used to mass-select the precursor ions, and the remaining EB portion provides the mass analysis of the product ions.

Four-sector instruments have the advantage of retaining the high-resolving power within MS1, while providing an improved mass resolution in MS2. However, the energy spread of the product ions limits the resolution of MS2 to <10,000. The linked-field scans discussed above can all be implemented on these multisector instruments. Another unique feature of multisector instruments is the feasibility of higher-order (MS^n) MS/MS reactions [24]. The use of multichannel array detectors for second-stage mass analysis has considerably improved detection sensitivity [25,26]. Four-sector mass spectrometers have been used primarily to obtain the amino acid sequence of peptides and proteins and to determine posttranslational modifications [27]. A major disadvantage of these instruments is their exorbitant cost. Also, the coupling of high-resolution separation devices is problematic.

4.3.2. Triple-Sector Quadrupole

A triple-sector quadrupole (TSQ) has established a unique position in the field of tandem mass spectrometry. The wide popularity of tandem mass spectrometry can be attributed largely to this instrumental development. Yost and Enke were the first to design a TSQ instrument for analytical applications in 1978 [28]. These researchers reported that ions could be effectively dissociated at low collision energies. A similar instrument was used earlier for photodissociation studies [29,30]. Since then, it has become the most widely used tandem mass spectrometer [4,31–33]. In this simple device, three quadrupoles are arranged sequentially (Figure 4.5). The first (Q_1) and the last (Q_3) quadrupoles function as normal mass filter devices. The direct current (dc) and radiofrequency (rf) potentials both control the operation of these quadrupoles. In contrast, only the rf potential is applied to the central quadrupole (Q_2). The rf-only mode of operation allows all ions to pass through Q_2. Because of the rapidly rotating saddle field that is created by the alternating rf potential, every ion within the mass range of interest can be focused within the boundaries of the rf-only quadrupole. Therefore, this device serves as a total ion containment region and an efficient collision cell. In order to perform CID, this cell is filled with a neutral gas at a pressure of 10^{-4}–10^{-2} torr. In TSQ, the low-energy fragmentation channels are the dominant reactions because only those ions that possess kinetic energies in the range of 0–200 eV can pass through quadrupole devices. An offset voltage between the ion source and the rf-only quadrupole can be adjusted to vary the collision energy.

All four types of scan discussed above (Section 4.1) can be implemented with a TSQ. The product ion spectrum is obtained by setting Q_1 to transmit a specified m/z value into Q_2 where it is fragmented via low-energy CID, and Q_3 is scanned to mass-analyze the products that are formed in Q_2. This procedure is reversed when the precursor ion spectrum is acquired. In other words, Q_3 is set to the mass value of a desired product ion, and Q_1 is scanned to provide the identity of all precursors of this selected product ion. In contrast to the magnetic sector–based tandem instruments, a simple scan law is used in a TSQ to monitor the loss of a neutral. The fields of Q_1 and Q_3 are offset by a factor that is related to the mass of the neutral, and scanned in tandem to maintain that mass difference.

Low cost, operational simplicity, straightforward scan laws, linear mass scale, and a unit mass resolution in the precursor mass selection and product mass analysis are the key advantages that have made a TSQ a popular tandem mass spectrometry instrument. The mass resolution of the precursor ion selection and product ion analysis, and the mass range (∼4000 Da) of a TSQ are much lower than those of the

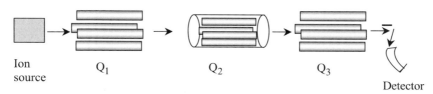

Figure 4.5. Schematic diagram of a triple-sector quadrupole tandem mass spectrometer.

four-sector instruments. A major disadvantage of this instrument is the gradual falloff in performance at $m/z > 1000$. Also, this instrument cannot perform certain charge permutation reactions discussed in Section 4.4.2.

4.3.3. Hybrid Tandem Mass Spectrometers

Hybrid tandem mass spectrometers are constructed with devices that use two different mass analysis principles [33]. The motivation is to benefit from the best performance features of both types of mass analyzers. Over the years, several different combinations have evolved. The combination of a magnetic sector mass spectrometer for MS1 and a quadruple for MS2 has been a popular concept for some time [34,35]. The EB and BE double-focusing magnetic sector instruments have both been utilized for this purpose. A schematic of an EB-qQ instrument is shown in Figure 4.6. Translational energies of the ions that emerge from MS1 must be reduced from kiloelectronvolts to electronvolts before their entry into the quadrupole section. A deceleration lens assembly is added to the intermediate region for this purpose. These instruments also contain an off-axis detector, which permits the first stage to be used as a standalone double-focusing mass spectrometer. Because of the double-focusing action of the MS1 section, the mass selection of the precursor is at a high resolution, and the product ion analysis occurs at a unit mass resolution. This hybrid combination is a much simpler and cheaper alternative to four-sector magnetic field–type instruments, and allows low- and high-energy CID studies to be performed on a single machine.

Another common design that has become popular since mid 1990s is the combination of a quadrupole and an orthogonal acceleration (oa) TOF instrument (Q-oaTOF) [36–39]. A schematic diagram of this instrument is shown in Figure 4.7. The quadrupole section consists of a normal quadrupole and an rf-only quadrupole, and serves as an MS1, and a TOF as an MS2. The mass-selected precursor is fragmented using low-energy collisions in the rf-only quadrupole. The product ions

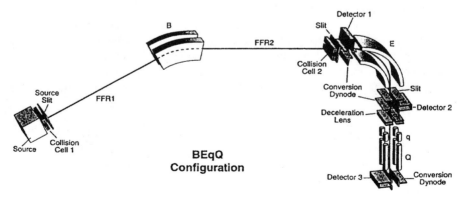

Figure 4.6. A BE-qQ hybrid tandem mass spectrometer. [Reproduced from Ref. 5 by permission of John Wiley & Sons (copyright 1996).]

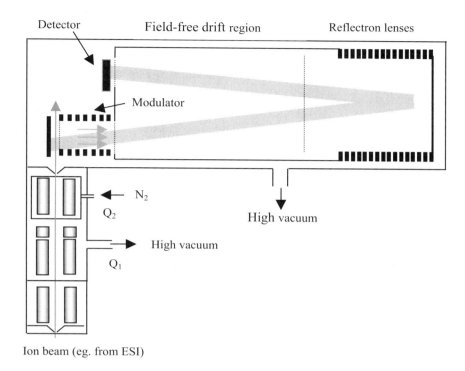

Figure 4.7. Schematic diagram of a quadrupole–orthogonal acceleration time-of-flight hybrid tandem mass spectrometer.

are mass-analyzed at a high resolution with TOF to provide their accurate mass. The TOF can also be used to acquire a normal mass spectrum. In this mode, the main quadrupole is also operated in the rf-only mode, where it serves as a wide-bandpass filter to allow the transmission of ions over a wide mass range. Microchannel plates are used for the parallel mode of detection of all ions. A reasonably good resolution, high transmission, ultrahigh sensitivity (low femtomole range for peptide sequencing), and good mass accuracy are the attributes of the Q-oaTOF design. A Q-TOF instrument, in which ions are injected along the axis of TOF, has also been reported for tandem MS applications [40].

Magnetic sector instruments have also been combined with an oaTOF analyzer [41–43]. A commercial instrument based on the EBE-oaTOF design is on the market (Micromass, Beverly, MA). With this design, precursor and products are both analyzed at a high resolution. Also, fragmentation of the selected precursors is via high-energy CID. As a consequence, the MS/MS spectra are similar to four-sector MS/MS instruments, but this hybrid instrument in contrast is much cheaper and easier to use. The coupling of MALDI has also been implemented with the EBE-oaTOF design [44,45]. Other hybrid designs that have been used for MS/MS applications include Q-FT-ICR [46], Q-QIT, BE-QIT, QIT-TOF [47,48], and QIT/FT-ICR [49].

4.3.4. Tandem Mass Spectrometry with Time-of-Flight Instrument

A substantial portion of excited ions after ejection from the ion source undergoes metastable dissociation in the FFR of the flight tube. The time dispersion of the products of metastable decay has been exploited to obtain the structural information. A technique termed *postsource decay* (PSD) has emerged for structural studies with TOF instruments [50–52]. Normally, a charged fragment and the corresponding neutral produced in the flight tube both continue to travel with the velocity of their precursor, but their kinetic energy is reduced in proportion to the change in their mass. As a consequence, neutral and charged fragments both arrive at the detector along with their precursor in a linear TOF instrument, with the result that the structural information is lost. However, by using a reflectron TOF (RTOF) instrument, those flight tube fragmentation reactions can be monitored. In the correlated reflex technique developed by LeBeyec and co-workers [53], a linear mass spectrum is first recorded to determine the flight times of the precursor ions. The reflectron is then turned on, and the products of metastable decay are recorded within a preset time interval only when neutrals arrive at the first detector. In the PSD approach developed by Spengler and co-workers (see Figure 4.8), the neutral products are detected as usual by placing the detector behind the reflectron [50–52]. This measurement provides the arrival time of the precursor. The precursor ion is mass-selected by a fast computer-controlled ion gating. The product ions are recorded by switching on the reflectron. Figure 4.9 demonstrates the ability of the PSD technique for the mass selection of precursor ions. Single-stage and two-stage reflectrons can both be used for this purpose, but the latter is preferred because it provides a better mass resolution in the product ion mass analysis. In a two-stage reflectron, the product ion spectrum is obtained in several segments by stepping up the reflectron voltage. Only those fragment ions that have enough energy will pass through the first stage of the two-stage reflectron. All ions below this kinetic energy will be detected as one peak.

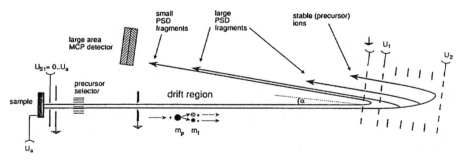

Figure 4.8. Postsource decay measurement with a reflectron time-of-flight mass spectrometer. [Reproduced from Ref. 51 by permission of Humana Press (copyright 1996).]

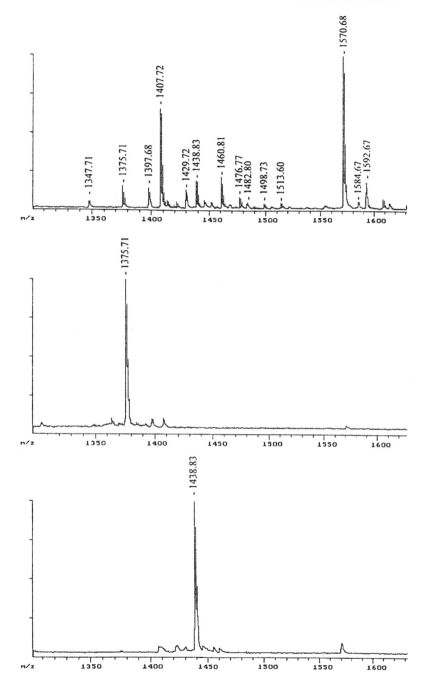

Figure 4.9. The gated mass selection of precursor ions in PSD-TOFMS: (*a*) normal spectrum (i.e., ion gate off); (*b*) ion gate on and set to transmit *m/z* 1375.8; (*c*) ion gating of *m/z* 1438.7. [Reproduced from Ref. 52 by permission of John Wiley & Sons (copyright 1997).]

The mass selectivity (i.e., $M/\Delta M$) of the gate-selected precursor ions in a single-reflectron TOF instrument is poor (typically of the order of 40–70 at FWHM). To improve the mass selectivity, true tandem mass spectrometers, in which the TOF principle is used in the first and second stages of mass analysis, have been designed. Various combinations of TOF and RTOF have been tested, including two linear TOFs (i.e., TOF-TOF) [54], one TOF and one reflectron TOF (i.e., TOF-RTOF) [55,56], and two reflectron TOFs (i.e., RTOF-RTOF) [57,58]. A fragmentation region is added between the two analyzers. The use of an RTOF for MS1 and MS2 has the advantage of greater mass resolution. In this design, the energy-focusing action of the RTOF1 permits mass selection of the isotopically resolved precursor. In practice, a desired precursor ion is usually selected by setting the appropriate delay between the ion extraction pulse and the laser-triggering event. The fragmentation of the mass-selected precursor is induced in the intermediate region by either collision activation or high-intensity laser flux. The mass dispersion and energy focusing of product ions is achieved by the RTOF2 to allow the analysis of product ions at a unit mass resolution.

4.3.5. Tandem Ion-Trap Mass Spectrometry

In the tandem instruments discussed above, different stages of MS/MS analysis are separated in space. On the other hand, the QITMS [59] and FTICR-MS [60] both act as tandem-in-time mass spectrometers. Both these instruments have proved to be powerful tools in the analysis of bimolecules, providing high-sensitivity and high-resolution capabilities. Although these two instruments differ in design, conceptually their MS/MS operation is similar. As mentioned above, all steps of MS/MS, namely, the precursor selection, excitation, and detection, are all performed in the same space, but with a temporal sequence. After the initial steps of ion formation and their confinement in the trapping field, all ions except the precursor are ejected out of the trap with the resonance ejection procedure [61]. An excitation pulse is applied to the precursor ion to facilitate collisions with an inert bath gas. Helium is usually present as a bath gas in QITMS. In the FTICR-MS, helium can be pulsed into the cell concurrent with the excitation pulse. The CID products are mass-analyzed as usual. A pictorial depiction of these steps is presented in Figure 4.10. Only the product ion scan is feasible with these instruments. One unique advantage of these instruments is the availability of higher-order MS/MS (MS^n) experiments [62]. After dissociation of the precursor ions, a particular product ion can be mass-selected by the ejection of other ions in the trap, and the activation process can be repeated for the new precursor ion. For MS^n, the sequential product spectra can be acquired until the ion population is too low for useful ion statistics. A practical limit of tandem mass spectrometry with ion traps is MS^9 or MS^{10}. This feature has been recently used to characterize oligosaccharides [63]. In single-stage MS/MS experiments, some structural features in glycans, such as N-acetyllactosamine antennae, neuraminic acids, and nonreducing terminal GlcNAc monosaccharides, usually suppress cross-ring and core saccharide cleavages. However, the removal of these structural features via CID allows the determination of branching patterns and

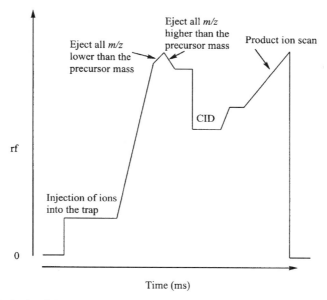

Figure 4.10. A pictorial representation of various steps of MS/MS analysis in an ion trap mass spectrometer.

linkage via higher-order MS/MS experiments. An example is the analysis of GlcNAc$_8$Man$_3$ sugar from chicken ovalbumin [63]. The MS^8 experiment in a QIT instrument revealed the branching patterns of the GlcNAc residues. A major disadvantage of MS/MS with trap instruments is the possibility of side reactions because all steps are performed within the same space. A significant enhancement of ion signal and dynamic range is realized by use of selected-ion accumulation and sustained off-resonance irradiation (SORI) experiments [64].

4.4. MODES OF ION ACTIVATION

4.4.1. Metastable Ions

During the ionization of a compound, ions are formed with a range of internal energies. As a consequence, some ions fragment instantaneously in the ion source, whereas others remain intact. Certain ions, called *metastable ions*, fragment after exiting the ion source, but before reaching the detector [14]. These ions have energies just above the threshold for the lowest dissociation channel, but have slow fragmentation rate constants (10^5–10^6 s^{-1}). In a spectrum that is acquired by a magnetic sector instrument without the aid of a data system, they appear as diffuse peaks at the nonintegral mass values given by Eq. 4.1. The peak broadening is the consequence of the partitioning of a fragmenting ion's internal energy into the translational energy of the products. The width of the peak and the magnitude of the kinetic energy release (KER) are diagnostic features of the fragmenting ion, and can be utilized to determine its structure [14]. These pieces of information, along with the ion abundances in the metastable spectrum, have been used to distinguish ionic structures [24].

4.4.2. Collision-Induced Dissociation

Collision-induced dissociation (CID) has played a major role in the development of tandem mass spectrometry. It was first used in 1968 for structural studies [1,65]. The motivation behind this development was twofold: (1) analytical applications of tandem mass spectrometry cannot rely on spontaneously fragmenting metastable ions because they usually are of low abundance, and fragment via a few selected pathways; and (2) ions that are generated by softer modes of ionization, such as FAB and ESI, exhibit a weaker fragmentation pattern, and yield a little structural information. An exhaustive review of this technique is beyond the scope of this chapter. A few reviews and books are worth reading on this subject [66–74]. Several studies have also appeared on the dynamics of collision process [75–77].

Collision-induced dissociation is a means by which the intact ions are activated in an intermediate region of a tandem instrument to produce a large population of fragment ions. These instruments are equipped with a collision cell, which is filled with a suitable neutral gas (helium or argon). The CID is a two-step process, namely, collision activation and unimolecular dissociation. During collision activation, the fast-moving mass-selected ions collide with atoms of an inert gas, and are excited to higher energy states (Eq. 4.4, where $m_p^{+\cdot}$ is the mass-selected ion and N and N' respectively are pre- and postcollision forms of a neutral target) [73]. During this excitation process, a part of the initial translational energy of the incident ions is converted into the excitation energy to cause its fragmentation (Eq. 4.5, where m_1^+ and m_2, are fragments). The maximum collision energy (E_{kin}) that can be converted into internal energy (E_{int}) in a single collision is given by Eq. 4.6, where m_p is the mass of the mass-selected ion and m_t the mass of the target atom:

$$m_p^+ + N \rightarrow [m_p^+]^* + N' \tag{4.4}$$

$$[m_p^+]^* \rightarrow m_1^+ + m_2 \tag{4.5}$$

$$E_{int} = E_{kin} \frac{m_t}{m + m_t} \tag{4.6}$$

Two collision energy regimes have been used for the fragmentation of the mass-selected ions: those in which ions have translational energies in the kiloelectronvolt range (3–10 keV) [69], and those in which ions have <200 eV energies [70]. The high-energy collisions are accessible in magnetic sector and TOF tandem instruments. In this energy regime, ions undergo a single collision, leading to a vertical electronic excitation of the incident ion. The excitation energy is rapidly converted statistically into various vibrational modes to effect fragmentation of the excited ions. During the unimolecular dissociation step, a part of the excess energy is released as translational energy to cause a broadening of peaks in the CID spectrum. A poor mass resolution in the MS2 stage is the result of this peak-broadening process. In high-energy CID, the products that result from consecutive fragmentations dominate the spectra. Also, the higher-energy fragmentation pathways, such as remote-site fragmentation [78] and side-chain fragmentations of peptides [79], can be accessed only in this energy regime.

The low-energy collisions are prevalent in quadrupole- and FTICR-based tandem instruments. Here, the interaction time during collisions between the ion and the target is somewhat longer to allow excitation mainly to higher vibrational states. Also, the low-energy ions spend more time in the reaction region to undergo a greater number of collisions. The low-energy CID spectra are dominated by the products formed via the low-activation-energy fragmentation pathways. In addition, the products of ion–molecule reactions may also appear in those spectra. Also, the target gas has a profound effect on the nature of the MS/MS spectrum. A gradual falloff in the efficiency of low-energy collisions is observed above 1000 Da mass.

A wide variety of reactions also compete with CID process during collision activation. Although these reactions have not been exploited for biological mass spectrometry, they are worth a passing note for their potential role in the structure elucidation of organic ions. Under some conditions, ion–molecule reactions and charge permutation reactions (Eqs. 4.7–4.10) can occur. Some of these reactions are endothermic and are observed only under high-energy collisions.

Charge exchange: $$m^{+\cdot} + N \rightarrow m + N^{+\cdot} \tag{4.7}$$

Charge inversion: $$m^{-\cdot} + N \rightarrow m^{+\cdot} + N + 2e^{-} \tag{4.8}$$

Charge stripping: $$m^{+\cdot} + N \rightarrow m^{2+} + N + e^{-} \tag{4.9}$$

Collision ionization: $$m + N \rightarrow m^{+\cdot} + N + e^{-} \tag{4.10}$$

4.4.3. Other Modes of Ion Activation

The energy required for an ion to dissociate increases as the mass of the ion increases. CID is limited to ions with mass below 3500 Da. Researchers have sought alternative modes of ion activation in an attempt to improve efficiency of internal energy deposition in the mass-selected ion. However, none has yet been used routinely for biological compounds. These modes are surface-induced dissociation (SID) [80,81], electron impact activation (a phenomenon similar to the EI process) [82], photodissociation (using high-energy photons, e.g., from a laser beam) [83] and infrared multiphoton dissociation (IRMPD) [84]. Of these, SID has been exploited by some investigators [80,81]. In SID, a solid surface is introduced in the path of the incident ion. The fragment ions are formed by the impact of collisions with the solid surface. The technique has been implemented on all mass analyzer types. The SID spectra of a number of peptides have been reported [85].

A newer mode of ion activation, called *electron-capture dissociation* (ECD), has been introduced for the ESI-produced multiply protonated biomolecules, $[M + nH]^{n+}$ in FT-ICRMS instruments [86–89]. The capture of a low-energy (<0.2-eV) electron by a mass-selected protonated ion produces the odd-electron ion $[M + nH]^{(n-1)+\cdot}$, which dissociate rapidly via energetic H$^\cdot$ transfer to the backbone carbonyl group to form amino acid sequence-specific ions (c and $z^\cdot$ types) in large abundance. It was shown that ECD produces far more backbone cleavages than CID, and thus can be used to sequence larger-sized peptides.

REFERENCES

1. K. R. Jennings, *Int. J. Mass Spectrom. Ion. Phys.* **1**, 227–235 (1968).

2. F. W. McLafferty, *Science* **214**, 280–287 (1981).

3. R. G. Cooks, *Chem. Eng. News* **59**, 40–52 (1981).

4. R. A. Yost and D. D. Fetterolf, *Mass Spectrom. Rev.* **2**, 1–45 (1983).

5. E. de Hoffmann, *J. Mass Spectrom.* **31**, 129–137 (1996).

6. F. W. McLafferty, *Tandem Mass Spectrometry*, Wiley-Interscience, New York, 1983.

7. K. L. Busch, G. L., Glish, and S. A. McLuckey, *Mass Spectrometry/Mass Spectrometry: Techniques and Applications of Tandem Mass Spectrometry*, VCH Publishers, New York, 1988.

8. R. K. Boyd, P. A. Bott, B. R. Beer, D. J. Harvan, and J. R. Haas, *Anal. Chem.* **59**, 189–193 (1987).

9. C. R. Bazemore Walker, N. E. Sherman, J. Shabanowitz, and D. F. Hunt, in B. S. Larsen and C. N. McEwen, eds., *Mass Spectrometry of Biological Materials*, Marcel Dekker, New York, 1998, pp. 115–135.

10. G. Neubauer and M. Mann, *Anal. Chem.* **71**, 235–242 (1999).

11. C. Dass and D. M. Desiderio, *Anal. Biochem.* **163**, 52–66 (1987).

12. A. P. Bruins, K. R. Jennings, and S. Evans, *Int. J. Mass Spectrom. Ion. Phys.* **26**, 395–404 (1978).

13. J. H. Beynon, R. M. Caprioli, and T. Ast, *Org. Mass Spectrom.* **5**, 229–234 (1971).

14. R. G. Cooks, J. H. Beynon, R. M. Caprioli, and G. R. Lester, *Metastable Ions*, Elsevier, Amsterdam, 1973.

15. C. Dass, K. N. Thimmaiah, B. S. Jayashree, R. Seshadiri, M. Israel, and P. J. Houghton, *Biol. Mass Spectrom.* **23**, 140–146 (1994).

16. K. R. Jennings and R. S. Mason, in F. W. McLafferty, ed., *Tandem Mass Spectrometry*, Wiley-Interscience, New York, 1983, pp. 197–222.

17. C. Dass and D. M. Desiderio, *Int. J. Mass Spectrom. Ion Proc.* **92**, 267–287 (1989).

18. C. Dass, in D. M. Desiderio, ed., *Mass Spectrometry of Peptides*, CRC Press, Boca Raton, FL, 1991, pp. 345–327.

19. C. Dass, J. J. Kusmierz, and D. M. Desiderio, *Biol. Mass Spectrom.* **20**, 130–138 (1991).

20. R. K. Boyd, C. J. Potter, and J. H. Beynon, *Org. Mass Spectrom.* **16**, 490–494 (1981).

21. M. L. Gross and D. H. Russell, in F. W. McLafferty, ed., *Tandem Mass Spectrometry*, Wiley-Interscience, New York, 1983, pp. 255–270.

22. F. W. McLafferty, P. J. Todd, D. C. McGilvery, and M. A. Baldwin, *J. Am. Chem. Soc.* **102**, 3360–3363 (1980).

23. M. L. Gross, in J. A. McCloskey, ed., *Methods in Enzymology*, Academic Press, San Diego, 1990, Vol. 193, pp. 150–152.

24. C. Dass and M. L. Gross, *Org. Mass Spectrom.* **25**, 24–32 (1990).

25. J. A. Hill, S. A. Martin, and K. Biemann, *Biomed. Environ. Mass Spectrom.* **17**, 141–151 (1988).

26. A. L. Burlingame, in T. Matsuo, R. M. Caprioli, M. L. Gross, and Y. Seyama, eds. *Biological Mass Spectrometry, Present and Future*, Wiley, New York, 1994, pp. 147–164.

27. K. Biemann, in J. A. McCloskey, ed., *Methods in Enzymology*, Academic Press, San Diego, 1990, Vol. 193, pp. 455–479.

28. R. A. Yost and C. G. Enke, *J. Am. Chem. Soc.* **100**, 2274–2275 (1978).

29. M. L. Vestal and J. H. Futrell, *Chem. Phys. Lett.* **28**, 559 (1974).

30. D. C. McGilvery and J. D. Morrison, *Int. J. Mass Spectrom. Ion Phys.* **25**, 81 (1978).

31. R. A. Yost and C. G. Enke, *Anal. Chem.* **51**, 1251A–1264A (1979).

32. V. H. Wysocki, in M. L. Gross, ed., *Mass Spectrometry in the Biological Sciences: A Tutorial*, Kluwer, Dordrecht, The Netherlands, 1992, pp. 59–77.

33. S. J. Gaskell, in M. L. Gross, ed., *Mass Spectrometry in the Biological Sciences: A Tutorial*, Kluwer, Dordrecht, The Netherlands, 1992, pp. 29–58.

34. S. A. McLuckey, G. L., Glish, and R. G. Cooks, *Int. J. Mass Spectrom. Ion Phys.* **39**, 219–230 (1981).

35. A. E. Shoen, J. W. Amy, J. D. Ciupek, R. G. Cooks, P. Dobberstein, and G. Jung, *Int. J. Mass Spectrom. Ion Proc.* **65**, 125–140 (1985).

36. A. V. Loboda, A. N. Krutchinsky, W. Ens, and K. G. Standing, *Rapid Commun. Mass Spectrom.* **14**, 1047–1057 (2000).

37. H. R. Morris, T. Paxton, A. Dell, J. Langhorne, M. Berg, R. S. Bordoli, J. Hoyes, and R. H. Bateman, *Rapid Commun. Mass Spectrom.* **10**, 889–896 (1996).

38. H. R. Morris, T. Paxton, M. Pannico, R. McDowell, and A. Dell, *J. Protein Chem.* **16**, 469–479 (1997).

39. A. Schevchenko, I. Chernuschevich, W. Ens, and K. G. Standing, *Rapid Commun. Mass Spectrom.* **11**, 1015–1024 (1997).

40. G. L. Glish and D. E. Goeringer, *Anal. Chem.* **56**, 2291–2295 (1984).

41. M. Guilhaus, D. Selby, and V. Mlynski, *Mass Spectrom. Rev.* **19**, 65–107 (2000).

42. R. H. Bateman, M. R. Green, G. Scott, and E. Clayton, *Rapid Commun. Mass Spectrom.* **9**, 1227–1233 (1995).

43. E. Clayton and R. H. Bateman, *Rapid Commun. Mass Spectrom.* **10**, 889–896 (1996).

44. K. F. Medzihradszky, G. W. Adams, R. H. Bateman, M. R. Green, and A. L. Burlingame, *J. Am. Soc. Mass Spectrom.* **7**, 1–10, (1996).

45. K. F. Medzihradszky, D. A. Maltby, Y. Qui, Z. Yu, S. C. Hall, Y. Chen, and A. L. Burlingame, *Int. J. Mass Spectrom. Ion Proc.* **160**, 357–369 (1997).

46. F. H. Strobel, L. M. Preston, K. S. Washburn, and D. H. Russell, *Anal. Chem.* **64**, 754–762 (1992).

47. I. V. Chernushevich, W. Ens, and K. G. Standing, *Anal. Chem.* **71**, 452A–461A (1999).

48. A. N. Krutchinsky, I. V. Chernushevich, V. L. Spicer, W. Ens, and K. G. Standing, *J. Am. Soc. Mass Spectrom.* **9**, 569–579 (1998).

49. S. M. Michael, B. M. Chien, and D. M. Lubman, *Anal. Chem.* **65**, 2614–2620 (1993).

50. R. Kaufmann, B. Spengler, and F. Lutzenkirchen, *Rapid Commun. Mass Spectrom.* **7**, 902–910 (1989).

51. B. Spengler, in J. R. Chapman, ed., *Protein and Peptide Analysis by Mass Spectrometry*, Humana Press, Totowa, NJ, 1996, pp. 43–56.

52. B. Spengler, *J. Mass Spectrom.* **32**, 1019–1036 (1997).

53. S. Della-Negra and Y. LeBeyec, in A. Benninghoven, ed., *Ion Formation from Organic Solids*, IFOS III, 1986, pp. 42–45.

54. D. R. Jardine, J. Morgan, D. S. Alderdice, and P. J. Derrick, *Org. Mass Spectrom.* **27**, 1077–1083 (1992).

55. K. L. Schey, R. G. Cooks, A. Kraft, R. Grix, and H. Wollnik, *Int. J. Mass Spectrom. Ion Proc.* **94**, 11–14 (1989).

56. U. Boesl, R. Weinkauf, and E. W. Schlag, *Int. J. Mass Spectrom. Ion Proc.* **112**, 121–166 (1992).

57. T. J. Cornish and R. J. Cotter, *Rapid Commun. Mass Spectrom.* **8**, 781–785 (1994).

58. D. J. Beussman, P. R. Vlasak, R. D. McLane, M. A. Seeterlin, and C. G. Enke, *Anal. Chem.* **67**, 3952–3957 (1995).

59. J. N. Louris, R. G. Cooks, J. E. P. Syka, P. E. Kelley, G. C. Stafford, Jr., and J. F. J. Todd, *Anal. Chem.* **59**, 1677–1685 (1987).

60. R. B. Cody and B. S. Freiser, *Anal. Chem.* **54**, 1431–1433 (1982).

61. R. G. Cooks and R. E. Kaiser, *Acc. Chem. Res.* **23**, 213–218 (1992).

62. B. D. Nourse, K. A. Cox, K. L. Morand, and R. G. Cooks, *J. Am. Chem. Soc.* **114**, 2010–2016 (1992).

63. A. S. Weiskopf, P. Vouros, and D. J. Harvey, *Anal. Chem.* **70**, 4441–4447 (1998).

64. J. W. Gauthier, T. R. Trautman, and D. B. Jacobson, *Anal. Chim. Acta* **246**, 211–225 (1991).

65. W. F. Haddon and F. W. McLafferty, *J. Am. Chem. Soc.* **90**, 4745–4746 (1968).

66. R. G. Cooks, ed., *Collision Spectroscopy*, Plenum Press, New York, 1978.

67. L. Levsen and H. Schwarz, *Mass Spectrom. Rev.* **2**, 77–148 (1983).

68. M. S. Kim, *Org. Mass Spectrom.* **26**, 565–574 (1991).

69. R. N. Hayes and M. L. Gross, in J. A. McCloskey, ed., *Methods in Enzymology*, Academic Press, San Diego, 1990, Vol. 193, pp. 237–267.

70. D. Despeyroux and K. R. Jennings, in T. Matsuo, R. M. Caprioli, M. L. Gross, and Y. Seyama, eds., *Biological Mass Spectrometry, Present and Future*, Wiley, New York, 1994, pp. 227–238.

71. A. K. Shukla and J. H. Futrell, *Mass Spectrom. Rev.* **12**, 211 (1993).

72. R. G. Cooks, *J. Mass Spectrom.* **30**, 1215 (1995).

73. S. A. McLuckey, *J. Am. Soc. Mass Spectrom.* **3**, 599–614 (1992).

74. M. S. Kim, *Int. J. Mass Spectrom. Ion Phys.* **50**, 189–203 (1983).

75. R. K. Boyd, *Int. J. Mass Spectrom. Ion Proc.* **75**, 243–264 (1987).

76. D. J. Douglas, *J. Am. Soc. Mass Spectrom.* **9**, 101–113 (1998).

77. P. O. Denis, C. Wesdemiotis, and F. W. McLafferty, *J. Am. Chem. Soc.* **105**, 7454–7456 (1983).

78. K. B. Tomer, F. W. Crow, and M. L. Gross, *J. Am. Chem. Soc.* **105**, 5487–5488 (1983).

79. R. S. Johnson, S. Martin, K. Biemann, J. T. Stults, and J. T. Watson, *Anal. Chem.* **59**, 2621–2625 (1987).

80. M. J. Dekrey, Md. A. Mabud, R. G. Cooks, and J. E. P. Syka, *Int. J. Mass Spectrom. Ion Proc.* **67**, 295–303 (1985).

81. V. H. Wysocki, J. L. Jones, A. R. Dongre, A. Somogyi, and A. L. McCormack, in T. Matsuo, R. M. Caprioli, M. L. Gross, and Y. Seyama, eds., *Biological Mass Spectrometry, Present and Future*, Wiley, New York, 1994, pp. 249–254.

82. R. B. Cody and B. S. Freiser, *Anal. Chem.* **54**, 1054–1056 (1982).

83. R. E. Tecklenburg, Jr., and D. H. Russell, *Mass Spectrom. Rev.* **9**, 405–451 (1990).

84. D. P. Little, J. P. Spier, M. W. Senko, P. B. O'Conner, and F. W. McLafferty, *Anal. Chem.* **66**, 2809–2815 (1994).

85. A. R. Dongre, A. Somogyi, and V. H. Wysocki, *J. Mass Spectrom.* **31**, 339–350 (1996).

86. E. Mirgorodskya, P. Roepstroff, and R. A. Zubarev, *Anal. Chem.* **71**, 4431–4436 (1999).

87. R. A. Zubarev, N. K. Kelleher, and F. W. McLafferty, *J. Am. Chem. Soc.* **120**, 3265–3266 (1998).

88. R. A. Zubarev, N. A. Kruger, E. K. Fridriksson, M. A. Lewis, D. M. Horn, B. K. Carpenter, and F. W. McLafferty, *J. Am. Chem. Soc.* **121**, 2857–2862 (1999).

89. N. A. Kruger, R. A. Zubarev, D. M. Horn, and F. W. McLafferty, *Int. J. Mass Spectrom. Ion Proc.* **185–187**, 787–793 (1999).

5

MOLECULAR MASS MEASUREMENT

One of the most important pieces of information that is required in the elucidation of the molecular structure of a biological compound is its mass. For small molecules (<500 Da), the value of the exact molecular mass can define their elemental composition. In biochemical applications, molecular mass can be used to confirm the DNA-derived sequence of proteins, to determine mutation in proteins, to ascertain posttranslational modifications in proteins, to confirm the base composition of oligonucleotides, to determine the carbohydrate content of proteins, and to confirm the correctness of chemical syntheses of peptides and other biomolecules. Therefore, the first essential step to analyze a compound is to measure its molecular mass.

5.1. DEFINITION OF MOLECULAR MASS

Because atoms and molecules are so tiny, expressing their mass in the normal unit of mass (i.e., in kilograms) is cumbersome. As an example, the mass of a single carbon atom is 1.99266×10^{-26} kg. To avoid this inconvenience, the mass of atoms and molecules is expressed in the units of dalton (Da), where by definition 1 Da is equal to $\frac{1}{12}$th the mass of a single atom of the ^{12}C isotope (i.e., $1\,Da = 1.66054 \times 10^{-27}$ kg). The unified atomic mass unit, abbreviated u, is also used alternatively in place of dalton. Several different molecular mass terms are in use. They are defined as follows:

> *Nominal mass*—the mass of the molecular ion calculated with the mass of the most abundant isotope of each element, and neglecting the mass defect (H = 1, C = 12, etc.).
>
> *Monoisotopic mass*—the mass of the molecular ion calculated with the mass of the most abundant isotope of each element, but including the mass defect (^{1}H = 1.007825, ^{12}C = 12.000000, etc.).
>
> *Most abundant mass*—the mass that corresponds to the most abundant peak in the isotopic cluster of the molecular ion.
>
> *Average mass*—the mass calculated with the average of the isotopic mass of each element, weighted for isotopic abundance. In other words, the average mass represents the centroid of the distribution of the isotopic peaks of the molecular ion.

These distinctions are made clear in Figure 5.1, which gives the theoretical molecular ion distribution of polystyrene with (*a*) *n* = 10, (*b*) *n* = 100, and (*c*) *n* = 1000 [1].

5.2. MONOISOTOPIC VERSUS AVERAGE MASS

Monoisotopic mass is meaningful for compounds of lower mass, because the elemental composition of such compounds can be determined from a well-defined isotopic pattern of the molecular ion (i.e., by using the relative abundances of [M + 1] and [M + 2] ions) [2,3]. The isotopic pattern of small molecules is simple, and can be observed even at low resolution (unit resolution). However, as the mass of a compound increases, the isotope pattern becomes very complex as is evident in Figure 5.1 [1]. Two changes are noticeable in the isotopic pattern of high-mass compounds. First, the isotopic pattern extends over many masses (e.g., over 15 at 10,000 Da, and over 50 at 100,000 Da). Second, it becomes more symmetric. For low-mass compounds, both nominal and monoisotopic masses can be identified with the most abundant peak in the isotopic cluster. However, for most compounds with a molecular mass above 1000 Da (unless they contain more of the mass-deficient elements, e.g., O, P), the nominal mass is meaningless. Also, the monoisotopic peak becomes difficult to identify. For example, above 2000 Da it is no longer the most abundant peak, and may not be visible in the spectrum for high-mass compounds (Figure 5.1*b,c*). Also, the gap between the nominal, monoisotopic, and average masses increases with increasing mass. As an example, in Figure 5.1*b*, the nominal mass is six mass units below the monoisotopic mass (10,458 vs. 10,464.34 Da), and the average mass is 15 and 9 units above the nominal and monoisotopic masses, respectively. This gap will be wider for compounds that contain a large number of hydrogens than for those that contain more elements with a larger negative mass

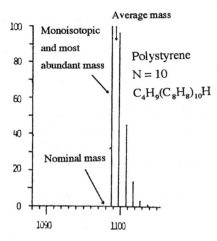

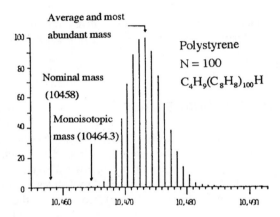

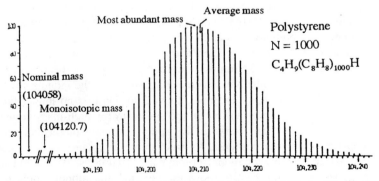

Figure 5.1. Theoretical molecular ion distribution of polystyrene: (*a*) *n* = 10; (*b*) *n* = 100; (*c*) *n* = 1000. [Reproduced from Ref. 1 by permission of American Chemical Society, Washington, DC (copyright 1983).]

defect. This characteristic behavior can be of help in distinguishing compound type. As an example, proteins will exhibit a larger gap than carbohydrates, which in turn will have a larger gap than will oligonucleotides [4].

Therefore, the need for high resolution becomes more critical in resolving the isotopic peaks of high-mass ions and to accurately analyze the isotopic composition of the most abundant peak. However, it becomes difficult to maintain a unit mass resolution for high-mass compounds, and the measured molecular ion profiles coalesce and become a single asymmetric peak. Therefore, in practice, the molecular mass of high-mass biopolymers is determined at low resolution and the value of the average mass is acceptable.

However, there are two valid concerns in the use of the average mass values. First, the measured average mass values always have some uncertainty, which is caused by natural variations in isotopic abundance of elements [5,6]. Second, the measured average mass value is always underestimated because of the inherent nonsymmetric nature of the isotopic distribution profile [7]. Zubarev et al. have shown that an uncertainty of ± 0.1 Da always remains for $<10,000$-Da mass compounds, and that above this mass the uncertainty can be as high as ± 10 ppm [8].

5.3. RESOLVING POWER

The resolving power of a mass spectrometer highly impacts the accuracy of the measured mass. By definition, the resolving power of a mass spectrometer is its ability to distinguish between ions that differ only slightly in their m/z ratio. The justifications for high resolving power are

- To improve accuracy in mass measurement. The exact mass values are required to determine the structure of an analyte, and to ascertain the fragmentation mechanism of an ion.
- To resolve isotopically labeled species when the percent incorporation of the label is to be determined.
- To resolve an isotopic cluster when the charge state of high-mass compounds is to be determined.
- To improve the accuracy of quantification.
- To unambiguously mass-select a precursor ion in tandem mass spectrometry (MS/MS) experiments.

Resolving power is loosely related to resolution (R), which is numerically expressed as

$$R = \frac{m}{\Delta m} \tag{5.1}$$

Resolution is defined in several different ways. The 10% valley definition is shown in Figure 5.2 [9]. As the name implies, in this definition, the two equal height peaks

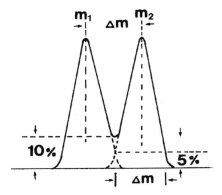

Figure 5.2. Mass resolution (10% valley definition) of two ions of mass m_1 and m_2. [Reproduced from Ref. 9 by permission of Plenum Press, New York (copyright 1994).]

of mass m and $m + \Delta m$ are considered resolved when the valley between the two ions is 10% of the height of either one (i.e., each contributes 5% to the valley). The value of R can also be determined in terms of the width of a single symmetric peak. The peak-width definition is pictorially depicted on the right-hand side of Figure 5.2, where m is the mass of the peak and Δm the width of the peak (at 5% height). The use of 10% valley definition is a usual practice with magnetic sector and quadrupole instruments. With some other types of mass analyzers, such as time-of-flight (TOF), quadrupole ion traps (QIT), and Fourier transform (FT)–ion cyclotron resonance (ICR), the peaks separated by 50% valley are considered as resolved. In such cases, resolution is reported in terms of the peak width measured at half of the maximum height (i.e., FWHM, full width at half maximum).

Depending on a particular application, the molecular mass can be measured at a low ($R < 1000$) or high ($R > 10,000$) resolution. A low-resolution mass measurement is a simple procedure, and can be accomplished with an ordinary instrument. A low resolution here implies that the ions are separated at a unit mass resolution. The low-resolution measurements may not provide the exact mass. Although not a necessary requirement, high resolution is a time-honored tradition for exact mass measurements. However, for this type of measurement, it is absolutely essential that the ion of interest is completely resolved from all other neighboring ions. A high resolving power can be an asset in this respect. If an interfering ion is not resolved, it would shift the position of the peak of interest, and introduce an error in mass measurement. The need for a higher resolution is not so critical for small size molecules, but it assumes a greater importance as the mass of the analyte ion increases. As an example, to mass-resolve ethylene$^{+\cdot}$ (mass $= 28.031300$) and $CO^{+\cdot}$ (mass $= 27.994915$), a resolution of 770 is needed. On the other hand, a resolution of 5500 is required for the separation of the two ions of the elemental composition $C_{13}H_{16}-C_2H_4^{+\cdot}$ and $C_{13}H_{16}-CO^{+\cdot}$, although the mass difference in both examples is 0.036385. Similarly, for low-mass compounds, the possible elemental compositions are few, but become exceedingly large as the mass increases.

$200/.036385 = 5500$

$28/.036385 = 770$

5.4. HIGH-RESOLUTION MASS SPECTROMETERS

As discussed in Chapter 3, a high resolution can be realized with double-focusing magnetic sector mass spectrometers. These instruments provide both velocity and direction focusing through the combined action of electrostatic and magnetic fields. Several factors contribute to a poor resolution, and include ion kinetic energy distribution, and focusing behavior and imperfections in ion optical devices. Efforts have been made to remove these limitations. As a result, a resolution of $>150,000$ can be realized with double-focusing instruments.

Strides have also been made to improve the resolving power of other instrument designs. As an example, the incorporation of delayed extraction (DE) [10] and reflectron (R) mirror [11] has pushed the resolution of TOF machines beyond 10,000. Through novel ideas (multiple pass, operating in higher stability regions, etc.), high-resolution measurements have also been made with quadrupole mass filters [12,13]. By scanning the electric fields at very slow rates, impressive gain in resolution of QIT can be made [14,15]. The field of high-resolution mass spectrometry has benefited immensely from developments in FT-ICR instrumentation [16]. It is now possible to resolve isotopic fine structures of certain midsize proteins.

5.5. EXACT MASS MEASUREMENTS

Many situations require high accuracy in the molecular mass measurement. Primarily, the exact mass data permit the assignment of an elemental composition to an ion of interest. This information is essential in the structure determination and interpretation of mass spectra of an unknown compound or of a component of a mixture. Chemists working with synthetic compounds and natural products rely heavily on the exact mass measurement data for structural assignment. This value is acceptable in lieu of the combustion or other elemental analysis data. An acceptable value of the measured mass should be within ± 5 ppm of the exact mass [17]. An error of 1 ppm corresponds to a deviation of 0.0002 Da for a mass of 200 Da. The exact mass measurement data are also required to distinguish mutations in peptides (e.g., between lysine and glutamine; the mass difference between them is 0.03638 Da).

5.5.1. High-Resolution Exact Mass Measurements by Magnetic Sector Instruments

Traditional methods to determine the exact mass of small molecules have relied on high-resolution mass spectrometry. Double-focusing magnetic sector mass spectrometers, in conjunction with electron ionization (EI), are the ideal instruments for such experiments. [18]. The following procedures are used to acquire the high-resolution mass measurement data.

Full-Scan Mode. One of the simplest ways to obtain the exact mass is to record the spectrum at high resolution ($>10,000$) in the presence of a reference compound. The magnet is scanned to cover the complete mass range of interest. With this technique, the exact masses of all ions present in the spectrum can be determined simultaneously. During the scan, the computer stores the centroid (the center of gravity) and area of each peak. The mass of the ion is exponentially related to the peak centroid time as shown in Eq. 5.2, where M_{max} is the highest mass in the scan, t is the time elapsed from the start of the scan, and τ is a time constant. The instrument is first mass-calibrated by fitting the reference peak mass centroids to the known mass values of the reference ions. From the known mass reference ions, the masses of the sample ions are determined with an accuracy of <5 ppm.

$$M = M_{max} \exp^{-t/\tau} \tag{5.2}$$

In the past, the Mattauch–Herzog geometry double-focusing mass spectrometers, which simultaneously detected all separated ions on a focal plane, were also used for exact mass measurements. A photographic plate was used to obtain a trace of all ions that reach the detector. By using a microdensitometer, the exact position of each mass was accurately measured with respect to the ions of known mass [19]. Currently, diode array detectors can perform a similar task for a limited mass range of the spectrum [20].

Peak-Matching Mode. One of the most precise methods of exact mass measurement is the peak-matching technique. The mass of an unknown is measured by comparing it against the known mass of a reference ion. A precision of <0.3 ppm can be achieved with this method, but with one major disadvantage that the mass of only one ion can be determined at a time. In a typical procedure, the magnetic field is held constant, and the accelerating voltage is changed in such a manner that a reference ion and the unknown ion alternatively follow the same trajectory. The lower of the two masses is first displayed on an oscilloscope screen, and the accelerating voltage is varied by a precision (within 1 ppm) voltage divider until the higher mass ion is superimposed on it. The unknown mass is measured from this ratio using Eq. 5.3. In order to achieve a higher accuracy in mass measurement by this method, it is essential that the mass of the reference ion be within 2% of the unknown mass. Computer programs have been written for the peak-matching procedure and implemented on the current generation of mass spectrometers. Computer algorithms perform signal averaging, smoothing, and peak-centroiding routines to provide an improved mass accuracy [21].

$$m_2 = m_1 \frac{V_2}{V_1} \tag{5.3}$$

The exact mass data can be easily obtained for samples that are accessible to EIMS analysis. In contrast, such measurements are difficult with fast-atom bombardment (FAB) because the alkali salts that are used as reference material tend to suppress the sample ionization/desorption processes. Also, the sample current does not remain steady during the measurement, and the reference ions are often not sufficiently close. For exact mass measurement in FAB, it is advisable to use the matrix ions as the reference ions, or to use some internal mass calibrants such as polyethylene or polypropylene glycols (PEGs or PPGs). Alternatively, a compound closely related in mass and chemical characteristics to the analyte can also be used as a reference compound. An automated exact mass measurement system that consists of ESI and double-focusing magnetic sector instrument has been developed for organic compounds [22]. This system can analyze 15–20 samples per hour with a mass measurement accuracy within the acceptable limit of 5 ppm.

5.5.2. High-Resolution Exact Mass Measurements by MALDI-TOFMS

Russell and Edmondson have discussed factors that affect exact mass measurements by matrix-assisted laser desorption/ionization (MALDI)-TOFMS [23]. These factors include the peak-shape profile, flight-time drift, and sample surface inhomogeneity. Abnormal peak shapes occur as a result of unresolved multiplets and poor signal-to-noise (S/N) ratio, both of which cause an error in centroiding the peak. The S/N ratio can be improved by signal averaging. The flight-time drift is caused primarily by a drift in the high-voltage power supply, and can be minimized by allowing enough time for warming the power supply prior to data acquisition. The sample surface inhomogeneity leads to variations in the sample thickness, which can affect the accuracy of the flight time and mass measurements. An improvement in sample preparation techniques [24,25], and the use of internal mass calibration can overcome the effect of the sample hetereogeneity problem. Another improvement that can be made is in the mass calibration equation shown below:

$$t - k_1 \left(\frac{m}{z}\right)^{1/2} + k_2 \tag{5.4}$$

where k_1 is the constant of proportionality between the arrival time of an ion and its m/z and k_2 is a time offset that arises from the difference in time between the ion extraction and the data acquisition start pulse. It is assumed in this equation that all ions that exit the source have the same kinetic energy; that assumption may not be true for the MALDI-formed ions. A third constant term is usually included into this equation to account for the initial velocity spread [26]. When all of these factors are controlled, a mass measurement accuracy of better than ± 3 ppm can be achieved with an internal mass calibration on a MALDI-DE/R-TOFMS. With an external mass calibration, the mass measurement accuracy is degraded somewhat. By use of a single lock mass, even a low-resolution TOFMS can be used to provide a mass accuracy better than 5 ppm.

5.5.3. High-Resolution Exact Mass Measurements by FT-ICRMS

Fourier transform–ICRMS currently yields the highest resolving power of any mass spectrometry instrument [27–29]. The achievable resolution in this instrument, however, is inversely proportional to the mass of an ion, and degrades when the pressure and population of ions inside the ICR cell increase. The ion-trapping voltage and a greater number of ions in the cell can cause shifts in the cyclotron frequencies of the trapped ions. These shifts are translated into errors in the measured mass values. Despite these reservations, impressive results have been obtained with respect to the resolution and accuracy in mass measurement. A resolution of several millions has been demonstrated with this instrument [30]. The use of an internal calibrant can minimize the effect of frequency instabilities. Wu et al. have shown that, by bracketing the unknown mass with two references, the acceptable accuracy (<3 ppm) for the elemental analysis can be achieved [31]. A mass measurement accuracy of <2 ppm is common with FT-ICRMS even for midsize molecules (~5000 Da) [32].

Because resolution is inversely proportional to mass, the multiple-charging feature of ESI is attractive for high-resolution measurements of high-mass biopolymers with FT-ICRMS. Marshall and coworkers have resolved the isotopic fine structure of p16 tumor suppressor protein (molecular mass = 15.8 kDa) at a resolving power of 8,000,000 by using a 9.4-T magnet [30]. They were able to count the number of sulfur atoms from the ratio of the ^{34}S abundance to the monoisotopic abundance.

5.5.4. Exact Mass Measurements at Low Resolution with a Quadrupole Mass Filter

Exact mass measurement can also be performed at low resolution, with the stipulation that the sample not contain any impurity, and that the interference from the background and reference ions be minimal. A high value of the sample ion current is also a prerequisite [33]. In this respect, ESI has an edge over other modes of ionization. By statistical evaluation of peak shapes, it is possible to ascertain the presence of interfering peaks [34]. An interference peak that is 4.4% as abundant as the analyte peak can cause mass measurement errors in excess of 5 ppm.

The use of ESI with a single-quadrupole mass spectrometer has been demonstrated for exact mass analysis [35]. Two improvements are made to the normal procedure when a low-resolution instrument, such as quadrupole, is used for exact mass measurement. First, the mass calibration is performed in the multichannel acquisition (MCA) mode, which is a signal-averaging routine, in which multiple scans are summed up. Because the S/N ratio increases as a function of the square root of the number of scans, a higher value of the S/N ratio is realized with the MCA routine, the consequence of which is a much improved peak profile that has less distortion at the peak top. As a result, an improvement in the peak assignment is ensured. Second, an improved data analysis procedure, such as a Gaussian curve-fitting routine, is used to calculate the average mass of the ion. By using this protocol, the mass of a protein, equine myoglobin (16,950.4 Da), was determined with a precision of 12 ppm [35].

A somewhat similar procedure was reported by another research group [33]. In this procedure, bracketing the sample ion with known mass reference ions provides the mass calibration, and the spectra are acquired in the MCA mode with 128 channels per m/z unit. The noise is removed by smoothing and subtraction routines. Knowledge of the correct position of the centroid of each peak is essential to improve the mass measurement accuracy. The inclusion of only those data points that have intensities greater than a predetermined threshold in the mass measurement routine leads to improvements in assigning the position of the centroid. A mass measurement accuracy of <4.5 ppm has been reported with this procedure. A few other reports of exact mass measurements at low resolution have also been described [36,37].

5.6. MASS MEASUREMENT OF BIOPOLYMERS

Traditionally, the molecular mass of biopolymers such as proteins has been estimated with sodium dodecyl sulfate–polyacrylamide gel electrophoresis (SDS-PAGE). For example, the migration pattern of a protein is compared to a set of known mass proteins. Many biochemistry laboratories still rely on this technique. However, the accuracy of this procedure is very poor (5–20%), and mitigates against its use in studies such as differentiation of protein variants. The exact mass values are required to distinguish a mutation between Asp and Asn, Asn and Ile/Leu, Glu and Gln, and Lys and Glu/Gln (for three-letter code and residue masses of amino acids, see Table 9.1). Currently, mass spectrometry is meeting this challenge with an eloquent success. The introduction of two highly sensitive ionization techniques, namely, ESI [38] and MALDI [39], has contributed to the routine use of mass spectrometry for molecular mass measurements of biopolymers with an accuracy of <0.03%.

The basis of molecular mass measurement by ESI is the formation of multiply charged ions of the analyte. For high-mass biopolymers, an envelope of a series of ions, each differing by one charge from its neighbor, is formed (see Figure 2.16). From the knowledge of the measured masses (m' and m'' given by Eqs. 5.5 and 5.6, respectively) of two neighboring ions and their charge states, the molecular mass of a biopolymer can be calculated by solving Eqs. 5.7 and 5.8.

$$m' = \frac{M + nH}{n} \tag{5.5}$$

$$m'' = \frac{M + (n + 1)H}{n + 1} \tag{5.6}$$

$$n = \frac{m'' - H}{m' - m''} \tag{5.7}$$

$$M = n(m' - H) \tag{5.8}$$

Here, m' is the mass of the $[M + nH]^{n+}$ ion (H = 1.007829 is the mass of the proton) and m'' ($m' > m''$) of the adjacent $[M + (n + 1)H]^{n+1}$ ion. A better accuracy in mass measurement is achieved if the value of M is first calculated from the successive pairs of adjacent ions, and then the average of these values is computed. The mass spectrum is obtained by scanning the narrow mass range that encompasses the multiply charged ion envelope in the MCA mode. The data are processed with usual smoothing, background subtraction, and peak-centroiding routines. The use of the maximum entropy (MaxEnt) method further improves the accuracy of mass measurements [40]. In favorable cases, the molecular mass of macromolecules can be calculated with an accuracy of better than $\pm0.005\%$. If ESI is combined to a high-resolution instrument, such as FT-ICRMS, then this figure can be improved further to $\pm0.001\%$.

5.7. MASS MEASUREMENTS OF MEGADALTON-SIZED MOLECULES

A new mass spectrometry technique, called *charge detection mass spectrometry* (CDMS), has been developed to measure the mass of megadalton-sized molecules such as nucleic acids [41,42]. The instrumentation is very simple, and consists of an ESI source and a charge detector that is made up of a small thin-walled brass tube and an associated charge amplifier. In principle, this technique measures the velocity of the ESI-produced ions, and from the relation given in Eq. 5.9, the mass of the ion can be calculated. Here, q is the ion charge, V is the bias voltage on the ion guide, v_m is the measured velocity, and v_g is the velocity due to the gas expansion. The value of v_g is measured by electrically grounding all electrostatic elements and measuring the time of passage of the residue particles through the detector. The resolution of this technique is very poor. The technique has been applied to determine the mass of double-stranded, circular DNA and single-stranded, circular DNA in the range of 2500–8000 base pairs (bp) (1.5–5.0 MDa) [42]:

$$m = \frac{2qV}{v_m^2 - v_g^2} \tag{5.9}$$

The mass of megadalton DNA molecules has also been determined with an ESI and FT-ICRMS combination [43,44].

5.8. MASS CALIBRATION STANDARDS

A reference compound is needed for the measurement of exact mass. The mass of an unknown is computed by comparing its signal with the signal of the known mass reference peaks. A computer establishes a mass scale from the known masses of reference ions to assign mass values to the sample ions. A reference compound is also required to calibrate the data system and to tune and check the performance of the instrument.

An ideal reference compound should have the following characteristics:

- It should provide a sufficient number of high relative intensity ions that are spaced regularly throughout the scan range.
- The reference ions should have a negative mass defect.
- The reference compound should be readily available.
- It should be chemically inert.
- It should be sufficiently volatile so that it can be pumped away readily from the ion source.

The compounds that are used commonly as calibration standards in EI, CI, FAB, and ESI modes of ionization are listed in Table 5.1.

Table 5.1. Mass calibration standards for EI, CI, FAB, and ESI

Compound	Structure	Ionization Mode	Mass Range (Da)
Perfluorokerosene (PFK)	$CF_3(CF_2)_nCF_3$	EI, CI	0–900
Perfluorotributylamine	$(C_4F_9)_3N$	EI, CI	0–600
Ultramark F series (Fomblin)	$CF_3O(CFCF_3CF_2O)_m-$ $(CF_2O)_nCF_3$	EI, CI	≤ 3500
Ultramark 1621 (perfluoroalkoxycyclo-triphosphazines)		EI, CI	≤ 3000
		ESI	≤ 6000
Tris(perfluoroalkyl)-s-triazine		EI	≤ 1500
Glycerol	$CH_2(OH)CH(OH)-CH_2OH$	FAB	20–1200
Cesium iodide	CsI	FAB	130–30,000
		ESI	≤ 3000
CsI/glycerol		FAB	90–3500
CsI/NaI/RbI		FAB	20–1000
LiI/NaI		FAB	0–1000
Polyethylene glycol	$H(OCH_2CH_2)_nOH$	FAB, ESI	50–2000
Polypropylene glycol	$H(OCH(CH_3)CH_2)_nOH$	FAB, ESI	50–2000
Myoglobin		ESI	600–2500
Lysozome		ESI	1000–2100
Polypropylene glycol sulfate		ESI (-ve)	300–1700
Sodium trifluoroacetate	CF_3COONa	ESI	100–4000
Tetraethyl ammonium iodide	$(C_2H_5)NI$	ESI (-ve)	≤ 6000
Water clusters	$H(H_2O)_n$	ESI	≤ 4000
Cesium tridecafluoroheptanoate	$C_7H_{13}O_2Cs$	ESI	$\leq 10,000$

Table 5.2. Mass calibration standards for MALDI

Compound	Mass
Gramicidin-S	1,141.5
ACTH (18–39)	2465.7
Bovine insulin	5733.5
Horse heart cytochrome C	12,360
Chicken lysozyme	14,307
Horse heart myoglobin	16,951
Bovine trypsinogen	23,957
Bovine serum albumin	66,431

The most widely used mass calibrants in EIMS are perfluorokerosene (PFK) and perfluorotributylamine [PFTBA; $(C_4F_9)_3N$; also known as *heptacosa*]. Both perform well for ≤ 900-Da mass compounds. For the high-mass (≤ 3000-Da) analysis, triazines and Ultramark (a mixture of fluorinated phosphazenes) are appropriate as reference mass markers.

Alkali-metal halides (e.g., CsI, LiI, NaI, RbI) are the most appropriate reference compounds for FABMS. CsI yields a series of ions of the compositions $Cs(CsI)_n^+$ and $I(CsI)_n^-$ in the positive- and negative-ion modes, respectively, with masses that extend beyond 30,000 Da. One difficulty with these standards is that the adjacent peaks are too far apart. To overcome this limitation, a mixture of NaI, CsI, and RbI can be used. The glycerol matrix can also serve as a mass calibrant up to ~ 1200 Da. Intense cluster ions of m/z $(92n + 1)$ and $(92n - 1)$ (where 92 is the mass of glycerol and n is an integer) are seen in the positive- and negative-ion analysis modes, respectively. A saturated solution of CsI in glycerol can be an effective mass calibrant up to 3000 Da.

Several calibration standards have been proposed for ESIMS. These compounds include PEG [45], PPG [45], CsI [46–48], sodium trifluoroacetate [49], tetraethyl ammonium iodide [48], and cesium salts of monobutyl phthalate, heptafluorobutyric acid, tridecafluoroheptanoic acid, and perfluorosebacic acid [51]. The multiple-charged ions of proteins [52] and ion clusters of water [53] have also been used as mass calibrants.

Mass calibration in MALDI-TOF is a very simple task, and requires only two ions, one of which can be from the matrix and the other from an internal mass calibrant. Certain peptides and proteins can be used as one of the internal calibrants. Common calibrants used in MALDIMS are listed in Table 5.2.

REFERENCES

1. J. Yergy, D. Heller, G. Hansen, R. J. Cotter, and C. Fenselau, *Anal. Chem.* **55**, 353–356 (1983).

2. J. H. Beynon, in R. M. Elliott, ed., *Advances in Mass Spectrometry*, Pergamon Press, London, 1962, Vol. 2; *Proc. Natl. Acad. Sci.* (USA) **38**, 667–678 (1952).

3. F. W. McLafferty and F. Turecek, *Interpretation of Mass Spectra*, University Science Books, Mill Valley, CA, 1993.

4. K. Biemann, in J. A. McCloskey, ed., *Methods in Enzymology*, Academic Press, New York, 1990, Vol. 193, pp. 295–395.

5. S. C. Pomerantz and J. A. McCloskey, *Org. Mass Spectrom.* **22**, 251–253 (1987).

6. R. C. Beavis, *Anal. Chem.* **65**, 496–497 (1993).

7. R. A. Zubarev, *Int. J. Mass Spectrom. Ion. Proc.* **107**, 17–27 (1991).

8. R. A. Zubarev, P. A. Damirev, P. Hakansson, and B. U. R. Sundqvist, *Anal. Chem.* **67**, 3793–3798 (1995).

9. C. Dass, in D. M. Desiderio, ed., *Mass Spectrometry: Clinical and Biomedical Applications*, Plenum Press, New York, 1994, Vol. 2, pp. 1–52.

10. R. S. Brown and J. J. Lennon, *Anal. Chem.* **67**, 3990–3999 (1995).

11. B. A. Mamyrin, *Int. J. Mass Spectrom. Ion Proc.* **131**, 1–19 (1994).

12. M. Amad and R. S. Houk, *Anal. Chem.* **70**, 4885–4889 (1998).

13. M. Amad and R. S. Houk, *J. Am. Soc. Mass Spectrom.* **11**, 407–415 (2000).

14. J. C. Schwartz, J. E. P. Syka, and I. Jardine, *J. Am. Soc. Mass Spectrom.* **2**, 198–204 (1991).

15. J. D. Williams, K. A. Cox, R. G. Cooks, and J. C. Schwartz, *Rapid Commun. Mass Spectrom.* **5**, 327–329 (1991).

16. J. A. Loo, J. P. Quinn, S. I. Ryu, K. D. Henry, M. W. Senko, and F. W. McLafferty, *Proc. Natl., Acad. Sci.* (USA) **89**, 286–289 (1992).

17. M. L. Gross, *J. Am. Soc. Mass Spectrom.* **5**, 57 (1994).

18. W. J. McMurray, B. N. Green, and S. R. Lipsky, *Anal. Chem.* **38**, 1194–1204 (1966).

19. K. Biemann, P. Bommer, and D. M. Desiderio, *Tetrahedron Lett.* **38**, 1725–1731 (1964).

20. A. L. Burlingame, in T. Matsuo, R. M. Caprioli, M. L. Gross, and Y. Seyama, eds. *Biological Mass Spectrometry, Present and Future*, Wiley, New York, 1994, pp. 147–164.

21. C. G. Hammer, G. Peterson, and P. T. Carpenter, *Biomed. Mass Spectrom.* **1**, 397–411 (1974).

22. G. Perkins, F. Pullen, and C. Thompson, *J. Am. Soc. Mass Spectrom.* **10**, 546–551 (1999).

23. D. H. Russell and R. D. Edmondson, *J. Mass Spectrom.* **32**, 263–276 (1997).

24. F. Xiang and R. C. Beavis, *Rapid Commun. Mass Spectrom.* **8**, 199–204 (1994).

25. O. Vorm, P. Roepstorff, and M. Mann, *Anal. Chem.* **66**, 3281–3287 (1994).

26. C. C. Vera, R. A. Zubarev, H. Ehring, P. Hakansson, and B. U. R. Sundqvist, *Rapid Commun. Mass Spectrom.* **10**, 1429–1432 (1996).

27. F. W. McLafferty, *Acc. Chem. Res.* **27**, 379–386 (1994).

28. Y. Li, R. T. McIver, Jr., and R. L. Hunter, *Anal. Chem.* **66**, 2077–2083 (1994).

29. I. Vidavsky and M. L. Gross, in F. Settle, ed., *Handbook of Instrumental Techniques for Analytical Chemistry*; Prentice-Hall PTR, Upper Saddle River, NJ, 1998, pp. 589–608.

30. S. D.-H. Shi, C. L. Hendrickson, and A. G. Marshall, *Proc. Natl. Acad. Sci.* (USA) **95**, 11532–11537 (1998).

31. J. Wu, S. T. Fannin, M. A. Franklin, T. F. Molinski, and C. B. Lebrilla, *Anal. Chem.* **67**, 3788–3792 (1995).

32. K. L. Walker and C. L. Wilkins, in F. Settle, ed., *Handbook of Instrumental Techniques for Analytical Chemistry*; Prentice-Hall PTR, Upper Saddle River, NJ, 1998, pp. 665–682.

33. A. N. Tyler, E. Clayton, and B. N. Green, *Anal. Chem.* **68**, 3561–3569 (1996).

34. K. F. Blom, *J. Am. Soc. Mass Spectrom.* **9**, 789–798 (1998).

35. R. Feng, Y. Konishi, and A. W. Bell, *J. Am. Soc. Mass Spectrom.* **2**, 387–401 (1991).

36. C.-G. Hammer and R. Hessling, *Anal. Chem.* **43**, 298–306 (1971).

37. J. Roboz, J. F. Holland, M. A. McDowell, and M. J. Hillmer, *Rapid Commun. Mass Spectrom.* **2**, 64–66 (1988).

38. J. B. Fenn, M. Mann, C. K. Meng, S. F. Wong, and C. M. Whitehouse, *Science* **246**, 64–71 (1989).

39. M. Karas and F. Hillenkamp, *Anal. Chem.* **60**, 2299–2301 (1988).

40. A. G. Ferrige, M. J. Seddon, B. N. Green, S. A. Jarvis, and J. Skilling, *Rapid Commun. Mass Spectrom.* **6**, 707–711 (1992).

41. S. D. Fuerstenau and W. H. Benner, *Rapid Commun. Mass Spectrom.* **9**, 1528–1538 (1995).

42. J. C. Schultz, C. A. Hack, and W. H. Benner, *J. Am. Soc. Mass Spectrom.* **9**, 305–313 (1998).

43. X. Chen, D. G. Camp II, Q Wu, R. Bakhtiar, D. L. Springer, B. J. Morris, J. E. Bruce, G. A. Anderson, C. G. Edmonds, and R. D. Smith, *Nucl. Acids Res.* **24**, 2183–2189 (1996).

44. R. Chen, X. Chen, D. W. Mitchell, S. A. Hofstadler, Q. Wu, A. L. Rockwood, M. G. Smith, and R. D. Smith, *Anal. Chem.* **67**, 1159–1163 (1995).

45. B. S. Larsen and C. N. McEwen, *J. Am. Soc. Mass Spectrom.* **2**, 205–211 (1991).

46. R. B. Cody, J. Tamura, and B. D. Musselman, *Anal. Chem.* **64**, 1561–1570 (1992).

47. S. Pleasance, P. Thibault, P. G. Sim, and R. K. Boyd, *Rapid Commun. Mass Spectrom.* **5**, 307–308 (1991).

48. C. E. C. A. Hop, *J. Mass Spectrom.* **31**, 1314–1316 (1996).

49. M. Moini, B. L. Jones, R. M. Rogers, and L. Jiang, *J. Am. Soc. Mass Spectrom.* **9**, 977–980 (1998).

50. M. Moini, *Rapid Commun. Mass Spectrom.* **8**, 711–714 (1994).

51. S. König and H. M. Fales, *J. Am. Soc. Mass Spectrom.* **10**, 273–276 (1999).

52. C. N. McEwen and B. S. Larsen, *Rapid Commun. Mass Spectrom.* **6**, 173–178 (1992).

53. D. W. Ledman and R. O. Fox, *J. Am. Soc. Mass Spectrom.* **8**, 1158–1164 (1997).

6

QUANTITATIVE ANALYSIS

Quantitative analysis is a cornerstone of many health-related fields such as clinical chemistry, pharmaceutical science, forensic science, and environmental science. The quantification of tissue levels of a biological compound provides information that leads to an understanding of its functional role in various neurological and pathophysiological events. The development of a drug and its clinical trial heavily rely on quantitative analysis. The management of various forms of illnesses requires quantitative analysis of endogenous biomolecules in the extracts of body tissues and fluids. The quantification of drugs of abuse and their metabolites is also essential in the management of the overdose patients.

Mass spectrometry has several desirable features that make it the most sought-after analytical technique for quantitative analysis, including the following:

- *Sensitivity*—mass spectrometry is one of the most sensitive analytical instrumental techniques. It has the capability to detect a single ion. With current instrumentation, detection limits in the attomole–femtomole range are common.
- *Specificity*—mass spectrometry provides structure-specific data in terms of the molecular mass and compound-related fragment ions. In addition, its combination with a separation device provides a high-resolution separation of the target compound(s), and imparts an additional compound-specific retention parameter.
- *Stable isotope*—labeled internal standards can be used.
- *High dynamic range.*

- *Speed of analysis.*
- *Applicability to a variety of compound types.*

Quantitative analysis must be performed in compliance with good laboratory practices (GLP) as defined by various regulatory agencies. This aspect has been discussed in a report in the *Journal of the American Society for Mass Spectrometry* [1]. Some of the essential aspects of quantitative analysis by mass spectrometry are discussed in this chapter. The field of quantitative analysis has been reviewed [2,3].

6.1. CALIBRATION

Like any other analytical instrument, the response of a mass spectrometer is not absolute. The instrumental response may show variation with time. In addition, the sample matrix may influence the signal. Therefore, calibration of the instrument is an essential part of a quantitative procedure. Calibration involves finding the correlation between a known concentration of the analyte and the resultant mass spectrometry signal. Calibration curves can be plotted manually, or calibration algorithms can be made part of the computer software.

Depending on the level of accuracy and precision desired, the calibration may be performed in one of the following ways. The first procedure involves the analysis of several standards of exactly known concentrations of the analyte, and plotting of the mass spectrometry response (signal intensity or the area underneath a chromatography peak) versus the concentration of the analyte. Three to five replicate standards of each concentration are analyzed. This procedure is the simplest form for preparation of a calibration. However, its accuracy and precision are limited because of the possibility that the mass spectrometry response may change from the time the calibration is drawn, and because of the difficulty to match the matrix of the calibration standards with the sample matrix. In the second method, known as the *standard addition method*, various concentrations of the standard solutions of the analyte are added to the fixed amount of the analyte, and the calibration curve is obtained as in the first method. The third method is the most accurate because it makes use of an internal standard. In this method, a constant amount of the internal standard is added to all calibration solutions that contain known concentrations of the analyte. The ratio of the response between the internal standard and the analyte (or the reverse) is plotted as a function of the analyte concentration.

6.2. INTERNAL STANDARD

In order to achieve a high level of accuracy and precision in quantitative analysis, the use of an *internal standard* is mandatory. An internal standard acts as a self-correcting system because it can account for any fluctuations in the mass spectrometry response and the sample losses that may occur in various sample-handling and chromatographic steps. An ideal internal standard is the one whose physical and chemical properties exactly match those of the analyte in terms of the extraction efficiency, chromatographic characteristics, ionization efficiency, and fragmentation behavior. Three different types of internal standards have found a niche in quantitative analysis. The first type includes any compound whose chemical and physical properties are similar to those of the analyte. This type of internal standard, however, does not provide the level of precision and accuracy that is required for many applications. The second type includes those compounds that are structurally homologous to the analyte. For example, [Ala]^2-leucine enkephalin was used as an internal standard in the quantification of leucine enkephalin from tissue samples [4]. It is a common practice to use this type of internal standard in the quantification of pharmaceutical drugs and their metabolites, primarily because of the cost-effectiveness. However, a homologous compound must be used only when a greater match is found between its chromatographic and mass spectrometry characteristics with those of the analyte molecules. The most ideal internal standard is a stable isotope-labeled analog of the analyte because this type of compound fulfills all the criteria listed above. However, its cost can be prohibitive. In addition, there is a danger that the stable isotope-labeled analog may contain unlabeled analyte as an impurity.

Several stable isotopes (^{2}H, ^{13}C, ^{15}N, ^{18}O, ^{34}S, and ^{37}Cl) can be incorporated in an isotope–labeled internal standard. However, ^{2}H is the most frequently used isotope because it is less expensive than other stable isotopes. Other factors that must be considered in the selection of a stable isotope–labeled internal standard are ease of synthesis, the position of the isotope label, and the required shift in the mass of the internal standard. The isotope label must be in nonexchangeable positions, and the mass of the internal standard should be away from the isotopic pattern of the analyte ion. A shift of 4–5 Da in the mass of the internal standard is reasonable for compounds of <500 Da mass.

The use of deuterated internal standards has been reported for the quantification of endogenous peptides, prostoglandins, pharmaceutical drugs, and the drugs of abuse. In the examples of quantification of methionine enkephalin [3,5–7] and β-endorphin [3,5–7], the deuterium was incorporated in the phenyl ring of the phenylalaline and Ile^{22} residues, respectively. The phenyl ring–deuterated internal standarad was also used in the quantification of μ-opioid receptor agonist Tyr–D–Arg–Phe–Lys–NH$_2$ from ovine plasma [8].

6.3. MASS SPECTROMETRY MEASUREMENTS

In a quantification experiment, the mass spectrometry data can be collected in one of the following ways:

- A wide-range mass scan
- A narrow-mass range scan

- A selected-ion monitoring (SIM)
- A selected-reaction monitoring (SRM) mode

The sensitivity, specificity, the type of information desired, and the cost of instrumentation are the criteria that influence the choice of a particular technique. The first two techniques are used less frequently because of the poor detection sensitivity, which is the consequence of the fact that a large amount of the sample ion current is wasted while the wide- and narrow-mass scans are acquired. The advantage of a wide-mass range scan is that one can retrieve from it a compound-specific spectrum, which provides additional data to confirm the identity of the analyte. Also, the presence of impurities, if any, can be ascertained from this spectrum. The scanning in the narrow-mass range improves the detection sensitivity somewhat, and still provides the isotopic pattern of the sample ion. An example of the use of a narrow-mass range scan is presented in Figure 6.1, which represents the molecular ion traces of NAIIK (Asn–Ala–Ile–Ile–Lys) and its deuterated analog [5]. This pentapeptide is a surrogate compound for the quantitative measurement of β-endorphin, which is a large polypeptide that contains 31 amino acids, and has a

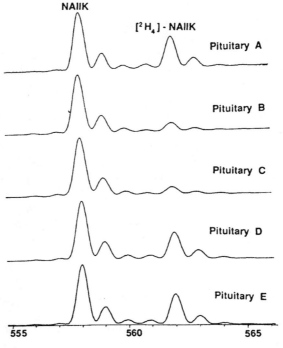

Figure 6.1. FABMS narrow-mass scan of the $[M + H]^+$ region (m/z 554–567) of NAIIK. The $[M + H]^+$ of the deuterated internal standard is also monitored. These scans are from the extracts of human pituitaries. [Reproduced from Ref. 5 by permission of John Wiley & Sons (copyright 1991).]

molecular mass of 3463 Da. The detection sensitivity of the intact β-endorphin by fast-atom bombardment (FAB)MS is very poor [9]. A significant increase in the detection sensitivity is observed when NAIIK, a smaller segment of β-endorphin$_{20-24}$, is chosen as the analyte. This pentapeptide is generated by trypsinolysis of β-endorphin. [^{2}H$_4$-Ile^{22}]-β-endorphin serves as the internal standard, which when digested with trypsin produces [^{2}H$_4$-Ile^{22}]-NAIIK.

6.3.1. Selected-Ion Monitoring

The SIM mode of ion detection, which has also been referred to as *mass fragmentography* [10], is the most frequently used approach for quantitative analysis. In this procedure, the ion current from only one or a few selected ions, rather than a complete mass spectrum, is repetitively recorded. In contrast to the acquisition of a full-scan spectrum, SIM exclusively monitors during the entire analysis time the selected m/z values. By this dedicated use of the mass spectrometry detection, the most of the sample ion current at the selected mass values can be recorded; therefore, a significant improvement in the detection sensitivity is realized as compared to the full-scan data acquisition. However, this scan mode results in massive loss of spectral information.

In practice, the switching the accelerating voltage (or of some other parameter) of a mass spectrometer between the desired values provides a means of monitoring the selected ions by a focal plane detector (FPD). Magnetic sector [11], quadrupole, quadrupole ion-trap (QIT) [12], and time-of-flight (TOF) [13] mass spectrometers all can be used for SIM measurements. In magnetic sector instruments, a convenient way to record the ion current at different m/z values is to change the accelerating voltage rather than the magnetic field strength. The monitoring of different mass-selected ions is even more convenient in quadrupoles and QIT instruments because only the direct current and radiofrequency fields are involved in the mass analysis.

Generally, the molecular ion, provided it is the most intense ion, is chosen for the analysis. In the absence of a prominent molecular ion, any other next most abundant high-mass fragment ion can serve the purpose. In certain situations, the molecular specificity of the SIM procedure may be at risk. In the selection of a specific ion for SIM measurements, it is presumed that this ion is exclusively derived from the analyte molecules. In practice, it is likely that some extraneous ions may have m/z values identical to the chosen analyte ion. The likelihood of this situation is more common with biological or environmental samples, which often lack the fidelity of a pure sample even after extensive chromatographic separation. Furthermore, matrices (as in FAB) or solvents used in some ionization techniques may also produce a background signal at the selected m/z values.

The detection selectivity of the SIM procedure can be improved in several ways: (1) a judicial selection of the ion to be monitored ensures that no background ion contributes to the chosen m/z value; (2) the monitoring of more than one m/z values will also reduce the contribution of the extraneous ions; (3) increasing the resolution may eliminate the contribution of the interfering ions, but at the cost of reduced detection sensitivity [14,15]; (4) the SRM technique can be used instead of SIM

[16]; and (5) finally, the coupling of SIM and SRM with chromatography separation devices pushes the detection selectivity to a much higher level. The last two procedures are discussed below.

6.3.2. Selected-Reaction Monitoring

The SRM is a tandem mass spectrometry technique that can be used in situations where SIM is beset with interference from other impurities or from the matrix background. In this technique, the two mass analyzers of a tandem mass spectrometer are set to monitor the ion current due to a specific precursor–product pair. The molecular ion of the analyte is first mass-selected by MS1, and is allowed to fragment unimolecularly or via collision-induced dissociation (CID). A specific product ion that is exclusively formed in this fragmentation reaction is monitored by MS2. The molecular specificity of the SRM technique is the result of the structural link that is maintained during the analysis between the mass-selected precursor ion and its product ions. Only those compounds that produce the preselected precursor ion and the chosen product ion will be detected by this procedure. The term *multiple-reaction monitoring* (MRM) refers to the use of more than one reaction from either the same precursor or more than one precursor. The following three criteria must be considered for the success of an SRM (or MRM) analysis:

- The fragmentation reaction chosen for quantification should be unique to the analyte.
- The precursor-product selection should provide strong ion current signal.
- An isotope-labeled internal standard, in which the stable isotope is present at a site that yields the labeled fragment ion, should be available.

The obvious advantage of SRM over SIM is the increased level of selectivity. In addition, because SRM virtually eliminates chemical noise, the confidence level of the measurement is increased further. However, the procedure is less sensitive than SIM because a large portion of the precursor ion current is distributed in its CID products. Also, the tandem instruments are more expensive than those used in the SIM analysis.

An SRM quantitative analysis is performed by a procedure identical to that used in the product ion MS/MS analysis (see Chapter 4) except that, instead of acquiring a full-scan spectrum, only a selected product(s) is monitored. A triple-sector quadrupole is ideally suited for SRM studies, although the magnetic sector and QIT instruments have also been used. A typical protocol that uses an SRM procedure for the quantification of biological compounds from tissue extracts is illustrated in Figure 6.2. This is an example of the quantification of β-endorphin in pituitary tumors. In this procedure, FAB is used as the ionization method, and a B/E linked-field scan on a magnetic sector tandem mass spectrometer is the SRM technique to monitor the transitions shown in Eqs. 6.1 and 6.2 [5]. For the reason mentioned above, NAIIK acts as a surrogate analyte. The product ion chosen is

NAI^+ because it is the most abundant product ion in the MS/MS spectrum of the $[NAIIK + H]^+$ ion [9]. The tissue is homogenized for 30 s in cold (4°C) acetic acid and equilibarted for 2 h with $[^2H_4\text{-}Ile^{22}]\text{-}\beta$-endorphin. The supernatant obtained after centrifugation of the homogenate is purified by C-18 solid-phase extraction (SPE) cartridge, followed by RP-HPLC. The lyophilized β-endorphin fraction is digested with trypsin, and NAIIK is isolated by RP-HPLC. The calibration curve is obtained from five standard solutions of synthetic β-endorphin. A fixed amount of the internal standard is added to the tissue sample and to each standard solution:

$$(NAIIK + H)^+ \rightarrow NAI^+$$

$$\text{558 Da} \qquad \text{299 Da} \tag{6.1}$$

$$([^2H_4\text{-}Ile^{22}]\text{-}NAIIK + H)^+ \rightarrow [^2H_4\text{-}Ile^{22}]\text{-}NAI^+$$

$$\text{562 Da} \qquad\qquad\qquad \text{303 Da} \tag{6.2}$$

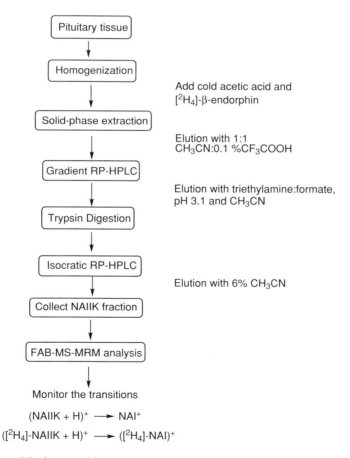

Figure 6.2. A protocol for the quantification of β-endorphin from human pituitary.

6.4. LIQUID CHROMATOGRAPHY/MASS SPECTROMETRY FOR QUANTITATIVE ANALYSIS

The development of a robust and reliable on-line combination with LC and mass spectrometry via atmospheric pressure ionization (API) [17,18] has greatly advanced the science of quantitative analysis of biological compounds [19,20]. The speed of analysis, simplicity, and enhanced selectivity are the high points of this combination. Despite its high cost as compared to the conventional UV and fluorescence detection systems, LCMS has become the most sought-after technique for the quantification of a variety of compound types. As is discussed in Chapter 7, HPLC is more conveniently coupled to mass spectrometry via atmospheric pressure chemical ionization (APCI) and electrospray ionization (ESI) interfaces [19,20]. Both ionization methods are complementary. APCI is applicable to those compounds that do not readily accept charges under standard ESI conditions. Several types of small organic molecules fall in this category. Although a common norm is to use narrow- and wide-bore columns with the APCI interface, the current ESI source designs can accept a wide range of solvent flows [nanoliter–milliliter (nL–mL) range]. Thus, a greater flexibility exists in the coupling LC columns of various dimensions, from capillary to conventional wide-bore, with ESI as an interface [21,22]. In both APCI and ESI, their high efficiency of ionization produces excellent detection sensitivities.

Gas chromatography (GC)MS continues to enjoy an excellent reputation for quantitative work. The technique is superior to LCMS because it provides excellent chromatographic resolution, and the background chemical noise is lower. However, it is applicable to small, thermally stable, and relatively volatile compounds only, and in many situations, it requires additional problematic chemical derivatization steps.

As enumerated in Chapter 7, the on-line combination of a chromatography system with mass spectrometry has several distinct advantages. Briefly, it provides enhanced sensitivity, selectivity, and speed of analysis. The opportunity for automation is a great asset of LCMS, thus allowing a large number of samples to be analyzed unattended. The feasibility of the quantification of coeluting components or compounds that have masses identical to those of the molecular ion or a chosen fragment ion, but elute at different times, is another useful feature of LCMS. As a consequence, the stable isotope–labeled analogs can be used as internal standards. Although the analyte and its stable isotope–labeled analog both coelute from the HPLC column, they can be readily distinguished because of their distinct m/z values.

Two instrumental systems are available for the detection of LC eluents. In one system, LC is connected to a single-stage mass spectrometer. Quadrupole and QIT are the two popular mass spectrometers for this instrumental configuration. Magnetic sector and TOF instruments can also be used, but their applications are very few. The mass spectrometry data are acquired in the SIM mode. The second system consists of the combination of LC with a tandem mass spectrometer (LC-MS/MS). Until the late 1990s, most of the LC-MS/MS instruments used triple-sector quadrupoles for this combination. Since then, the increased use of QIT instruments has been reported. In the LC-MS/MS system, the analyte signal is recorded in the SRM mode. Although LCMS in the SIM mode provides better detection sensitivity than does the LC/MS/MS combination, the later exhibits an extra dimension of specificity over the SIM procedure. LC-MS/MS instrumentation, however, is more expensive. The introduction of QIT as an MS/MS detector in place of a triple-sector quadrupole has reduced the cost of LC-MS/MS systems somewhat. The extra selectivity afforded by the SRM procedure helps minimize the sample preparation steps, and puts less stringent requirements on the HPLC separation step. As a consequence, fast-flow LC with shorter chromatographic run times can be used to speed up the analysis. Also, LC can serve merely as a means of sample introduction.

The basis of quantification by LCMS is the fact that the area beneath a peak in an ion chromatogram is proportional to the sample ion concentration. The literature contains extensive examples of the use of LCMS for the quantification of biological compounds, pharmaceutical drugs, and their metabolites. It is beyond the scope of this book to cover all the examples. However, a few examples are mentioned here. These examples illustrate how typically a quantitative analysis is performed with LCMS and LC-MS/MS techniques.

An LC-ESIMS method for quantitative determination of a peptide drug Ac–Arg– Pro–Asp–Pro–Phe–NH_2 in human and rabbit plasma has been described [23]. In this procedure, after the addition of the desired deuterated internal standard, the sample is desalted, proteins are precipitated, and the drug is extracted by SPE. The HPLC separation is performed in the isocratic mode with a mobile phase that consists of 1% methanol and 20% acetonitrile in 10 mM ammonium formate (pH 5.2). A flow splitter is used to allow only 5% of the flow to enter the ESI source.

The use of LC-ESIMS/MS for quantification is exemplified by a procedure reported for the determination of a potent 5-HT_{2a} receptor antagonist (373 Da) and its desmethyl metabolite (359 Da) in rat brain extracellular fluid (ECF) and plasma [24]. A homologous compound (343 Da) was used as an internal standard. ECF was collected via microdialysis sampling. The protonated molecue of 5-HT_{2a} receptor antagonist, its desmethyl analog, and the internal standard all, on CID, yield intense signal at m/z 123. Thus, the monitoring of the ion currents due to the transitions $374 \rightarrow 123$ and $360 \rightarrow 123$ provides a means of quantification of the 5-HT_{2a} receptor antagonist and its desmethyl analog, respectively.

Another example of quantification with the LC-MS/MS system is the determination of mevinolinic acid at the 50-pg/mL level [25]. This compound is a major metabolite of lavastatin in human blood. Clinical trials have shown that lavastatin lowers the serum level of cholesterol by acting as a potent inhibitor of 3-hydroxy-3-methylglutoryl-coenzyme A. Mevinolinic acid is the active form of lavastatin in this functional role. In this method, a microbore HPLC column is coupled to ESI-MS/MS. The fragmentation of the $[M - H]^-$ ions of the analyte and methylmevinolinic acid (used as an internal standard) produces a common fragment ion of m/z 319. The negative-ion MRM acquisition mode is used to monitor the precursor–product reactions of the analyte and internal standard.

The HPLC/ESI-MS/MS procedure has been successful in the quantification of the drug Ro 48-6791 {3-(5-dipropylaminomethyl-1,2,4-oxadiazol-3-yl)-8-fluoro-5-

methyl-5,6-dihydro-4H-imidazol[1,5-α][1,4]benzodiazepin-6-one} and its major metabolite Ro 486792 [26]. The quantification limit of 1 pg/mL in serum is achieved for both compounds by monitoring the reactions 413 → 114 and 371 → 300, respectively, in the MRM mode. Finasteridein has been quantified in human plasma at the pg/mL concentration [27]. Nanobore HPLC coupled to ESI-MS/MS has been used for the sensitive and specific determination of cocaine and its metabolites in human hair [28].

6.4.1. Nanoelectrospray Ionization for Quantitative Analysis

An unprecedented low level of detection sensitivity can be achieved with nanoelectrospray (nanoESI) mode of ionization [29]. Another unique advantage of this mode of ionization is the need of extremely low sample volumes. These two advantages have been utilized in a method developed for the detection of steroids, which are first converted to their sulfated conjugates [30]. This method uses precursor ion scanning, in which the instrument monitors the precursor molecules that fragment to yield a sulfate group marker ion at m/z 97. The detection sensitivity in the amol range is achieved, and only 20% of the 1 µL sample volume is consumed in the analysis.

6.4.2. High-Throughput LCMS and LC-MS/MS Analyses

A significant advantage of the coupling of LC with mass spectrometry is the automation of the quantitative protocol. Automation of the sample preparation and analysis steps increases the sample throughput. A high-throughput quantification of steroid estrogens in human urine has been reported [31]. Analysis of these compounds is of interest because of their significance in estrogen replacement therapy and their putative involvement in breast cancer risk. High-throughput quantification methods were developed for two representative estrogens, equilenin and progesterone, with estrone-d_4 as an internal standarad. A 96-well plate is used for the SPE step. The analytes are ionized in the APCI mode, and their response is measured by monitoring the precursor to product ion transitions 267 → 209 for equilenin, 315 → 97 for progesterone, and 275 → 135 for estrone-d_4. A total of 384 samples and standards can be analyzed in a 24-h period. The same group has also developed a high-throughput quantification method for an anticancer–anti-inflamatory drug methotrexate and its major metabolite, 7-hydroxymethotrexate [32]. Liquid–liquid extraction (LLE) in the deep-well 96-well plate format is used for sample preparation. The analytes, along with the deuterated internal standards, are monitored in the SRM mode. A total of 820 samples could be analyzed in a 24-h period.

6.5. VALIDATION STUDIES

Issues involved in the validation of a mass spectrometry method for quantitative analysis are similar to those in any other analytical technique. The method developed should provide the limit of detection, lower limit of quantification, within-day reproducibility and accuracy, and day-to-day reproducibility and accuracy. All of these parameters must be determined with commonly accepted GLP criteria that are used in the validation of bioanalytic methods.

REFERENCES

1. R. K. Boyd, J. D. Henion, M. Alexander, W. L. Budde, J. D. Gilbert, S. M. Musser, C. Palmer, and E. K. Zurek, *J. Am. Soc. Mass Spectrom.* **7**, 211–218 (1996).

2. D. M. Desiderio and X. Zhu, *J. Chromatogr. A* **794**, 85–96 (1998).

3. X. Zhu and D. M. Desiderio, *Mass Spectrom. Rev.* **15**, 213–240 (1996).

4. D. M. Desiderio, S. Yamada, F. S. Tanzer, J. Horton, and J. Trimble, *J. Chromatogr.* **217**, 437–452 (1981).

5. C. Dass, J. J. Kusmierz, and D. M. Desiderio, *Biol. Mass Spectrom.* **20**, 130–138 (1991).

6. J. J. Kusmierz, C. Dass, J. T. Robertson, and D. M. Desiderio, *Int. J. Mass Spectrom. Ion Proc.* **111**, 247–262 (1991).

7. D. M. Desiderio, J. J. Kusmierz, X. Zhu, C. Dass, D. H. Hilton, J. T. Robertson, and H. S. Sacks, *Biol. Mass Spectrom.* **22**, 89–97 (1993).

8. O. O. Grigoriants, J.-L. Tseng, R. R. Becklin, and D. M. Desiderio, *J. Chromatogr. B* **695**, 287–298 (1997).

9. C. Dass, G. H. Fridland, P. W. Tinsley, J. T. Killmar, and D. M. Desiderio, *Int. J. Peptide Prot. Res.* **34**, 81–87 (1989).

10. C.-G. Hammer, B. Holmstedt, and R. Ryhage, *Anal. Biochem.* **25**, 532–548 (1968).

11. C. C. Sweely, W. H. Elliott, I. Fries, and R. Ryhage, *Anal. Chem.* **38**, 1549–1553 (1966).

12. R. M. Caprioli, W. F. Fries, and M. S. Story, *Anal. Chem.* **46**, 453A–462A (1974).

13. R. S. Gohlke, *Anal. Chem.* **38**, 1332–1333 (1962).

14. D. S. Millington, in S. Facchetti, ed., *Applications of Mass Spectrometry, to Trace Analysis*, Elsevier, Amsterdam, 1981, pp. 189–202.

15. G. C. Thorne and S. J. Gaskell, *Biomed. Mass Spectrom.* **11**, 415–420 (1984).

16. S. J. Gaskell and D. S. Millington, *Biomed. Mass Spectrom.* **5**, 557–558 (1978).

17. *The API Book*, PE SCIEX (1990).

18. D. I. Carroll, I. Dzdic, R. N. Stillwater, and E. C. Horning, *Anal. Chem.* **47**, 1956–1959 (1975).

19. E. C. Huang, T. Wachs, J. J. Conboy, and J. D. Henion, *Anal. Chem.* **62**, 713A–725A (1990).

20. T. Wachs, J. J. Conboy, F. Garcia, and J. D. Henion, *J. Chromatogr. Sci.* **29**, 357–366 (1991).

21. C. Dass, in F. Settle, ed., *Handbook of Instrumental Techniques for Analytical Chemistry*; Prentice-Hall PTR, Upper Saddle River, NJ, 1998, pp. 647–664.

22. C. Dass, *Curr. Org. Chem.* **3**, 193–209 (1999).

23. D. A. Volmer and J. P. M. Hui, *Rapid Commun. Mass Spectrom.* **11**, 1926–1934 (1997).

24. T. G. Heath and D. O. Scoot, *J. Am. Soc. Mass Spectrom.* **8**, 371–379 (1997).

25. R. E. Calaf, M. Carrascal, E. Gelpi, and J. Abian, *Rapid Commun. Mass Spectrom.* **11**, 75–80 (1997).

26. M. Zell, C. Husser, and G. Hopfgartner, *Rapid Commun. Mass Spectrom.* **11**, 1107–1114 (1997).

27. B. K. Matuszewski, M. L. Constanzer, and C. M. Chavez-Eng, *Anal. Chem.* **70**, 882–889 (1998).

28. K. M. Clauwaert, J. F. van Bocxlaer, W. E. Lambert, E. G. van den Eeckhout, F. Lemiere, E. L. Esmans, and A. P. De Leenheer, *Anal. Chem.* **70**, 2336–2344 (1998).

29. M. Wilm and M. Mann, *Anal. Chem.* **68**, 1–8 (1996).

30. K. Chatman, T. Hollenbeck, L. Hagey, M. Vallee, R. Purdy, F. Weiss, and G. Siuzdak, *Anal. Chem.* **71**, 2358–2363 (1999).

31. G. Rule and J. Henion, *J. Am. Soc. Mass Spectrom.* **10**, 1322–1327 (1999).

32. S. Steinborner and J. Henion, *Anal. Chem.* **71**, 2340–2345 (1999).

7

COUPLING OF SEPARATION TECHNIQUES WITH MASS SPECTROMETRY

In many situations, mass spectrometry must deal with real-world samples, such as extracts from biological tissues and fluids, which are complex in nature and contain hundreds and thousands of components. The major drawback of mass spectrometry is its limitation in handling a complex mixture of compounds. An exhaustive protocol must be followed to purify such tissue extracts before mass spectrometry analysis can be performed. An off-line purification procedure is one of the solutions to this situation, but it is laborious and time-consuming, and entails losses of precious samples. To meet the challenge of increasing sample complexities and small sample volumes, the on-line combination of separation techniques with mass spectrometry is a sensible approach. This combination is an ideal marriage between the two most powerful standalone analytic techniques; chromatography and mass spectrometry.

Several separation devices, such as gas chromatography (GC) [1], high-performance liquid chromatography (HPLC) [2,3], and capillary electrophoresis (CE) [4,5], occupy an established position in analytical chemistry as standard techniques for high-resolution separation of a complex mixture. However, they do not provide an unequivocal identity of the analyte when used with conventional detection systems. A compound-specific detector is an essential adjunct for the unambiguous characterization of eluting components from any chromatographic separation device. In this respect, mass spectrometry offers unique advantages of high molecular

143

specificity, detection sensitivity, and dynamic range. It can distinguish closely related compounds on the basis of the molecular mass and structure-specific fragment ion information. In addition, mass spectrometry comes very close to being a universal detector. The coupling of a separation device with mass spectrometry thus benefits mutually. The result is a powerful two-dimensional analysis approach, where high-resolution separation and highly sensitive and structure-specific detection are both realized simultaneously. The primary advantages of combining a separation device with mass spectrometry are as follows:

- The capabilities of both techniques are synergistically enhanced, with the result that both instruments can be operated at lower than their optimal performance.
- The molecular specificity of the analysis is further optimized because the analyte is identified by chromatographic retention time, as well as by the structure-specific mass spectrometry data.
- Mixtures of varying complexities can be analyzed.
- The sensitivity of analysis is improved because the sample enters the mass spectrometer in the form of a narrow focused band.
- A number of laborious off-line experimental steps are avoided, which results in a minimal sample loss and reduction in overall analysis time.
- Less sample is required compared to an off-line analysis by two techniques.
- The mutual signal suppression in the mass spectrometry analysis is minimized and the quality of mass spectral data is improved.
- Stable isotope analogs of the analyte can be used as internal standards in quantitative analysis.

In order to accrue the benefits of these advantages, various separation techniques, including GC [6,7], LC [8], CE [9] supercritical fluid chromatography (SFC) [10,11], size-exclusion chromatography [12], and thin-layer chromatography (TLC) [13], have been combined with mass spectrometry. Some of these important developments are highlighted in this chapter. To further increase the degree of chemical information, GC and LC have been combined with mass spectrometry in parallel with other well-developed analytic techniques such as UV–visible spectroscopy with diode array detector, Fourier transform (FT) infrared spectroscopy [14,15], and nuclear magnetic resonance (NMR) [16–18].

7.1. GENERAL CONSIDERATIONS

Instrumentally, such a combination consists of three major components: a separation device to resolve a complex mixture into individual components, an interface to transport the separated components into the ion source of a mass spectrometer, and a mass spectrometer to analyze those individually resolved components.

7.1.1. Types of Mass Analyzers

To achieve the full benefits of the combination of a separation device with mass spectrometry, the mass spectrometry system should be inexpensive and have high scan speed, adequate mass range, reasonable mass resolution, high sensitivity, and useful dynamic range. Although several types of mass analyzers are available (see Chapter 3), a quadrupole mass spectrometer offers most of these desirable features. A quadrupole ion trap (QIT) mass spectrometer presents an attractive alternative as a coupling device for GC and HPLC. A QIT is more compact, cheaper, and sensitive than its cousin quadrupole is. In addition, it can be operated in the tandem mass spectrometry mode. As a consequence, ion trap-based GCMS and LCMS systems are gaining more acceptance. For special applications, magnetic-sector-, FT–ion cyclotron resonance (ICR)-, and time-of-flight (TOF)-based mass spectrometry systems have also been coupled with GC, LC, and CE separation techniques [19–22]. These analyzers possess certain unique features that are not available with quadrupoles. For exact mass measurements, double-focusing magnetic sector and FT-ICRMS instruments are the ideal choice as ion detectors. The fast spectrum acquisition rate and multiplex detection capability of TOFMS and FT-ICRMS instruments are attractive for the GCMS and HPLCMS combinations. However, the coupling of LC (or CE) with these mass spectrometry systems is complicated. The high voltages required for the ion extraction in the magnetic sector and TOFMS pose risks of electrical discharge when used with atmospheric pressure ion sources. High pressure is also detrimental for the optimal performance of FT-ICRMS. Another problem with FT-ICRMS and TOFMS is that the ions must enter the mass analyzers as a short pulse, whereas the effluents from a separation device emerge in a continuous fashion.

7.1.2. Types of Mass Spectrometry Signal

Determination of the molecular mass, identification of compounds, and quantification are three general areas of applications of the combination of a separation device with mass spectrometry. Depending on the end use, the mass spectrometry data can be recorded by scanning the mass analyzer either in a wide mass range or by monitoring a few selected ions. The mass spectrometry signal is plotted as an ion chromatogram or a mass spectrum. Three different types of ion chromatograms can be obtained from the mass spectrometry data: a total ion current (TIC) chromatogram, a mass chromatogram (or ion extraction chromatogram), and a selected-ion monitoring (SIM) chromatogram. The TIC chromatogram is obtained by recording the signal due to all ions in a certain mass range. It is similar to a conventional UV chromatogram, and can be used to identify compounds on the basis of their retention time. In addition, a mass spectrum can be retrieved for any chosen scan in the TIC chromatogram. A mass chromatogram is a plot of the ion current due to a characteristic ion versus the scan number. It is a useful means to distinguish coeluting components and to identify a homologous series of compounds. The SIM chromatogram is obtained by recording the signal due to only one or more compound-specific ions, and is useful for quantification of a target compound and for selective detection of a homologous series of compounds. Improved detection limits can be achieved in SIM, but at the cost of extensive loss of spectral information.

7.1.3. Mass Spectral Data Acquisition

The signal output from GCMS, LCMS, and CEMS poses a challenge for the mass spectrometry data acquisition system; first, the volume of data is huge, and second, the peak width usually is very narrow. In addition, the partial pressure of the sample that enters the mass spectrometry ion source continuously changes. A rapid scanning mass spectrometry system is very essential to ensure that no useful data point is missed during the scan. Rapid scanning of a mass spectrometer ensures that there is no distortion in the relative peak intensities and that the chromatographic peak profile is preserved. The mass scan rate is critically dependent on the functional features of a mass spectrometer. Therefore, for certain types of mass spectrometers, ion-counting statistics will be compromised at fast scan speeds. In this respect, magnetic sector instruments have a limited utility, whereas quadrupole, QIT, TOF, and FT-ICR mass spectrometers are ideally suited for fast-scanning applications. Magnetic sector instruments with laminated magnets can acquire a spectrum at the repetition rate of 2–3 scans/s (scans per second). A slightly higher repetition rate (5–10 scans/s) is realized with quadrupole-based mass analyzers. Scan speeds that approach 10 scans/s are also attainable with FT-ICRMS. TOFMS, on the other hand, acquires data at much higher repetition rates (100 scans/s).

7.2. GAS CHROMATOGRAPHY/MASS SPECTROMETRY

Gas chromatography/mass spectrometry has acquired a unique position as a modern analytical technique for the sensitive and selective identification and quantification of relatively volatile organic compounds [6,7,23]. GCMS is considered a standard technique for confirmation of the presence of a compound. Methods based on the GCMS protocol are readily acceptable in medical–legal defense cases. Although GCMS is a standard protocol for the analysis of organic compounds, some compounds of biological relevance can also be made accessible to this technique after derivatization with a suitable reagent. Therefore, a brief discussion of GCMS is warranted in this chapter. The combination of GC with mass spectrometry was inevitable because of the benefits of the high-resolution separation capability of GC and a unique capability of mass spectrometry to provide a means of highly structure-specific detection of the GC eluents. In addition, GC and mass spectrometry are both highly compatible with respect to the sample size, and both deal with the gaseous samples.

7.2.1. Basic Principles of Gas Chromatography

In gas chromatography, the sample is vaporized by injecting it through a heated injector, and is placed onto the head of a chromatographic column that contains a liquid stationary phase [1,23,24]. The components of a mixture are separated on the basis of their varying affinity for and solubility in the stationary phase, and elution of the separated components is effected by the flow of an inert carrier gas (usually helium). A GC separation is conducted at elevated temperatures (150–300°C) to bring many not so volatile, but thermally stable, compounds under its domain. In the past, a vast majority of GC separation was carried out on wide-bore (1–3-mm-i.d.) packed columns. Modern analytical GC is practiced with more efficient and faster long capillary columns. The inside wall of these fused-silica open tubular (FSOT) columns is coated with a thin layer of a stationary phase, usually an organosiloxane polymer. Polydimethyl siloxane is a general-purpose nonpolar stationary phase. The replacement of methyl groups with polar functional groups, such as phenyl, trifluoropropyl, and cynopropyl, increases the polarity of this phase. Polyethylene glycol is another common stationary phase.

7.2.2. Coupling of Gas Chromatography with Mass Spectrometry

The coupling of gas chromatography with mass spectrometry is very straightforward because both operate in the gas phase. The only incompatible factor is the pressure mismatch. The separated components exit from the GC column at atmospheric pressure, whereas the mass spectrometry source operates at 10^{-6}–10^{-5} torr. This problem, however, is not severe with capillary columns. The columns of up to 320 μm internal diameter (i.d.) can be directly coupled to mass spectrometry by inserting the end of the column through a heated sheath into the ion source of a mass spectrometer (Figure 7.1*a*) [25]. The high-capacity pumping systems of modern mass spectrometers can effectively handle the gas load (usually 1–2 mL/min) from a capillary column. Larger-diameter open tubular and packed columns are coupled to mass spectrometry via an interface. An open-split interface, shown in Figure 7.1*b*, uses a high flow of a purge gas to remove most of the carrier gas that emerges from the GC column. The end of the column is at atmospheric pressure rather than at a reduced pressure as is the case with the direct coupling of a capillary column. Therefore, degradation of the chromatographic resolution is minimized.

The jet separator (Figure 7.1*c*) is the most common interface for packed columns [26]. It operates on the principle that the fast-moving lighter-mass components effuse at a faster rate than do the slower, heavier-mass ones. In practice, a jet of GC effluents is sprayed through a small nozzle into a partially evacuated chamber. The lighter carrier-gas molecules effuse faster at wider angles, whereas heavier organic compounds travel over a narrow angle around the central axis. By collecting the middle portion, a skimmer permits an enriched portion of the sample to enter into the mass spectrometry ion source.

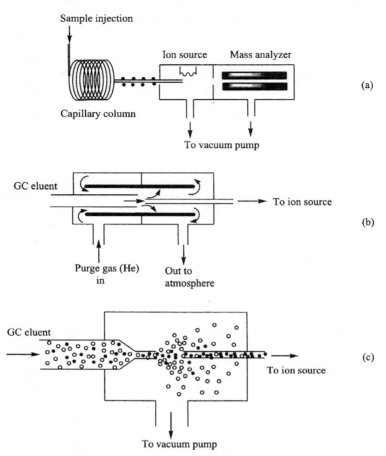

Figure 7.1. Various configurations for the coupling of GC with mass spectrometry: (*a*) direct coupling of capillary columns; (*b*) open-split interface; (*c*) jet separator. The latter two are used with higher-diameter columns.

Electron ionization and chemical ionization methods, both of which are compatible with gas-phase samples, are used for mass spectrometry analysis of the eluting components.

7.2.3. Applications

A complete list of applications of GCMS would be staggering, and beyond the scope of this volume. Its use is highly mentioned in the fields of forensic drug analysis [27,28], structural characterization of biomolecules [29,30], toxicology [31], medicinal chemistry [32], clinical science [33], and environmental chemistry [25,34,35].

7.3. HIGH-PERFORMANCE LIQUID CHROMATOGRAPHY/MASS SPECTROMETRY

As mentioned above, GCMS has long enjoyed the status of a proven and dependable analytical technique for the analysis of complex mixture of relatively volatile and thermally stable compounds. Researchers and analysts have felt a strong need for a similar combination of HPLC and MS for the analysis of polar, nonvolatile, and thermally labile compounds. Although attempts to combine LC with MS were initiated in the early 1970s [36,37], a robust and dependable combination that could be adapted for a wide range of applications has emerged only very recently. Currently, the practice of LCMS arguably has become the single most widely useful analytical technique. It is used for a variety of applications such as qualitative and quantitative analysis of complex mixtures of biochemical, inorganic, and organic compounds.

7.3.1. Basic Principle of HPLC Separation

HPLC analysis is performed by applying a sample onto a column filled with an appropriate stationary phase [2,3]. The components of a mixture are differentially partitioned on the sorbent material and are separated when eluted by a stream of a liquid mobile phase. Unlike GC, the stationary and mobile phases both play critical roles to optimize separation. The elution is accomplished in either the isocratic mode (i.e., at a fixed solvent composition) or the gradient mode (i.e., at a changing solvent composition). Although not a highly characteristic property, the basis of the identification of a compound is its retention time. The more commonly used approach for the separation of biological compounds is the reverse-phase (RP) mode of HPLC, in which the stationary phase consists of a nonpolar matrix and the mobile phase is a polar solvent (water mixed with a polar organic modifier such as methanol, isopropanol, and acetonitrile). The basic principle of RP-HPLC separation is the hydrophobic interaction between the nonpolar hydrocarbonaceous matrix of the stationary phase and the hydrophobic group of the analyte. A typical stationary phase is prepared by chemically bonding a long-chain hydrocarbon, such as *n*-octadecyl (C-18), *n*-octyl (C-8), and *n*-butyl (C-4), to porous silica. In practice, relatively hydrophilic compounds elute earlier with the aqueous mobile phase. Increasing the concentration of the organic modifier in the mobile phase elutes in sequence, the more strongly retained hydrophobic compounds. The selectivity of the separation is achieved by the addition of ion-pairing reagents in the mobile phase [37]. Although a large number of buffer systems have been used in a conventional LC/UV–visible detection system, only volatile ion-pairing reagents are suitable for the LCMS operation.

7.3.2. Coupling of Liquid Chromatography with Mass Spectrometry

One of the major hurdles in combining HPLC with mass spectrometry has been the fundamental difference in their operating conditions. The separation in conventional wide-bore analytical columns is accomplished at liquid flow rates of 0.5–1.5 mL/min. Unlike GCMS, this liquid flow is too large for safe operation of the mass spectrometry vacuum system (10^{-5}–10^{-8} torr). The common solvents and nonvolatile additives used in HPLC separations are also not compatible with the mass spectrometry operation. In addition, an LCMS interface must be able to handle the aqueous content, organic and ionic modifiers, and buffers of the mobile phase. In the early years of LCMS development, researchers also faced difficulty in the ionization of thermally labile and nonvolatile compounds. Sustained research by several groups has resulted in a variety of different LCMS interfaces. Some of the not-so-common approaches are discussed briefly below. The more popular current approaches are discussed in Section 7.3.3.

Direct Liquid Introduction Probe. As the term implies, the liquid effluent from HPLC is directly introduced into the ionization chamber of a mass spectrometer as a stream of droplets via a narrow-bore capillary or through a pinhole in a metal disk [38,39]. A flow splitter is needed to divert most of the LC solvent from conventional columns because only 10–50 µL/min liquid flow rates can be accommodated into the mass spectrometry ion source without breakdown of the vacuum. The nebulization takes place in a reduced-pressure desolvation chamber. The eluents are ionized via solvent-mediated CI. In some designs, preheating the solvent before it enters the CI source facilitates the nebulization [40]. With this interface, thermally labile compounds can be transported into the ion source without any thermal degradation. The method, however, is applicable only to low-mass compounds. This interface, however, has fallen out of favor mainly because of the limited utility and practical operational difficulty of frequent blockages of the small orifice.

Moving-Belt Interface. A moving-belt interface consists of a moving wire or ribbon to transport the column effluent into the mass spectrometry ion source [41–44]. This interface is restricted mainly to compounds of appreciable volatility and thermal stability. In practice, the LC effluent is deposited continuously over this moving belt, and is heated gently under an infrared (IR) heater to vaporize the solvent. After traveling through two differentially pumped vacuum chambers, the solvent is completely removed from the analyte. Finally, the sample components are transported into the ion chamber by flash evaporation. This interface has the advantage of choosing EI or CI as the ionization mode, and the composition of CI reagent gases. In addition, not easily volatilized components can be ionized by focusing a FAB beam onto the end of the moving belt [44]. Another significant advantage of the belt interface is that it can be used with nonvolatile buffers of moderate concentration. Depending on the aqueous content, solvent flow rates of 0.5–1.5 mL/min can be accepted by the moving-belt system. However, with the advent of more convenient LCMS systems, the use of this interface has almost declined.

The Particle Beam Interface. A particle beam interface was introduced with the aim to access "true EI spectra" of samples from the LC effluent so that the spectrum of an unknown can be searched against libraries of known compounds [45,46]. Its operating principle is similar to that used in the GCMS jet separator interface (Figure 7.2). The original device was called *monodisperse aerosol generation interface for chromatography* (MAGIC) [47]. The particle beam interface can accept 0.5– 1.5 mL/min flow rates, and has a preference for a normal-phase LC over reverse phase.

Three steps—the nebulization, desolvation, and momentum separation of the analyte and solvent molecules—are involved in the conversion of analytes from LC effluents into the gas phase. The LC effluent is nebulized by forcing the liquid, together with a flow of helium, through a small orifice. The jet of liquid that emerges breaks into a stream of uniform droplets under the influence of Rayleigh forces. In the original design, a high-velocity stream of helium was introduced at a right angle to the flow of a liquid stream to disperse the particles. The dispersed droplets travel at a very high velocity through a desolvation chamber where vaporization of the solvent occurs at atmospheric pressure. The desolvation chamber is heated ($\sim$40– 50°C) to provide thermal energy just sufficient to compensate for the latent heat of vaporization of the solvent. The solvent vapors, along with the sample molecules and helium, pass through a two-stage momentum separator. On entering the low-pressure region, the mixture expands into a supersonic jet, whereas the heavy sample particles remain confined in the central axis. The low-mass solvent molecules diffuse away at a wider angle, and are efficiently removed by the pumping system. The remaining high-mass analyte molecules continue to travel as a beam of uncharged particles toward a conventional mass spectrometry ion source. Although a common approach to ionize the constituents of the particle beam is to use either EI or CI, the use of FAB ionization has also been discussed for this purpose [48,49]. In another novel arrangement, the particle beam interface has been used to produce desolvated analytes, which are deposited on a solid target and ionized by bombardment with a primary beam of massive, multiply charged glycerol clusters with energies of $\leq$1 MeV [50].

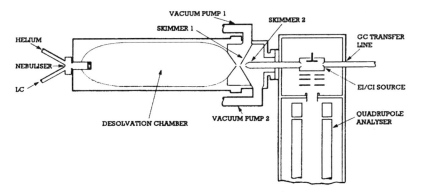

Figure 7.2. Particle beam LCMS interface. (Courtsey Micromass.)

The ion–solvent complexes often pose problems in LCMS analysis. In order to facilitate rapid desolvation, a particle beam interface has been designed to operate at eluent flow rates down to 1 µL/min [51,52]. In this design, the conventional nebulizer is replaced with a microflow nebulizer that is made of a 50-µm-i.d., 180-µm-o.d. (outer-diameter) fused-silica capillary tubing.

Thermospray Interface. The thermospray interface represents a first serious attempt to combine analytical columns with mass spectrometry [53–55]. The basic principle of thermospray ionization was discussed in Section 2.4.1; briefly, it makes use of a heated capillary to generate a mist of fine droplets of HPLC effluents, and to transport them directly into the ion source. Unlike the interfaces discussed above, no external mode of ionization is required. There is, however, a downward trend in the use of this interface in favor of a more versatile and convenient electrospray ionization (ESI) interface, discussed in Section 7.3.3.

Thermospray is confined mostly to RP-HPLC, but it is incompatible with nonvolatile solvents. A solvent flow of 2.0 mL/min can be effectively nebulized by this device. Therefore, it is an ideal interface for coupling of conventional wide-bore columns with mass spectrometry. Significant concentrations of volatile buffers [usually millimolar (mM) amounts of ammonium acetate] must be maintained in the mobile phase. The dissociation of this buffer produces gaseous ammonium and acetate ions, which are effective as CI reagent ions. A decrease in sensitivity is usually observed at high concentrations of organic solvents. This phenomenon can pose problems at the upper end of an RP-HPLC gradient run. One way to circumvent this problem is to combine the filament-on mode of ionization with the usual buffer-ion-initiated CI.

Liquid chromatography/thermospray-MS has proved to be extremely popular in the pharmaceutical industry for the characterization of drugs and metabolites. Qualitative and quantitative measurements have both been effectively performed. Other applications include the analysis of peptides [56,57], pesticides [58], drugs [59], and environmental samples [60].

Continuous-Flow Fast-Atom Bombardment. As discussed in Section 2.2.3, a continuous-flow (CF)-FAB probe can be used to maintain a steady flow of the matrix–sample solution over the atom gun target area in. This arrangement also has been exploited as an LCMS interface with much success [61–66]. The eluents from an HPLC column can be connected directly to the CF-FAB probe. Because for stable operation the solution reaching the target area must continuously evaporate under the influence of the ion-source vacuum, a maximum flow rate of only 1–10 µL/min is permissible with this probe. As a consequence, only microbore and capillary columns are compatible with the CF-FAB interface. An improved version of this probe for LCMS operation uses a wick made of porous cellulose fiber to absorb excess liquid. In addition, the probe is heated slightly (50°C). Although LCMS with CF-FAB has produced impressive results, its success-ful operation requires a high degree of skill. Because of this operational difficulty, its use is being phased out in favor of the more successful LC-ESIMS combination.

In an LC/CF-FABMS configuration, a microcolumn typically operates at a flow rate of 20–50 µL/min. However, the reliable and reproducible gradient is formed at much higher flow rates (~500 µL/min). Therefore, a precolumn flow splitter is employed before the injector. A postcolumn flow splitter is also used to reduce the flow that enters the CF-FAB probe to <10 µL/min. Conventional wide-bore analytical columns have also been used with a CF-FAB interface, but with a large (100 : 1) flow split. Glycerol, because it has the same eulotropic strength as methanol, is used invariably as the matrix additive. Two arrangements are used to add the FAB matrix to the LC eluents. A convenient approach is to add 5% glycerol to each solvent of the gradient. A possibility of peak tailing, however, exists when glycerol is added to the eluting buffer [67,68]. Another approach is to use a coaxial probe for the postcolumn addition of glycerol [69]. This modified probe contains two concentric fused-silica capillaries. The LC solvent is delivered through the inner capillary, and the liquid matrix, which is added via a tee-junction, flows through the outer capillary. The inner and outer capillaries both terminate at the probe tip, where mixing of the two liquid channels takes place. CF-FAB has also played the role of an interface for the coupling of LC with tandem mass spectrometry [70–73].

At one time, LC/CF-FAB was a very popular technique for the analysis of peptides [65,72–76], protein digests [71–73,77], drugs [78], and other classes of compounds [73].

7.3.3. Coupling of Liquid Chromatography with Electrospray Ionization Mass Spectrometry

The development of an atmospheric pressure ionization (API) source [79,80] is a significant breakthrough that has simplified the coupling of LC with MS [81–83]. API, when operated in the ESI format (the other format is atmospheric pressure chemical ionization), is unique because it has a great potential for the analysis of a variety of small and large molecules at fmol sensitivity [84,85]. ESI is viewed as the most general ionization technique. Although several other options were discussed above, ESI appears to provide the best approach to the on-line LCMS, where it serves at the same time as an LCMS interface and a means of generating ions. Thus, in a short time, it has gained prominence as the most versatile interface for HPLC.

The composition and flow rate of the solvent must be controlled for an optimum operation of the ESI system. A flow rate of 1–10 µL/min is required to maintain a stable spray. At higher flow rates, larger droplets are formed, that may lead to electrical breakdown. A fluid with a high surface tension such as pure water is difficult to electrospray. However, many polar solvents commonly used in HPLC (e.g., methanol, ethanol, isopropanol, acetonitrile) are suitable for the electrospray operation. In contrast, nonpolar solvents are difficult to disperse. As a consequence, HPLC in the normal-phase mode is not compatible with the ESI process unless a polar solvent is admixed with the nonpolar mobile phase. In order to make the ESI process more desirable for higher flow rates of a conventional HPLC, several modified designs of the ESI interface have emerged. A sheath of a nebulizing gas is added in pneumatically assisted electrospray, to enable higher liquid flows (100–

300 µL/min) to be electrosprayed [86,87]. This interface allows the use of larger-bore columns or smaller postcolumn split ratios. By supplying heat around the tip of the capillary, the flow rate acceptable for a stable ES operation can be extended beyond 300 µL/min [88]. The direct coupling of wide-bore columns is readily achieved with this type of ESI interface. In another design that can handle mL/min flow rates, a grounded liquid shield is introduced after the ESI needle, and the ion-sampling capillary is heated [89]. An ESI device suitable for nL/min flow rates has also been introduced [90,91]. Several applications of the combination of nanoscale chromatography with this design have been reported.

An interface similar in concept to ESI, called *sonic spray interface* (SSI), has also emerged for LCMS applications [92,93]. In this design, the solution is introduced through a fused-silica capillary, and sprayed with the help of a coaxial high-speed nitrogen gas flow. The intensity of the ion current increases with the gas flow, and is maximum near to the sonic gas velocity. If the electric field is applied to the solution, then a dramatic increase in the ion current is noted. In addition, the formation of multiply charged ions of proteins is facilitated. Normally, this system is applicable to solution flow rates of <0.2 mL/min, but by placing a multihole plate in front of the sampling cone of the mass spectrometer, flow rates as high as 1 ml/min can be accepted. The SSI has been combined with semimicro and conventional LC columns. Figure 7.3 depicts an SSI that is suitable for mL/min flow rates.

Liquid Chromatography–ESIMS Configurations. One of the most important factors to be considered in designing a successful on-line HPLC-ESIMS system is the rate of liquid flow that emerges from the HPLC column. To a large measure, the mobile-phase flow rate is dictated by the i.d. of the column. A variety of column sizes that range in i.d. from 0.1 to 4.6 mm are in common use. The sensitivity, efficiency, sample loading capacity, and sample size are important criteria in the selection of column dimensions. The smaller-diameter columns provide increased

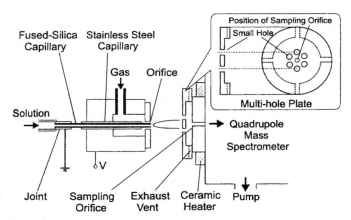

Figure 7.3. Sonic spray interface for LCMS combination. [Reproduced from Ref. 93 by permission of American Chemical Society, Washington, DC (copyright 1998).]

sensitivity, and are used when the amount of sample is limited; a fourfold increase in sensitivity can be realized by reducing the column diameter by one-half. However, with the proper technique and hardware, a column of any dimension can be combined with an ES interface. The choice of an actual experimental configuration will depend on the sample amount, column dimensions, and the available ion source. The common types of columns and their characteristics are given in Table 7.1, and a few on-line LC-ESIMS configurations with column dimensions and flow rates are illustrated in Figure 7.4 [94,95].

Because of their high efficiency and large sample loads, narrow-bore (2.1 mm i.d.) and wide-bore (standard analytical columns of 4.6 mm i.d.) are used in most HPLC applications. However, the transfer of the methods that are developed with these columns to MS-based procedures is not an easy task. The inability of a conventional ES source to handle larger liquid flow rates is the major obstacle. Large-size droplets are formed during electrospraying the solvent flows (0.2–2.0 mL/min) that are commonly employed in these columns. At these high flow rates, the probability of the ejection of ions from larger droplets is lower, and results in reduced ionization efficiency. One solution to this problem is to use postcolumn flow splitting, as shown in Figure 7.4a. Another alternative is to use an ESI interface that can handle large flow rates such as a megaflow ESI source (Figure 7.4b). The pneumatically assisted ESI can be operated at up to 300 μL/min flow rates, and is suitable for the direct coupling with narrow-bore columns [96]. At higher flow rates, however, there is a danger of an increased background due to the formation of solvent clusters, which results in lower detection sensitivity. In some applications, flow splitting is a desirable option because a large portion of the precious peak material can be recovered for other experiments.

In many biochemical studies, the sample amounts and volumes are limited. Researchers use packed-capillary columns (100–300 μm i.d.) for the HPLC separation of low-volume samples. The combined use of small-i.d. columns with an ES ion source has a further advantage of optimal detection sensitivity because of its concentration-dependent response. Because these columns operate in the nL–μL/min flow range, they can be directly connected to nanoES or pure ES

Table 7.1. Typical RP-HPLC columns

Column	Dimensions (length × i.d., in mm)	Flow rate (μL/min)	Injection volume (μL)	ESI* configuration	Flow-split
Analytical	150 × 4.6	500–2000	10–500	a	Postcolumn
				b	No split
Narrow-bore	150 × 2.1	50–500	5–50	a	Postcolumn
				b	No split
Microbore	150 × 1.0	20–100	1–20	d	No split
Capillary	300 × 0.3	1–10	0.05–2	c	Precolumn
Nanoscale	500 × 0.05	0.05–0.5	0.01–1	c	Makeup sheath liquid

* Refers to Figure 7.4.

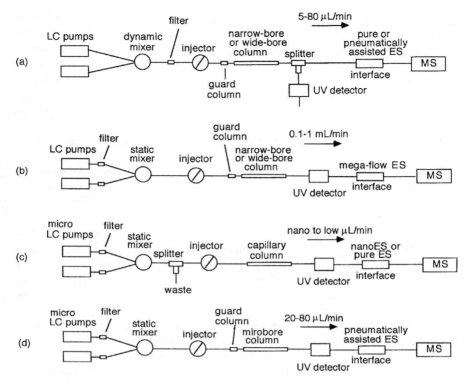

Figure 7.4. Various configurations for the coupling of LC with ESI–mass spectrometry.

sources (Figure 7.4c). These columns can also be coupled to a conventional or pneumatically assisted ES source, if an additional sheath liquid is added to raise the flow to a range that is acceptable by these mass spectrometry sources. Although capillary columns have the potential of a reduced sample load and a high sensitivity, they also have limitations with respect to reproducible flow rates and gradients and uniformity in packing. Also, they require unduly long equilibration times between analyses. Therefore, optimum efficiency may not be achieved with capillary columns. One way to achieve reliable gradients is to use precolumn flow splitting (Figure 7.4c) [97]. MicroLC pumps deliver the solvents at higher flow rates (200–500 μL/min). After mixing the solvents in a static mixer, the required flow is diverted to the capillary column via a flow splitter. Currently, special-purpose LC pumps for capillary columns have become commercially available, and thus can aid in direct coupling of these columns with an appropriate ESI interface. The use of a commercial pneumatic flow splitter capable of providing steady split flow in the low nL/min range has been described [98]. With this flow controller, conventional narrow- and wide-bore columns can be conveniently combined with a nanoES source.

Microbore columns (0.8–1 mm, i.d.) offer a compromise between wide-bore and capillary columns. Chromatography separation by these columns is carried out at a flow rate of 20–100 µL/min. HPLC hardware is now available that can deliver a precise volume and solvent composition in this low flow range. A microbore column can be coupled to mass spectrometry via a pneumatically assisted ESI source with no precolumn or postcolumn flow splitting (Figure 7.4d) [99].

Fast-Flow Liquid Chromatography–Mass Spectrometry. An important area of research in LCMS is to reduce the analysis times. This innovation can improve productivity through high-throughput analysis, reduce solvent waste, and reduce total method development and quality analysis cost. The process of drug discovery and development can benefit highly from this innovation. Several imaginative attempts have been made to develop fast LC. One approach is to use short LC columns (typically <50 mm long) [100–102]. A decrease in the analysis time by a factor of ~7.5 is realized when the column length is reduced from 250 to 33 mm. In addition, by packing these columns with 3-µm particles rather than standard 5-µm particles, a further improvement in the analysis time is realized because the optimal linear mobile-phase velocity is inversely proportional to the particle diameter. Although the separation efficiency increases with a decrease in the particle size, overall resolution is somewhat degraded because of a decrease in column length. The loss in resolution is compensated for when fast LC columns are combined with tandem mass spectrometry detection. Another successful attempt in this direction is to use a turbulent flow through short columns [103]. In this approach, the column is operated at high mobile-phase flow rates (4 mL/min). A considerable improvement in column efficiency is realized because the solvent moves through the column as a plug to provide a planar profile rather than the usual parabolic profile. To compensate for the increase in pressure that results due to the increased flow rate, the column needs to be packed with larger diameter particles.

Another approach to fast chromatography is the use of perfusion columns [104,105]. These columns are packed with polymer particles that have through-pores of 6000–8000 Å diameter, each is connected by 300–1500 Å diffused pores. This network of pores allows 10–20 times higher solvent flow rates through it compared to conventional HPLC columns. Nonporous silica columns have also been used for fast chromatograpy [106]. To increase the mass measurement accuracy across the fast-eluting peaks, it is important to couple fast-flow LC with a high-acquisition data collection system such as TOF instruments [107].

Types of Mass Analyzer. Although a variety of mass analyzers are in common use [108], the coupling of HPLC with mass spectrometry is more conveniently achieved with quadrupole and QIT mass spectrometers [109], mainly because of their high tolerance to low vacuum and the absence of high potentials in the ion source. In addition, they are inexpensive and provide high-scan-speed data acquisition. The coupling of LC with FT-ICRMS instruments provides the benefits of high-resolution and accurate mass measurements [110–112].

Although multiplex detection and high scan speeds are welcome features of a TOF mass spectrometer, its coupling with LC-ESI is not straightforward. Several imaginative approaches have been devised to couple LC-ESI with TOFMS. The most common approach uses the orthogonal ion extraction concept [113], in which ESI-produced ions are stored between each duty cycle, and are pushed into the flight tube by applying a high-voltage pulse [114,115]. An improvement in the original concept is to incorporate a QIT [116] or an rf-only quadrupole [115] between the ES interface and the TOF mass analyzer. These devices act as a source of ion storage between each duty cycle of TOF [117,118]. Another convenient way to introduce the effluents from LC is a continuous-flow probe interface [119] similar to that used for CF-FAB analysis [74]. The matrix solution, which is composed of 3-nitrobenzoyl alcohol, 0.1% TFA, 1-propanol, and ethylene glycol (3 : 3 : 5 : 9 by volume), is mixed with the LC flow postcolumn with a mixing tee. A pulse beam of fast ions or a laser beam ionizes the eluents that emerge from the tip of the probe. A pulsed sample introduction (PSI) interface and laser-induced multiphoton ionization have also been used to couple LC with TOF instruments [120]. The aerosol generator also holds the potential of being a successful LC interface for MALDI-TOF mass spectrometers [121,122]. In this design, the LC eluent is mixed with a matrix solution, and allowed to flow through a stainless-steel tube into a pneumatic nebulizer. The coupling of MALDI to liquid separation devices has also been reviewed [123].

Applications. The range of potential LC-ESIMS applications is too broad to cover in this chapter. In general, the technique has found applications in several areas of research such as the sequencing of proteins; identification of mixtures of compounds; tryptic maps; posttranslational modifications in proteins; structure elucidation of metabolic products; analysis of drugs, pesticides, and toxins; and quantification of a variety of compounds. A few selected examples are briefly mentioned below; a detailed discussion is presented in the following chapters.

Developments in the field of on-line LC-ESIMS have helped devise high-speed, high-sensitivity peptide mapping and protein sequencing techniques [124–126]. Automation of all the steps of the peptide-mapping procedure has been achieved [127]. The LC-ESIMS combination has also found utility in the selective detection of covalently modified peptides in the enzymatic digest of a protein [128–130]. Similarly, the selective detection of glycopeptides is accomplished by monitoring the carbohydrate-specific oxonium ions [131–133].

LCMS has been used to establish the purity of peptides and to identify impurities in the crude product of solid-phase peptide synthesis (SPPS) [95,134,135]. The rapid screening of combinatorial libraries is another area where LCMS has made important contributions [136,137]. Many studies have been reported in which HPLC coupled to ESIMS is used to identify the degradation products of bioactive compounds. Ayyoub et al., have studied degradation mechanisms of MHC class I–presented tumor antigenic peptides [138]. The LCMS technique has been used by us to study mechanisms of proteolytic degradation of phosphorylated and acetylated methionine and leucine enkephalins against the action of aminopeptidase M [139], and to study the long- and short-term stability of deslorelin, an active ingredient of a drug formulation [140].

The role of the LC-ESIMS combination for solution of many forensic [141] and environmental [142–144] problems has been demonstrated. A typical example is the characterization of conjugated metabolites of polycylic aromatic hydrocarbons (PAHs) from biological extracts [142]. LC-ESIMS has been used to determine the adducts of 1,3-butadiene with DNA [143]. LCMS has also found applications in the analysis of herbicides. The simultaneous determination of several imidazolinone herbicides from the soil and natural water samples is another example of its use in the environmental field [144].

Essential to many biochemical studies is an accurate determination of the concentration of the target compound. The development of LC-ESIMS has greatly advanced the science of quantification. This development is the consequence of the inherent high detection sensitivity and specificity of ESIMS, and its applicability to polar and thermally labile compounds [145]. Despite the high cost of LC-ESIMS and LC/ESI-MS/MS instrumentation compared with the conventional UV and fluorescence detection, they have become the methods of choice for the quantification of a variety of compound types. More details of quantitative analysis and a list of applications can be found in Chapter 6.

7.3.4. Atmospheric Pressure Chemical Ionization Interface

Atmospheric pressure chemical ionization (APCI) is another robust and dependable interface for the LCMS combination, but it is applicable only to relatively small compounds [146]. Similar to thermospray interface, APCI can be readily coupled to wide-bore analytical columns with a flow rate of $\leq 2\,mL/min$. A corona discharge generates reagent ions. A heated nebulizer inlet probe is used to couple LC with APCIMS [147,148]. An APCI interface suitable for semimicroLC has also emerged [149]. The interface used in this design consists of a stainless-steel capillary that is inserted into a stainless-steel block, and the vaporizer and nebulizer blocks are combined into one. The usefulness of LC-APCIMS has been demonstrated for a wide range of applications in the bioanalytical, pharmaceutical, and environmental fields [150]. Applications of this interface to the biomedical field have been summarized [151]. The detection of β-blockers at the fmol level has been reported from the author's laboratory [152].

7.4. CAPILLARY ELECTROPHORESIS/MASS SPECTROMETRY

Capillary electrophoresis (CE) is another separation device that has benefited from a highly structure-specific and sensitive detection capability of mass spectrometry. In the 1990s, CE emerged as a powerful technique for the rapid and efficient separation of biological compounds. The basis of separations in CE is the electrically driven flow of ions in free solutions. The hallmarks of CE are unparalleled resolving power, ultrahigh sensitivity, low-sample volumes, and speed of analysis. Currently, CE is meeting the challenge of analyzing the sample volumes as low as that available from a single biological cell. The mechanism of separation in CE is different from that in chromatographic techniques. Thus, it offers an orthogonal and complementary approach to HPLC separation. A few reviews and books on this subject are worth reading [4,5,153,154].

7.4.1. Basic Principle of Capillary Electrophoresis

The analytical selectivity in CE is achieved by the differential migration of electrically charged particles or ions in solution under the influence of an applied electric field. CE embodies the concept of the traditional polyacrylamide gel electrophoresis (PAGE) technique and the on-line detection and automation concepts of chromatography. The separation efficiency of PAGE is poor because of the heat generation at the high applied voltages. Jorgenson and Lucas demonstrated that effective heat dissipation could be achieved in small-diameter, open tubular, fused-silica capillaries; thus, providing a means of highly efficient separations [155]. The most common and simple mode of CE is capillary-zone electrophoresis (CZE; referred in this text simply as *CE*). It is performed when a narrow-bore (25–75 μm-i.d.) buffer-filled capillary is placed between two buffer reservoirs and a very high potential (200–500 V/cm) is applied between the two ends (Figure 7.5). The analyte is transported through the capillary by the combined action of electrostatic force and electroosmosis, the latter is the dominating factor. This phenomenon is a consequence of the surface charge on the wall of the silica capillary. The association of the positively charged counter ions from the solution with the negatively charged silanol groups on the capillary wall creates a zeta potential, ζ, which forces the movement of the solution toward the cathode. This flow enables simultaneous analysis of all solutes, regardless of their charge. The migration velocity of the analyte is the sum of the electrophoretic mobility, μ_e, and electroosmotic mobility, μ_{EO}, (i.e., $\mu = \mu_e + \mu_{EO}$). The migration time, t, is given by

$$t = \frac{l}{\mu E} = \frac{lL}{\mu V}$$

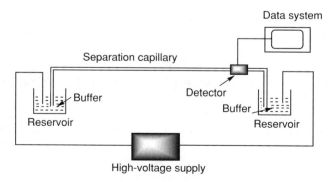

Figure 7.5. A schematic of a CE separation system.

where l is the length of the capillary to the detector, L is the total capillary length, E is the electric field, and V is the applied voltage.

Other modes of CE are miceller electrokinetic chromatography (MEKC), capillary gel electrophoresis (CGE), capillary isoelectric focusing (CIEF), and capillary isotachophoresis [3,4]. Another mode of CE, dubbed as *capillary electrochromatography* (CEC), is gaining fast acceptance for special applications [156]. In this mode, the concept of CE and LC are combined together. The separation is accomplished in a CE capillary packed with an LC stationary phase, but the liquid flow is due to electroosmosis. Because the solvent flow inside the capillary has a planar profile, the technique provides a high-resolution separation than HPLC does, and also retains the advantage of high sample load.

7.4.2. Coupling of Capillary Electrophoresis with Mass Spectrometry

Capillary electrophoresis in a variety of operational modes has been combined with mass spectrometry. CF-FAB [157–159], and ESI interfaces [160–162], both of which are appropriate for the ionization of samples present in solutions, are ideal for the coupling of CE with mass spectrometry. The coupling of CE with MALDIMS has also been described [163,164]. The current trend is in favor of the use of ESI for CEMS applications because of the obvious advantage of it being an atmospheric ionization technique. In this interface, because the terminating end of the CE capillary is also at atmospheric pressure, the planar profile and the electroosmotic flow of the CE buffer is not disturbed. As a consequence, the integrity of the CE separation is preserved. A CEMS interface must address the problem of the low liquid flow rates of the CE process, buffer compositions, and proper electrical contact between the separation buffer and the interface. As with the LCMS systems, only volatile buffers are tolerated in a CEMS operation.

Electrospray Ionization Interface. A conventional ESI source is ideally suited for coupling CE with MS. The coupling of CE with ESIMS has been achieved via a sheathless, a sheath-flow, or a liquid junction interface (Figures 7.6 and 7.7). Smith's group described in 1987 the first successful CEMS combination [160]. The electrical contact for the CE circuit at the capillary terminus, as well as for the ESI electric field, has been established in several ways. In the prototype design of Smith et al., the cathode end of the separation capillary terminated within a stainless-steel capillary, which acted as the CE cathode and the ESI needle [160]. An improvement in this design was the use of a metallized capillary end for electrical contact with the buffer (Figure 7.6*a*) [161]. In another version, the electrical contact is established by inserting a gold electrode into the outlet of the CE capillary (Figure 7.6*b*). A further improvement in the CE/MS interface was the adoption of a sheath-flow design of the LC/CF-FAB coupling (Figure 7.6*c*) [165,166]. The sheath liquid serves to establish the electrical contact at the cathode end of the capillary as well as provide a suitable solvent composition for a stable electrospray. For positive-ion analysis, the sheath liquid consists of a 1 : 1 (v : v) mixture of water–methanol that contains either 0.1% formic or acetic acid. Another successful design uses the liquid junction interface, in

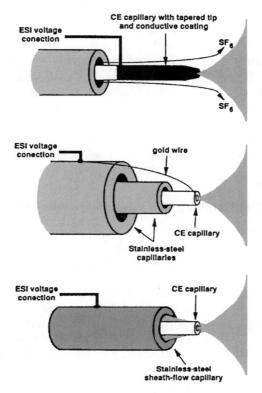

Figure 7.6. Schematic diagram of CE-ESIMS interfaces: (*a*) interface with conductive tip; (*b*) interface with a gold wire electrode; (*c*) coaxial sheath-flow configuration. [Reproduced from Ref. 198 by permission of Elsevier Science B. V. (copyright 1995).]

which a stainless-steel tee is used to introduce the makeup liquid (Figure 7.7) [162,167]. The cathode end of the CE capillary and the ESI needle are introduced from the opposite ends of the tee. The gap between the two terminals is adjusted to 10–25 μm. The electrical contact between the CE capillary and the ESI needle is established through the makeup liquid that surrounds the junction of the two capillaries.

The performance characteristics of the sheath flow and liquid junction interfaces have been evaluated, and the former appears to have an edge [168]. It is more robust and reproducible and provides enhanced signal-to-noise ratios and improved separations compared to those with liquid junction coupling. Also, it is ideal for flow-injection experiments. In contrast, the liquid junction interface is easy to assemble and operate. The additional liquid flow in both designs often degrades the detection sensitivity. In this respect, the sheathless design is superior. Other advantages of sheathless design are low flow rates, long-term stability, and the absence of any possible interference from the sheath liquids.

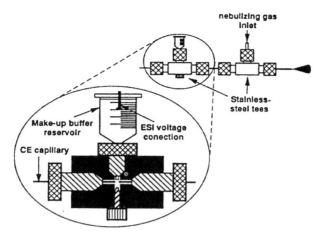

Figure 7.7. A diagram of a liquid junction interface for a CEMS combination. [Reproduced from Ref. 198 by permission of Elsevier Science B. V. (copyright 1995).]

An exciting development in CEMS methodology is the coupling of CE with microspray ionization [169,170]. In this method, the microsprayer tip and the CE column are connected and sheathed by a length of polysulfone microdialysis tubing. Electrolytic transfer of the buffer components provides the electrical connection to the CE and ESI processes across this membrane. No sheath liquid is required in this design, as a result of which 10–100-fold enhancement in the detection sensitivity is realized. Another rugged design that does not require any sheath liquid for the electrical connection makes use of a single-piece CE column tapered to a tip with an i.d. of 20 μm for microspraying [171]. This combination results in a high ionization efficiency and low-mass detection limit. The detection limit of 0.1–5 fmol for peptide standards has been reported. Enhancement in the signal intensity of the peaks has been achieved by a novel concept of voltage stepping, in which, by reducing the voltage during the time an analyte peak appears, the eluting time of the analyte is prolonged. In another simple version of a CEMS interface, inserting a platinum wire near the tapered capillary makes the electrical connection [172]. Smith and colleagues have applied another ingenious concept to couple a CE capillary with the ESI interface [173]. This interface makes use of a microdialysis membrane to provide the electrical connection. The separation capillary and ESI capillary are butted together inside a short dialysis tubing, which is surrounded by the background electrolyte. The sonic spray interface has also been coupled to CE [174].

Continuous-Flow Fast-Atom Bombardment Interface. The low flow rates (1–2 μL/min) of the solution that emerges from a CE capillary are not compatible with the liquid flow rates required for a stable CF-FAB operation. A makeup flow is required to overcome this deficiency. Sheath-flow and liquid junction designs discussed above have both been successfully used to couple CE with CF-FABMS (Figure 7.8) [157–159]. In fact, the coupling of CE with MS was first achieved via a CF-FAB probe. In that configuration, a liquid junction was used to introduce the makeup liquid and glycerol matrix. The sheath-flow design has the advantages that the composition and the flow rates of the CE effluents and of the FAB matrix solution can be optimized independently, and that the separation efficiency is higher. However, the benefit of high separation efficiency can be realized only with narrow (<15 μm-i.d.) separation capillaries. With larger capillaries, a significant vacuum-induced liquid flow is observed. This phenomenon degrades the separation because

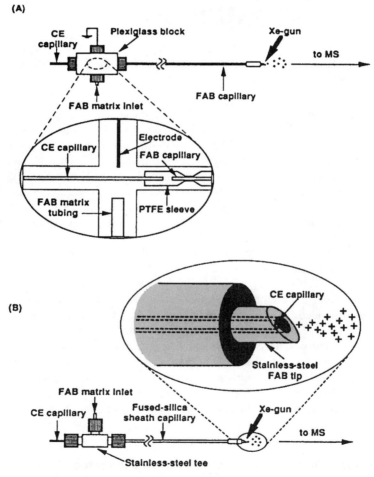

Figure 7.8. Schematic diagram of CE/CF-FAB interfaces: (*a*) liquid junction interface; (*b*) coaxial sheath-flow configuration. [Reproduced from Ref. 198 with permission from Elsevier Science B. V. (copyright 1995).]

the vacuum-induced flow is parabolic rather than of the plug type that is encountered in normal CE operation. The lower separation efficiency of the liquid junction coupling is due to the additional long transfer capillary. This design is easier to handle and operate.

Practical Consideration. For a successful coupling of CE with mass spectrometry, a few problems that are inherent to the CE operation must be surmounted. First, the ionization mode must be compatible with the charge of the eluting components. In practice, the CE separation is best achieved for the anionic form of analytes, whereas the best mass spectrometry response is obtained for cationic analytes. This problem can be alleviated with the postcolumn addition of an appropriate sheath buffer [175,176]. The second problem is the intrinsic narrow time width (<1 s) of CE peaks, similar to the ones that are encountered in a capillary GC analysis. This problem requires an ultrarapid scanning of the spectral data. An alternate solution to this problem is to use a programmable field-strength gradient [177]. This step increases the elution time of the analyte peak. The third problem is the poor concentration sensitivity of CE that is caused by the limited sample volumes, which usually are less than 1–2% of the total capillary volume. A solution to this problem requires a preconcentration step. One approach has used an on-line sample-stacking step via the capillary isotachophoresis mode of CE [178–181]. Another approach uses an on-line RP-HPLC packing [182,183] or an impregnated membrane [184,185] for the preconcentration step.

Coupling of Other Modes of CE with MS. For certain special applications other modes of CE have also been coupled with ESIMS. CIEF separates proteins on the basis of differences in isoelectric points (pI). To take advantage of its high-resolution capability, several researchers demonstrated the coupling of CIEF with ESIMS [186–188]. This combination can be effectively used for protein mapping [186]. Applications of CECMS [189,190], CGEMS [191], and MEKCMS [192] have also been reported.

Mass Spectrometry Instrumentation. Because of the obvious reasons of their tolerance of relatively higher pressures and voltages, quadrupole instruments are the preferred choice for the coupling of CE with mass spectrometry. Applications of CEMS with FT-ICRMS [193], TOF [194,195], and QIT [196] have also been reported.

Applications. Several areas of biochemical research have benefited from the emergence of the CEMS combination. Its applications include the analysis of peptides, proteins, enzymatic digests, noncovalent complexes, DNA adducts, pharmaceutical drugs and their metabolites, and compounds of environmental concern. A few excellent reviews can be consulted for these applications [197–200].

7.5. AFFINITY CHROMATOGRAPHY COMBINED WITH MASS SPECTROMETRY

Although the liquid chromatography–mass spectrometry combination provides a high-resolution separation and the specific detection of analytes in a complex mixture, in many situations it is necessary to provide purification, concentration, and isolation of the desired component in a biological matrix. This step is performed by affinity chromatography. Practical examples are the selective capture of a specific antigen by the corresponding antibody and isolation of phosphorylated proteins and peptides [201–204]. Immunoaffinity extraction in combination with mass spectrometry has been used for the mapping of epitopes [205].

7.6. OTHER SEPARATION TECHNIQUES COMBINED WITH MASS SPECTROMETRY

The coupling of several less common forms of separation techniques has also received attention. Several researchers have interfaced supercritical fluid chromatography (SFC) with mass spectrometry [10,11,206–209]. This combination exploits the liquid-like properties of supercritical fluids, and the ease of coupling of GC with mass spectrometry. Packed and capillary chromatography columns have both been utilized. Mass spectrometry interfaces include thermospray [210-212], particle beam [213], moving belt [214,215], and API interface [216,217]. Although thin layer chromatography is a low-resolution separation technique, its on-line combination has also been utilized for certain biological applications [13,218,219]. The polar and labile compounds separated by TLC can be analyzed by directly scanning the TLC plate with FAB ionization. TLC has also been coupled with MALDI through a hybrid TLC–MALDI plate, in which a silica layer and a MALDI layer are configured adjacently on a common backing [219]. The on-line coupling of SEC with ESIMS has been exploited to assess the molecular mass distribution of nonpolar surfactants [11], and to study proteolytic processing of prodynorphin-derived peptides in rat brain [220]. Countercurrent chromatography, which separates analytes through a distribution between two adjacent layers of a moving liquid, has also been coupled to mass spectrometry via moving-belt [221] and frit-FAB interfaces [222].

REFERENCES

1. R. L. Grob, *Modern Practice of Gas Chromatography*, Wiley, New York, 1995.
2. V. R. Meyer, *Practical High-Performance Liquid Chromatography*, Wiley, New York, 1999.
3. E. D. Katz, ed., *High-Performance Liquid Chromatography: Principles and Methods in Biotechnology*, Wiley, New York, 1996.
4. P. Camilleri, ed., *Capillary Electrophoresis—Theory and Practice*, CRC Press, Boca Raton, FL, 1998.

5. P. G. Righetti, ed., *Capillary Electrophoresis in Analytical Biotechnology*, CRC Press, Boca Raton, FL, 1998.

6. F. W. McLafferty, *Chemtech* **22**, 182–189 (1992).

7. F. G. Kitson, B. S. Larsen, and C. N. McEwen, *A Practical Guide to Gas Chromatography and Mass Spectrometry*, Academic Press, San Diego, 1996.

8. R. Willoughby, E. Sheehan, and S. Mitrovich, *A Globel View of LC/MS*, Global View Publishing, Pittsburg, PA, 1995.

9. J. H. Wahl, H. R. Udseth, and R. D. Smith, in W. S. Hancock, ed., *New Methods in Peptide Mapping for the Characterization of Proteins*, CRC Press, Boca Raton, FL, 1996, pp. 143–179.

10. D. Thomas, P. G. Sim, and F. M. Benoit, *Rapid Commun. Mass Spectrom.* **8**, 105–110 (1994).

11. P. Arpino and P. Haas, *J. Chromatogr. A* **703**, 479–488 (1995).

12. W. J. Simonsick and L. Prokai, *Adv. Chem. Ser.* **247**, 41–56 (1995).

13. K. L. Busch, *Trends Anal. Chem.* **11**, 314–324 (1992).

14. R. L. White and C. L. Wilkins, *Anal. Chem.* **54**, 2443–2447 (1982).

15. M. J. Tomlinson, T. A. Sasaki, and C. L. Wilkins, *Mass Spectrom. Rev.* **15**, 1–14 (1996).

16. A. Lommen, M. Godejohann, D. P. Venema, P. C. H. Hollman, and M. Spraul, *Anal. Chem.* **72**, 1793–1797 (2000).

17. F. S. Pullen, A. G. Swanson, M. J. Newman, and D. S Richards, *Rapid Commun. Mass Spectrom.* **9**, 1003–1006 (1995).

18. R. M. Holt, M. J. Newman, F. S. Pullen, D. S Richards, and A. G. Swanson, *J. Mass Spectrom.* **32**, 64–70 (1997).

19. J.-T. Wu, M. G. Qian, M. X. Li, L. Liu, and D. M. Lubman, *Anal. Chem.* **68**, 3388–3396 (1996).

20. R. W. Purves and L. Li, *J. Am. Soc. Mass Spectrom.* **8**, 1085–1093 (1997).

21. E. B. Ledford, Jr., R. L. White, S. Ghaderi, C. L. Wilkins, and M. L. Gross, *Anal. Chem.* **52**, 2450–2451 (1980).

22. J. F. Holland, B. Newcombe, R. E. Tacklenburg, Jr., M. Devenport, J. Allison, J. T. Watson, and C. G. Enke, *Rev. Sci. Instrum.* **62**, 69–76 (1991).

23. M. C. McMaster and C. McMaster, *GC/MS: A Practical User's Guide*, Wiley, New York, 1998.

24. H. M. McNair and J. M. Miller, *Basic Gas Chromatography*, Wiley, New York, 1997.

25. R. A. Hites, in F. Settle, ed., *Handbook of Instrumental Techniques for Analytical Chemistry*; Prentice-Hall PTR; Upper Saddle River, NJ, 1998, pp. 609–626.

26. R. Ryhage, *Anal. Chem.* **36**, 759–764 (1964).

27. E. J. Cone and W. D. Darwin, *J. Chromatogr. Biomed. Appl.* **580**, 43–61 (1992).

28. Y. Nygren, S-A. Fredriksson, and B. Nelsson, *J. Mass Spectrom.* **31**, 267–274 (1996).

29. R. E. Clement, ed., *Gas Chromatography: Biochemical Biomedical, and Clinical Applications*, Wiley, New York, 1990.

30. C. Koppel and J. Tenczer, *J. Am. Soc. Mass Spectrom.* **6**, 995–1003 (1995).

31. J. Farrario, C. Byrne, D. McDaniel, A. Dupuy, Jr., and R. Harless, *Anal. Chem.* **68**, 647–652 (1996).

32. J. Roboz, E. Nieves, and J. Holland, *J. Chromatogr.* **500**, 413–426 (1990).

33. A. M. Lawson, *Mass Spectrometry in Clinical Biochemistry*, Water de Gruyter, Berlin, 1988.

34. P. M. Buszka, D. L. Rose, G. B. Ozuna, and G. E. Groschen, *Anal. Chem.* **67**, 3659–3667 (1995).

35. V. L. Tal'rose, R. G. Skurat, I. G. Gorodetskii, and N. B. Zoltoai, *Russ. J. Phys. Chem.* **46**, 456 (1972).

36. E. C. Horning, D. I. Carroll, I. Dzidic, K. D. Haegele, M. G. Horning, and R. N. Stillwell, *J. Chromatogr.* **99**, 13–21 (1974).

37. C. Dass, P. Mahalakshmi, and D. Grandberry, *J. Chromatogr. A* **678**, 249–257 (1994).

38. M. A. Baldwin and F. W. McLafferty, *Org. Mass Spectrom.* **7**, 1111–1112 (1973).

39. P. Arpino, in M. L. Gross, ed., *Mass Spectrometry in the Biological Sciences*, Kluwer, Dordrecht, The Netherlands, 1992, pp. 253–267.

40. P. Arpino and C. Beaugrand, *Int. J. Mass Spectrom. Ion Proc.* **64**, 275–298 (1985).

41. P. Arpino, *Mass Spectrom. Rev.* **8**, 35–55 (1989).

42. W. H. McFadden, H. L. Schwartz, and D. C. Bradford, *J. Chromatogr.* **122**, 389–396 (1976).

43. M. J. Hayes, E. P. Lankmayer, P. Vouros, and B. L. Karger, *Anal. Chem.* **55**, 1745–1752 (1983).

44. J. G. Stroh, J. C. Cook, R. M. Milberg, L. Brayton, T. Kihara, Z. Huang, K. L. Rinehart, Jr., and I. A. S. Lewis, *Anal. Chem.* **57**, 985–991 (1985).

45. R. D. Voyksner, C. S. Smith, and P. C. Knox, *Biomed. Environ. Mass Spectrom.* **19**, 523 (1990).

46. C. S. Creaser and J. W. Stygall, *Analyst* **118**, 1467–1480 (1993).

47. R. C. Willoughby and R. F. Browner, *Anal. Chem.* **56**, 2626–2631 (1984).

48. J. D. Kirk and R. F. Browner, *Biomed. Environ. Mass Spectrom.* **18**, 355–357 (1989).

49. P. E. Sanders, *Rapid Commun. Mass Spectrom.* **4**, 123–124 (1990).

50. T.-C. Wang, L. J. Cornio, and S. P. Markey, *J. Am. Soc. Mass Spectrom.* **7**, 293–297 (1996).

51. A. Cappiello and F. Bruner, *Anal. Chem.* **65**, 1281–1287 (1993).

52. A. Cappiello, G. Famiglini, A. Lombardozzi, A. Massari, and G. G. Vadala, *J. Am. Soc. Mass Spectrom.* **7**, 753–758 (1996).

53. C. R. Blakley and M. L. Vestal, *Anal. Chem.* **55**, 750–754 (1983).

54. P. Arpino, *Mass Spectrom. Rev.* **9**, 631–669 (1990).

55. M. L. Vestal, *Science* **226**, 275–281 (1984).

56. H. Y. Kim, D. Pilosof, D. F. Dyckes, and M. L. Vestal, *J. Am. Chem. Soc.* **106**, 7304–7309 (1984).

57. C. Fenselau, D. J. Librato, J. A. Yargey, R. J. Cotter, and A. L. Yargey, *Anal. Chem.* **56**, 2759–2762 (1984).

58. D. Volmer and K. Levsen, in H. J. Stan, ed., *Analysis of Pesticides in Ground and Surface Waters II*, Springer-Verlag, New York, 1995, pp. 133–179.

59. D. A. Catlow, *J. Chromatogr.* **523**, 163–170 (1985).

60. R. D. Voyksner, J. T. Bursey, and E. D. Pellizzari, *Anal. Chem.* **56**, 1507–1533 (1984).

61. Y. Ito, T. Takeuchi, D. Ishii, and M. Goto, *J. Chromatogr.* **346**, 161–166 (1985).

62. R. M. Caprioli, W. T. Moore, B. DaGue, and M. Martin, *J. Chromatogr.* **443**, 355–362 (1988).

63. R. M. Caprioli, B. DaGue, and K. Wilson, *J. Chromatogr. Sci.* **26**, 640–644 (1988).

64. M. A. Moseley, L. J. Deterding, J. S. M. deWit, K. B. Tomer, and J. W. Jorgenson, *Anal. Chem.* **61**, 1577–1584 (1989).

65. M. A. Moseley, L. J. Deterding, K. B. Tomer, and J. W. Jorgenson, *Anal. Chem.* **63**, 1467–1473 (1991).

66. J. E. Coutant, T.-M. Chen, and B. L. Ackerman, *J. Chromatogr.* **529**, 265–275 (1990).

67. J. P. Gagne, A. Carrier, and M. Bertrand, *J. Chromatogr.* **554**, 61–71 (1991).

68. J. P. Gagne, A. Carrier, L. Varfalvey, and M. Bertrand, *J. Chromatogr.* **647**, 21–29 (1993).

69. J. S. M. deWit, M. A. Moseley, L. J. Deterding, K. B. Tomer, and J. W. Jorgenson, *Rapid Commun. Mass Spectrom.* **2**, 100–104 (1988).

70. B. L. Ackerman, J. E. Coutant, and T.-M. Chen, *Biol. Mass Spectrom.* **20**, 431–440 (1991).

71. A. Cappiello, P. Palma, A. Papayannopoulos, and K. Biemann, *Chromatographia* **30**, 477–483 (1990).

72. D. B. Kassel, B. D. Musselman, and J. A. Smith, *Anal. Chem.* **63**, 1091–1097 (1991).

73. L. J. Deterding, M. A. Moseley, K. B. Tomer, and J. W. Jorgenson, *Anal. Chem.* **61**, 2504–2511 (1989).

74. R. M. Caprioli, *Continuous-Flow Fast Atom Bombardment Mass Spectrometry*, Wiley, New York, 1990.

75. R. M. Caprioli and W. T. Moore, *Anal. Chem.* **63**, 1978–1983 (1991).

76. M. P. Knadler, B. L. Ackerman, J. E. Coutant, and G. H. Hurst, *Drug Metab. Dispos.* **20**, 89–95 (1992).

77. W. J. Henzel, J. H. Bourell, and J. T. Stults, *Anal. Biochem.* **187**, 228–233 (1990).

78. P. S. Kokkonen, W. M. A. Niessen, U. R. Tjaden, and J. van der Greef, *Rapid Commun. Mass Spectrom.* **5**, 19–24 (1991).

79. The API Book, PE SCIEX (1990).

80. D. I. Carroll, I. Dzdic, R. N. Stillwater, and E. C. Horning, *Anal. Chem.* **47**, 1956–1959 (1975).

81. E. C. Huang, T. Wachs, J. J. Conboy, and J. D. Henion, *Anal. Chem.* **62**, 713A–725A (1990).

82. T. Wachs, J. J. Conboy, F. Garcia, and J. D. Henion, *J. Chromatogr. Sci.* **29**, 357–366 (1991).

83. W. M. A. Niessen and A. P. Tinke, *J. Chromatogr. A* **703**, 37–57 (1995).

84. J. B. Fenn, M. Mann, C. K. Meng, S. F. Wong, and C. M. Whitehouse, *Mass Spectrom. Rev.* **9**, 37–70 (1990).

85. R. D. Smith, J. A. Loo, R. R. Ogorzalek Loo, M. Busman, and H. R. Udseth, *Mass Spectrom. Rev.* **10**, 359–451 (1991).

86. A. Bruins, T. R. Covey, and J. D. Henion, *Anal. Chem.* **59**, 2642–2646 (1987).

87. T. R. Covey, R. F. Bonner, B. I. Shushan, and J. D. Henion, *Rapid Commun. Mass Spectrom.* **2**, 249–256 (1988).

88. E. D. Lee and J. D. Henion, *Rapid Commun. Mass Spectrom.* **6**, 727–733 (1992).

89. G. Hopfgartner, T. Wachs, K. Bean, and J. D. Henion, *Anal. Chem.* **65**, 439–446 (1993).

90. M. S. Wilm and M. Mann, *Int. J. Mass Spectrom. Ion. Proc.* **136**, 167–180 (1994).

91. M. R. Emmet and R. M. Caprioli, *J. Am. Soc. Mass Spectrom.* **5**, 605–613 (1994).

92. A. Hirabayashi, Y. Hirabayashi, M. Sakairi, and H. Koizumi, *Rapid Commun. Mass Spectrom.* **10**, 1703–1705 (1996).

93. Y. Hirabayashi, A. Hirabayashi, M. Sakairi, and H. Koizumi, *Anal. Chem.* **70**, 1882–1884 (1998).

94. C. Dass, in F. Settle, ed., *Handbook of Instrumental Techniques for Analytical Chemistry*, Prentice-Hall PTR; Upper Saddle River, NJ, 1998, pp. 647–664.

95. C. Dass, *Current Org. Chem.* **3**, 193–209 (1999).

96. T. R. Covey, E. C. Huang, and J. D. Henion, *Anal. Chem.* **63**, 1193–1200 (1991).

97. E. C. Huang and J. D. Henion, *Anal. Chem.* **63**, 732–739 (1991).

98. J. C. Le, J. Hui, V. Katta, and M. F. Rohde, *J. Am. Soc. Mass Spectrom.* **8**, 703–712 (1997).

99. E. C. Huang and J. D. Henion, *J. Am. Soc. Mass Spectrom.* **1**, 158–165 (1990).

100. D. A. Volmer, *Rapid Commun. Mass Spectrom.* **10**, 1615–1620 (1996).

101. A. C. Hogenboom, J. Slobodnik, J. J. Vreuls, J. A. Rontree, B. L. M. van Baar, W. M. A. Niessen, and U. A. Th. Brinkman, *Chromatographia* **42**, 506–514 (1996).

102. H. Zhang, K. Heinig, and J. Henion, *J. Mass Spectrom.* **35**, 423–431 (2000).

103. J. Ayrton, G. J. Dear, W. J. Leavens, D. N. Mallett, and R. S. Plumb, *Rapid Commun. Mass Spectrom.* **11**, 1953–1958 (1997).

104. D. B. Kassel, B. Shushan, T. Sakuma, and J.-P. Salzmann, *Anal. Chem.* **66**, 236–243 (1994).

105. H.-Y. Lin and R. D. Voyksner, *Rapid Commun. Mass Spectrom.* **8**, 333–338 (1994).

106. T. J. Barder, P. J. Wohlman, C. Thrall, and P. D. DuBois, *LC-GC*, **15**, 918–926 (1997).

107. M. J. Lee, S. Monté, S. Sanderson, and N. J. Haskins, *Rapid Commun. Mass Spectrom.* **13**, 216–221 (1999).

108. C. Dass, in D. M. Desiderio, ed., *Mass Spectrometry: Clinical and Biomedical Applications*, Plenum Press, New York, 1994, Vol. 2, pp. 1–52.

109. G. J. Van Berkel, S. A. McLuckey, and G. L. Glish, *Anal. Chem.* **62**, 1284–1295 (1990).

110. J. E. Bruce, S. A. Hofstadler, B. E. Winger, and R. D. Smith, *Int. J. Mass Spectrom. Ion Proc.* **132**, 97 (1994).

111. M. W. Senko, C. L. Hendrickson, M. R. Emmett, S. D.-H. Shi, and A. G. Marshall, *J. Am. Soc. Mass Spectrom.* **8**, 970–986 (1997).

112. M. R. Emmett, F. M. White, C. L. Hendrickson, S. D.-H. Shi, and A. G. Marshall, *J. Am. Soc. Mass Spectrom.* **9**, 333–340 (1998).

113. M. Guilhaus, D. Selby, and V. Mlynski, *Mass Spectrom. Rev.* **19**, 65–107 (2000).

114. A. N. Krutchinsky, A. V. Loboda, V. L. Spicer, R. Dworschak, W. Ens, and K. G. Standing, *Rapid Commun. Mass Spectrom.* **12**, 508–518 (1998).

115. I. V. Chernushevich, W. Ens, and K. G. Standing, *Anal. Chem.* **71**, 452A–461A (1999).

116. B. M. Chien, S. M. Michael, and D. M. Lubman, *Rapid Commun. Mass Spectrom.* **7**, 837–843 (1993).

117. J.-T. Wu, L. He, M. X. Li, S. Parus, and D. M. Lubman, *J. Am. Soc. Mass Spectrom.* **8**, 1237–1246 (1997).

118. R. W. Purves and L. Li, *J. Am. Soc. Mass Spectrom.* **8**, 1085–1093 (1997).

119. D. Nagra and L. Li, *J. Chromatogr. A* **711**, 235–245 (1995).

120. A. P. L. Wang, X. Guo, and L. Li, *Anal. Chem.* **66**, 3664–3675 (1994).

121. K. K. Murray and D. H. Russell, *J. Am. Soc. Mass Spectrom.* **5**, 1–9 (1994).

122. K. K. Murray, T. M. Lewis, M. D. Beeson, and D. H. Russell, *Anal. Chem.* **66**, 1601–1609 (1996).

123. K. K. Murray, *Mass Spectrom. Rev.* **16**, 183–299 (1997).

124. W. S. Hancock, ed., *New Methods in Peptide Mapping for the Characterization of Proteins*, CRC Press, Boca Raton, FL, 1996.

125. C. Dass, in B. S. Larsen and C. N. McEwen, eds., *Mass Spectrometry of Biological Materials*, Marcel Dekker, New York, 1998, pp. 247–280.

126. T. Covey, in J. R. Chapman, ed., *Protein and Peptide Analysis by Mass Spectrometry*, Humana Press, Totowa, NJ, 1996, pp. 83–99.

127. D. B. Kassel, R. K. Blackburn, and B. Antonsson, in B. S. Larsen and C. N. McEwen, eds., *Mass Spectrometry of Biological Materials*, Marcel Dekker, New York, 1998, pp. 247–280.

128. M. J. Huddleston, R. S. Annan, M. F. Bean, and S. A. Carr, *J. Am. Soc. Mass Spectrom.* **4**, 710–717 (1993).

129. J. Ding, W. Burkhart, and D. B. Kassel, *Rapid Commun. Mass Spectrom.* **8**, 94–98 (1994).

130. X. Zhu and C. Dass, *J. Liq. Chromatgr. Rel. Technol.* **22**, 1635–1647 (1999).

131. M. J. Huddleston, M. F. Bean, and S. A. Carr, *Anal. Chem.* **65**, 877–884 (1993).

132. S. A. Carr, M. J. Huddleston, and M. F. Bean, *Prot. Sci.* **2**, 183 (1993).

133. A. W. Guzzetta and W. S. Hancock, in W. S. Hancock, ed., *New Methods in Peptide Mapping for the Characterization of Proteins*, CRC Press, Boca Raton, FL, 1996, pp. 181–217.

134. L. Rovatti, B. Masin, S. Catinella, and M. Hamdan, *Rapid Commun. Mass Spectrom.* **11**, 1223–1229 (1997).

135. S. Li and C. Dass, *Anal. Biochem.* **270**, 9–14 (1999).

136. R. Wieboldt, J. Zweigenbaum, and J. D. Henion, *Anal. Chem.* **69**, 1683–1691 (1997).

137. Y. M. Dunayevskiy, J.-J. Lai, C. Quinn, F. Talley, and P. Vouros, P. *Rapid Commun. Mass Spectrom.* **11**, 1178–1184 (1997).

138. M. Ayyoub, B. Monsarrat, H. Mazarguil, and R. Gairin, *Rapid Commun. Mass Spectrom.* **12**, 557–564 (1998).

139. D. Jayawardene and C. Dass, *Peptides* **20**, 963–970 (1999).

140. G. C. Wood, C. Dass, M. Iyer, A. M. Fleischner, and B. B. Sheth, *PDA J. Pharm. Technol.* **51**, 176–180 (1997).

141. Y. Xia, P.-P. Wang, M. G. Bartlett, H. M. Solomon, and K. L. Busch, *Anal. Chem.* **72**, 764–771 (2000).

142. Y. Yang, W. J. Griffiths, J. Sjövall, J.-Å. Gustafsson, and J. Rafter, *J. Am. Soc. Mass Spectrom.* **8**, 50–61 (1997).

143. N. Y. Tretyakova, S.-Y. Chiang, V. E. Walker, and J. A. Swenberg, *J. Mass Spectrom.* **33**, 363–376 (1998).

144. A. Lagana, G. Fago, and A. Marino, *Anal. Chem.* **70**, 121–130 (1998).

145. C. Dass, J. J. Kusmierz, D. M. Desiderio, S. A. Jarvis, and B. N. Green, *J. Am. Soc. Mass Spectrom.* **2**, 149–156 (1990).

146. T. R. Covey, E. D. Lee, A. Bruins, and J. D. Henion, *Anal. Chem.* **58**, 1451A–1463A (1986).

147. H. Kambara, *Anal. Chem.* **54**, 143–146 (1982).

148. M. S. Sakairi and H. Kambara, *Anal. Chem.* **60**, 774–780 (1988).

149. T. Nabeshima, Y. Takada, and M. S. Sakairi, *Rapid Commun. Mass Spectrom.* **11**, 715–718 (1997).

150. R. D. Voyksner, *Environ. Sci. Technol.* **28**, 118A–127A (1994).

151. E. Gelpi, *J. Chromatogr. A* **703**, 59–80 (1995).

152. R. Kendal and C. Dass, *S. Assoc. Agr. Scient. Bulletin: Biochem. Biotechnol.* **12**, 1–6 (1999).

153. A. G. Ewing, R. A. Wallingford, and T. M. Olefirowicz, *Anal. Chem.* **61**, 292A–303A (1989).

154. H. Carchon and E. Eggermont, *Am. Lab.* **24**, 67–72 (1992).

155. K. D. Lucas and J. W. Jorgenson, *J. High Res. Chromatogr.* **8**, 407–411 (1985).

156. M. M. Dittmann, K. Wienand, F. Bek, and G. P. Rozing, *LC-GC*, **13**, 800–814 (1995).

157. M. A. Moseley, L. J. Deterding, K. B. Tomer, and J. W. Jorgenson, *Rapid Commun. Mass Spectrom.* **3**, 87–93 (1989).

158. R. M. Caprioli, W. T. Moore, M. Martin, B. DaGue, K. Wilson, and S. Moring, *J. Chromatogr.* **480**, 247–257 (1989).

159. N. J. Reinhoud, W. M. A. Niessen, U. R. Tjaden, L. G. Gramberg, E. R. Verheij, and J. van der Greef, *Rapid Commun. Mass Spectrom.* **3**, 348–357 (1989).

160. J. A. Olivares, N. T. Nguyen, C. R. Yonker, and R. D. Smith, *Anal. Chem.* **59**, 1230–1232 (1987).

161. R. D. Smith, J. A. Olivares, N. T. Nguyen, and H. R. Udseth, *Anal. Chem.* **60**, 436–441 (1988).

162. E. D. Lee, W. Muck, J. D. Henion, and T. R. Covey, *J. Chromatogr.* **458**, 313–321 (1989).

163. H. Zhang and R. M. Caprioli, *J. Mass Spectrom.* **31**, 1039–1046 (1996).

164. J. Preisler, F. Foret, and B. L. Karger, *Anal. Chem.* **70**, 5278–5287 (1998).

165. R. D. Smith, C. J . Barinaga, and H. R. Udseth, *Anal. Chem.* **60**, 1948–1952 (1988).

166. H. R. Udseth, J. A. Loo, and R. D. Smith, *Anal. Chem.* **61**, 228–232 (1989).

167. I. M. Johansson, E. C. Huang, and J. D. Henion, *J. Chromatogr.* **559**, 515–528 (1991).

168. S. Pleasance, P. Thibault, and J. Kelly, *J. Chromatogr.* **559**, 197–208 (1991).

169. J. C. Severs, A. C. Harms, and R. D. Smith, *Rapid Commun. Mass Spectrom.* **10**, 1175–1178 (1996).

170. E. Rohde, A. J. Tomlinson, D. H. Jonson, and S. Naylor, *Electrophoresis*, **19**, 2361 (1998).

171. J. F. Kelly, L. Ramaley, and P. Thibault, *Anal. Chem.* **69**, 51–60 (1997).

172. P. Cao and M. Moini, *J. Am. Soc. Mass Spectrom.* **8**, 561–564 (1997).

173. J. Severs and R. D. Smith, *Anal. Chem.* **69**, 2154–2158 (1997).

174. Y. Hirabayashi, A. Hirabayashi, and H. Koizumi, *Rapid Commun. Mass Spectrom.* **13**, 712–715 (1999).

175. M. A. Moseley, L. J. Deterding, K. B. Tomer, and J. W. Jorgenson, *Anal. Chem.* **63**, 109–104 (1991).

176. P. Thibault, C. Paris, and S. Pleasance, *Rapid Commun. Mass Spectrom.* **5**, 484–490 (1991).

177. D. R. Goodlett, J. H. Wahl, H. R. Udseth, and R. D. Smith, *J. Microcolumn. Sep.* **5**, 57–62 (1993).

178. R. D. Smith, S. M. Fields, J. A. Loo, C. J. Baringa, H. R. Udseth, and C. G. Edmonds, *Electrophoresis* **11**, 709–717 (1990).

179. A. P. Tinke, N. J. Reinhoud, W. M. A. Niessen, and J. van der Greef, *Rapid Commun. Mass Spectrom.* **6**, 560–563 (1992).

180. N. J. Reinhoud, A. P. Tinke, U. R. Tjaden, W. M. A. Niessen, and J. van der Greef, *J. Chromatogr.* **627**, 263–271 (1992).

181. T. J. Thompson, F. Foret, P. Vouros, and B. L. Karger, *Anal. Chem.* **65**, 900–906 (1993).

182. D. Figeys, A. Durect, and R. Aebersold, *J. Chromatgr. A,* **763**, 295–306 (1997).

183. A. J. Tomlinson and S. Naylor, *J. Capillary Electrophor.* **2**, 225 (1995).

184. Q. Yand, A. J. Tomlinson and S. Naylor, *Anal. Chem.* **71**, 183A–189A (1999).

185. A. J. Tomlinson, M. L. Benson, S. Jameson, D. H. Johnson, and S. Naylor, *J. Am. Soc. Mass Spectrom.* **8**, 15–24 (1997).

186. J. Ding and P. Vouros, *Anal. Chem.* **71**, 378A–385A (1999).

187. Q. Tang, A. K. Harata, and C. S. Lee, *Anal. Chem.* **69**, 3177–3182 (1997).

188. N. J. Clarke, A. J. Tomlinson, and S. Naylor, *J. Am. Soc. Mass Spectrom.* **8**, 743–748 (1997).

189. M. R. Taylor and P. Teale, *J. Chromatogr. A,* **768**, 89–95 (1997).

190. J. Ding, T. Barlow, A. Dipple, and P. Vouros, *J. Am. Soc. Mass Spectrom.* **9**, 823–829 (1998).

191. A. Harsch and P. Vouros, *Anal. Chem.* **70**, 3021–3027 (1998).

192. M. H. Lamoree, U. R. Tjaden, and J. van der Greef, *J. Chromatogr. A* **712**, 219–225 (1995).

193. J. C. Severs, R. D. Smith, S. A. Hofstadler, F. D. Swanek, and A. G. Ewing, *J. High Res. Chromatgr.* **19**, 617–621 (1996).

194. J.-T. Wu, M. G. Qian, M. X. Li, K. Zheng, P. Huang, and D. M. Lubman, *J. Chromatogr. A* **794**, 377–389 (1998).

195. M. E. McComb, A. N. Krutchinsky, W. Ens, and K. G. Standing, *J. Chromatogr. A* **800**, 1–11 (1998).

196. R. Shepperd and J. Henion, *Anal. Chem.* **69**, 2901–2907 (1997).

197. W. M. A. Niessen, U. R. Tjaden, and J. van der Greef, *J. Chromatogr.* **636**, 3–19 (1993).

198. J. Cai and J. Henion, *J. Chromatogr. A* **703**, 667–692 (1995).

199. J. H. Wahl, H. R. Udseth, and R. D. Smith, in W. S. Hancock, ed., *New Methods in Peptide Mapping for the Characterization of Proteins*, CRC Press, Boca Raton, FL, 1996, pp. 143–179.

200. J. F. Banks, *Electrophoresis* **18**, 2255 (1998).

201. W. C. Brumely and W. Winnik, *J. Chromatogr. Librr.* **59**, 481–527 (1996).

202. S. Li and C. Dass, *Eur. Mass Spectrom.* **5**, 279–284 (1999).

203. G. S. Rule and J. D. Henion, *J. Chromatogr.* **582**, 103–112 (1992).

204. J. Cai and J. Henion, *Anal. Chem.* **66**, 3723–3726 (1994).

205. Y. Zhao and B. T. Chait, *Anal. Chem.* **68**, 72–78 (1996).

206. R. D. Smith, H. T. Kalinoski, H. R. Udseth, and C. G. Edmonds, *Mass Spectrom. Rev.* **6**, 445–496 (1987).

207. D. E. Games, A. I. Berry, I. C. Mylchreest, J. R. Perkins, and S. Pleasance, *Anal. Proc.* **24**, 371–372 (1987).

208. P. Arpino, in M. L. Gross, ed., *Mass Spectrometry in the Biological Sciences*, Kluwer, Dordrecht, The Netherlands, 1992, pp. 269–280.

209. D. M. Sheely and V. N. Reinhold, *J. Chromatogr.* **474**, 83–96 (1989).

210. R. D. Smith and H. R. Udseth, *Anal. Chem.* **54**, 13–32 (1987).

211. A. I. Berry, D. E. Games, I. C. Mylchreest, J. R. Perkins, and S. Pleasance, *Biomed. Environ. Mass Spectrom.* **15**, 105–109 (1988).

212. J. R. Chapman, *Rapid Commun. Mass Spectrom.* **15**, 105–109 (1988).

213. P. O. Edlund and J. D. Henion, *J. Chromatogr. Sci.* **27**, 274–282, (1989).

214. A. I. Berry, D. E. Games, and J. R. Perkins, *J. Chromatogr.* **363**, 147–158, (1986).

215. J. R. Perkins, D. E. Games, J. R. Startin, and J. Gilbert, *J. Chromatogr.* **540**, 239–256, 257–270 (1991).

216. E. Huang, J. Hanion, and T. R. Covey, *J. Chromatogr.* **511**, 257–270 (1990).

217. J. F. Ancleto, L. Ramaley, R. K. Boyd, S. Pleasance, M. A. Quilliam, P. G. Sim, and F. M. Benoit, *Rapid Commun. Mass Spectrom.* **5**, 149–155 (1994).

218. P. Martin, W. Moden, P. Wall, and I. D. Wilson, *J. Planar Chromatogr.* **5**, 255–258, (1992).

219. J. T. Mehl and D. M. Hercules, *Anal. Chem.* **72**, 68–73 (2000).

220. I. Nylander, K. Tan-No, A. Winter, and J. Silberring, *Life Sci.* **57**, 123–129 (1995).

221. J. N. McGuire, M. L. Proefke, W. D. Conway, and K. L. Rinehart, *ACS Symp. Ser.* (Chapter 12, 'Modern Countercurrent Chromatography") **593**, 129–142 (1995).

222. H. Oka, in *High-Speed Countercurrent Chromatography*, Y. Ito and W.D. Conway, eds., Chemical Analysis Series, Wiley, New York, 1996, Vol. 132, pp. 73–91.

8

STRUCTURAL ANALYSIS
OF PROTEINS

8.1. BIOLOGICAL FUNCTIONS AND STRUCTURES OF PROTEINS

Proteins are complex macromolecules, and by far the most important of all biological compounds. The word *protein* has its root in the Greek word *proteios*, which means "of first importance." Proteins play crucial roles in virtually all biological processes. Some of the many important functions of proteins are as enzymes, receptors, antibodies (for immune protection), and hormones; for control of gene expression; as major components of muscles, bones, skin, and other organs; and as a means of transport and storage of small molecules.

Although the human body contains over 100,000 different proteins, all are formed from 20 naturally occurring amino acids (see Figure 8.1 for their structure, three-letter abbreviations, and one-letter code) that are covalently bonded together in a head-to-tail arrangement via an amide linkage (also called the *peptide bond*). A shorter chain of 30–40 amino acids is termed a *peptide*. A detailed discussion of peptides is presented in the next chapter. A longer chain is called a *polypeptide*. Individual proteins are composed of discrete polypeptide chains that are linked together either covalently or by the local structure of the surrounding solvent molecules.

Proteins are organized into four structure levels [1]. The basic level is the *primary structure*, which describes the linear arrangement (sequence) of the constituent amino acids. The primary structure represents what is directly encoded by an organism's genome (the elaborate structure of DNA in a cell). In addition, it includes

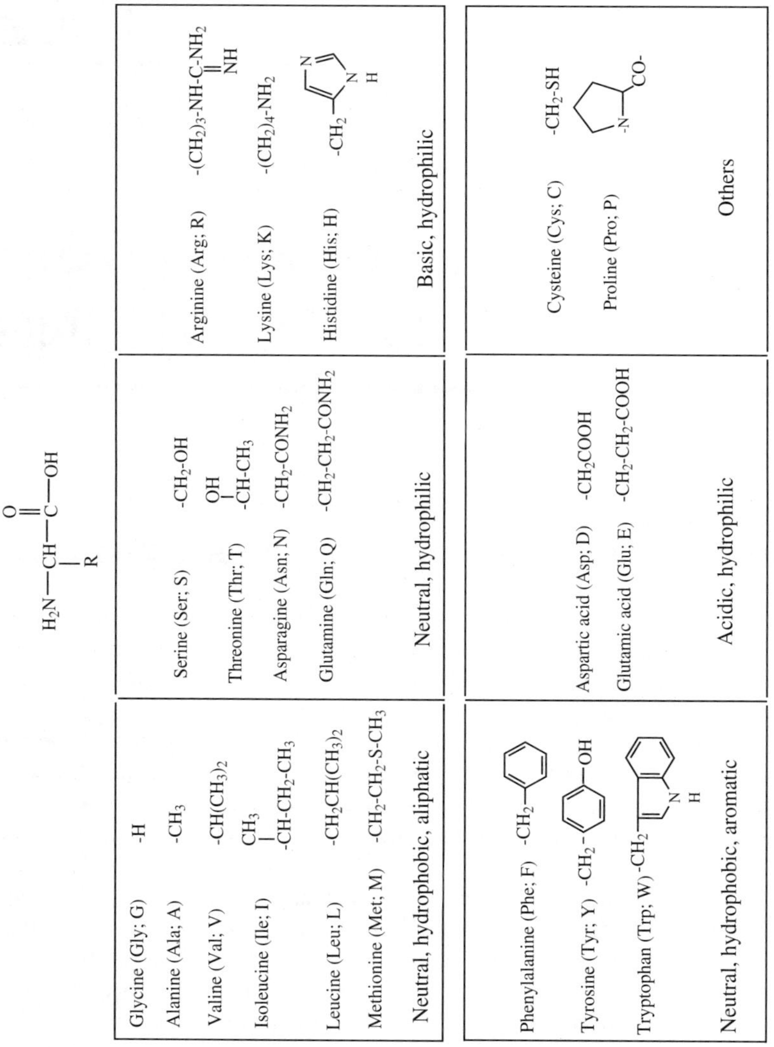

$$H_2N-\overset{\displaystyle \ \ \ \ \ O}{\underset{\displaystyle R}{CH}}-\overset{\displaystyle \|}{C}-OH$$

Glycine (Gly; G) -H

Alanine (Ala; A) -CH$_3$

Valine (Val; V) -CH(CH$_3$)$_2$

Isoleucine (Ile; I) -CH-CH$_2$-CH$_3$ (CH$_3$)

Leucine (Leu; L) -CH$_2$CH(CH$_3$)$_2$

Methionine (Met; M) -CH$_2$-CH$_2$-S-CH$_3$

Neutral, hydrophobic, aliphatic

Serine (Ser; S) -CH$_2$-OH

Threonine (Thr; T) -CH-CH$_3$ (OH)

Asparagine (Asn; N) -CH$_2$-CONH$_2$

Glutamine (Gln; Q) -CH$_2$-CH$_2$-CONH$_2$

Neutral, hydrophilic

Arginine (Arg; R) -(CH$_2$)$_3$-NH-C-NH$_2$ (NH)

Lysine (Lys; K) -(CH$_2$)$_4$-NH$_2$

Histidine (His; H) -CH$_2$-

Basic, hydrophilic

Phenylalanine (Phe; F) -CH$_2$-

Tyrosine (Tyr; Y) -CH$_2$- -OH

Tryptophan (Trp; W) -CH$_2$-

Neutral, hydrophobic, aromatic

Aspartic acid (Asp; D) -CH$_2$COOH

Glutamic acid (Glu; E) -CH$_2$-CH$_2$-COOH

Acidic, hydrophilic

Cysteine (Cys; C) -CH$_2$-SH

Proline (Pro; P)

Others

Figure 8.1. Structures of the naturally occurring amino acids.

any covalent modifications of certain amino acid residues (e.g., glycosylation, phosphorylation, acylation, amidation, lipidation, sulfation). Each protein in the body has its own unique sequence, which defines its biological function and chemical behavior. The *secondary structure* represents the folding of short segments of proteins such that certain patterns repeat themselves. An α helix and a β pleated sheet are the examples of the repeating pattern. The orderly alignment of these protein chains is maintained as a result of hydrogen bonds between the backbone $-C=O$ and $H-N-$ groups; the helices are stabilized by intrachain hydrogen bonds, whereas the pleated sheets are formed via hydrogen bonding between two or more chains. Proteins can also fold in the form of a random coil. In some proteins all three types of secondary structures are manifested. The *tertiary structure* describes how the secondary structural elements of a single protein chain interact with each other to fold into three-dimensional conformation of protein molecules. Tertiary structures are stabilized via disulfide bridges, hydrogen bonding, salt bridges, and hydrophobic interactions. The highest level of protein organization is the *quaternary structure*, which determines how individual polypeptide chains (called *subunits*) interact to form a multimeric complex. Subunits are held together via polar and hydrophobic interactions. This structure defines the stoichiometry and spatial arrangement of each subunit.

Proteins are synthesized in the cell body via a long chain of events that start with the transcription of the genetic information from the peptide gene to the mRNA, followed by the translation of this information into protein synthesis. The nascent proteins are then transported to various intracellular and extracellular locations. During their transport through the endoplasmic reticulum, Golgi complex, and secretory vesicles, proteins undergo a variety of posttranslational modifications, including acylation (acetyl, formyl, and myristyl), carboxylation, glycosylation, lipidation, amidation, phosphorylation, and sulfation. Consensus sequences have been defined for many posttranslational modifications. Proteins also undergo proteolytic cleavage before they are rendered fully functional. In addition, a protein that participates in a reaction may change its molecular form as a result of a physiological condition.

This chapter outlines the mass spectrometry–based procedures that are used for the determination of the primary structure of proteins. Methods for higher-order structures are discussed in Chapter 11. Determining the primary structure of a protein here means defining the sequence of amino acids and the covalent modifications to the side chains. In the 1990s there were dramatic developments in instrumental design and refinements in mass spectral analysis techniques. Concurrent with these developments, protein chemistry has also undergone profound changes. The older methods of analysis are being replaced with newer, more sensitive, and convenient methods.

Knowledge of the amino acid sequence of a protein is paramount for understanding biological events and the molecular basis of its biological activity, and predicting its three-dimensional structure. Identification of the exact molecular form of a protein is also important from the viewpoint of human health. Any aberration in the protein structure is caused by a particular diseased state. Furthermore, the aberration itself can also lead to a diseased state. In addition, the amino acid sequence of a protein is useful in unraveling evolutionary events. All proteins from a common ancestor have a greater sequence homology.

8.2. TRADITIONAL BIOCHEMICAL APPROACHES

Traditionally, biochemists have used appropriate antibodies and photoaffinity labeling for the identification of proteins, and subsequent Edman sequence analysis for the characterization of proteins [2]. Edman sequencing involves the sequential degradation of a peptide from the N terminus, and the identification of the released amino acids one at a time. The Edman technique has several limitations. It is time-consuming and less sensitive. Modern microEdman sequencing instruments have detection limits only in the low picomole range, and require 30 min per amino acid. Because sensitivity decreases and interference increases with the number of steps of the Edman cycle, the complete sequencing of polypeptides is seldom achieved. In addition, it is not applicable to N-terminally blocked peptides and proteins, and does not allow the direct identification of many posttranslational modifications. Antibodies are used for Western blots, and other immunology-based methods such as enzyme-linked immunoassay (ELISA) and radioimmunoassay (RIA). The use of antibodies, although rapid and sensitive, requires the ready availability of an extensive library of suitable antibodies. In addition, the identification of a protein is often uncertain because of cross-reactivity of antibodies with other compounds that may contain a similar structure. Similarly, photoaffinity labeling followed by autoradiography is also very sensitive, but requires the use of hazardous radioactive labels, and its specificity is also questionable.

Advances in microbiology techniques have made it possible to derive the amino acid sequence of a protein from a cDNA sequence [3,4]. From the mRNA that codes a protein, cDNA is made and cloned; the cDNA sequence provides the corresponding amino acid sequence of the protein. Although this procedure requires less effort and is more sensitive than the Edman procedure, errors in DNA-derived sequence may occur because a protein expressed by recombinant DNA techniques is not necessarily identical to the native protein. Information about posttranslational modifications also remains obscure.

Mass spectrometric methods can offer needed alternative strategies for the analysis of proteins. The 1980s and 1990s witnessed some impressive developments in ionization techniques for biomolecules, including fast-atom bombardment (FAB) [5] and its variation, liquid secondary ionization mass spectrometry (liquid–SIMS) [6], ^{252}Cf-plasma desorption (PD) ionization [7], electrospray ionization (ESI) [8,9], and matrix-assisted laser desorption/ionization (MALDI) [10,11]. These novel ionization methods have made the mass spectrometric analysis of macromolecules a reality. Currently, biopolymers with a molecular mass of $> 100,000$ Da are analyzed routinely. The interfacing of mass spectrometry with high-resolution separation devices such as high-performance liquid chromatography (HPLC) [12–14] and capillary electrophoresis (CE) has matured [15,16]. Another impressive development is the ability to correlate mass spectrometry data with protein databases [17–19]. New strategies have been developed that combine mass spectrometry techniques with two-dimensional gel electrophoresis, peptide mass mapping, and a computer protein database search for the rapid and sensitive analysis of proteomes [20,21]. As a result of these developments, the once difficult and time-consuming task of protein identification has become routine. Mass spectrometry is now playing an ever-increasing role in solving research problems in biological and medical sciences. The complete sequencing of peptides, the determination of the molecular mass of proteins and peptides, the verification of the primary structure of proteins predicted from the cDNA sequence, the characterization of natural mutants, and the identification of posttranslational modifications are now routinely accomplished by mass spectrometry techniques [22–25]. An extensive list of applications of mass spectrometry to the characterization of the primary structure of proteins has been compiled in several published scholarly reviews [26–28].

8.3. A GENERAL PROTOCOL FOR DETERMINATION OF THE PRIMARY STRUCTURE OF PROTEINS

Mass spectrometry-based approaches for the structure determination of proteins have been applied under different situations that include the verification and identification of protein structure or mutation, identification of proteomes, and characterization of a protein of unknown sequence. Figure 8.2 illustrates a general protocol that can be tailored for a specific situation. This protocol heavily relies on peptide mapping, and encompasses the following steps [29]:

- Homogenization of the target tissue, fluid, or cell culture, and the separation of the target proteins by using sodium dodecyl sulfate (SDS)–polyacrylamide gel electrophoresis (PAGE) or gel-permeation chromatography (GPC).
- Measurement of molecular mass of the intact protein. This task is currently accomplished readily via ESIMS and MALDIMS techniques.
- Reduction–alkylation of the protein. The molecular mass measurement of the protein before and after this step provides the information regarding the number of cysteine residues and disulfide bonds.
- Site-specific cleavage of the reduced and alkylated protein into smaller manageable peptide fragments by chemical degradation or proteolysis.
- Mass spectrometry analysis of the peptide fragments directly, or after fractionation (via RP-HPLC) or by an on-line combination of HPLC (or CE) and ESIMS.
- Sequence determination of each peptide fragment by tandem mass spectrometry (MS/MS) or via ladder sequencing.
- Finally, identification of the protein by correlating the mass spectrometry data with the data contained in a protein or DNA database. Alternatively, the amino

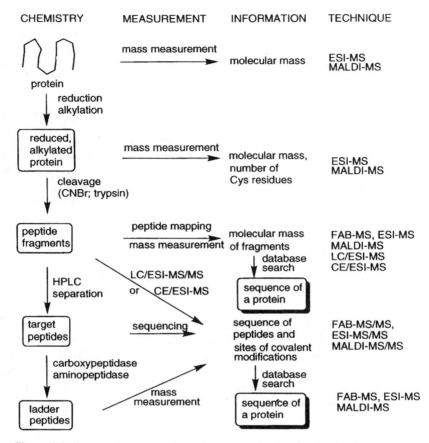

Figure 8.2. A general mass spectrometry protocol for the characterization of proteins.

acid sequence of each peptide fragment is determined for the de novo protein sequencing.

Mass spectrometry–based methods have the advantage of short analysis times, high absolute sensitivity, straightforward sample preparation steps, and optimum specificity, As little as 10 fmol will suffice to identify a newly isolated protein.

8.4. PURIFICATION OF PROTEINS

Mass spectrometry techniques for the identification of proteins do not work well on a mixture of proteins, and require their isolation into individual components. A multitude of techniques is available for the purification of proteins. SDS-PAGE, RP-HPLC, ion-exchange chromatography, and GPC are the methods of choice for purification of proteins. It is beyond the scope of this book to provide a detailed coverage of those methods. A brief account of SDS-PAGE is given in Section 8.7. Several volumes have provided an in-depth coverage of these techniques [30–33]. Special attention, however, must be paid to a particular method because certain reagents or chemicals (salts, detergents, surfactants, nonvolatile buffers, etc.) may interfere in the mass spectrometry detection by suppressing the signal or even causing structural modifications in proteins and peptides.

8.5. MOLECULAR MASS MEASUREMENT OF PROTEINS

The mass measurement of an intact protein is the first essential step in the characterization of that protein. The mass measurement alone can be used with a high degree of certainty to verify the correctness of the translated sequence. A similar application of the molecular mass information is the identification of point mutation (see Section 8.12). The mass measurement can also determine whether the protein has undergone posttranslational modifications. The molecular mass of the intact protein will increase by an amount equal to the mass of the posttranslational moiety (e.g., by 80 Da for each phospho group; see Section 8.14). To determine whether a protein contains sugar residues, the protein is treated with N-glycosidase F (PNGase F) and endo-β-N-acetylglucosaminidase H (endo H), and the molecular mass is measured again. Another use of mass measurement data is to determine the number of cysteine residues and disulfide bonds in a protein (see Section 8.13). The mass measurement is also essential because it provides a frame within which the final structure must fit. The emergence of ESIMS and MALDIMS has greatly facilitated the determination of molecular mass of intact proteins with an accuracy of better than $\pm 0.01\%$. A detailed discussion of molecular mass measurement can be found in Chapter 5. Procedures have been developed to measure the molecular mass of proteins separated by gel electrophoresis directly by MALDIMS (see Section 8.7.2).

8.6. PEPTIDE MAPPING

Peptide mapping is a powerful technique for defining the primary sequence of a protein [29,30,34,35]. Often, this approach is used for quality control of genetically engineered proteins. On other occasions, peptide maps are used to confirm the sequence of a native protein for which the cDNA sequence is known. In general, purified protein is degraded into smaller discrete segments, which are differentiated via a suitable mass spectrometry technique to create a distinctive peptide map or fingerprint. This approach is equally applicable to proteins that contain posttranslationally modified residues. Other uses of peptide mapping techniques in protein chemistry are to evaluate the higher-order structures (discussed in Chapter 11), and to elucidate noncovalent interactions of proteins (see Chapter 14).

8.6.1. Reduction and Carboxymethylation

Prior to cleaving, proteins may be subjected to a reduction–alkylation process [33]. This step is optional and required only when a protein contains disulfide bonds or free sulfhydryl groups because cysteine residues are often prone to autooxidation, leading to a variety of unwanted products and random formation of disulfide bridges (see Section 8.13) during the sample preparation procedure. Alkylation converts sulfhydryl groups to stable derivatives. Reduction–alkylation also facilitates the cleavage of a protein by chemical–proteolytic digestion. During this procedure, the three-dimensional structure of proteins is disrupted to allow more cleavage sites to be accessible. Proteins are first reduced by treatment with 2-mercaptoethanol or dithiothreitol to convert disulfide bonds to free sulfhydryl groups (Eq. 8.1), which are subsequently converted to an *S*-carboxymethyl derivative by reacting with iodoacetic acid (Eq. 8.2). The reduction is carried out in alkaline pH (> 8.0) to generate a reactive intermediate (thiolate anion). Other widely used alkylating agents are 4-vinylpyridine, iodoacetamide, and acrylamide.

$$\text{S-S} \xrightleftharpoons{\text{RSH}} \text{S-SR, SH} \xrightleftharpoons{\text{RSH}} \text{SH, SH} + \text{RS-SR} \qquad (8.1)$$

$$\text{SH, SH} + 2\text{ICH}_2\text{COOH} \longrightarrow \text{S-CH}_2\text{COOH, S-CH}_2\text{COOH} + 2\text{HI} \qquad (8.2)$$

8.6.2. Cleavage of Proteins

Two methods are widely prevalent for cleaving proteins at specific sites: (1) digestion with a selected endoprotease and (2) reaction with a suitable chemical reagent [30,34,35]. A list of these cleaving reagents, along with the conditions used for cleavage, is given in Table 8.1. A protocol for cleavage of proteins via vapor-phase acid hydrolysis has also emerged [36,37].

Digestion with Chemical Reagents. Cyanogen bromide (CNBr) is the most commonly used chemical reagent. It cleaves the amide bond on the C-terminal side of methionine. Under the acidic environment, most of the C-terminal homoserine residues that result are in the lactone form (Scheme 8.1), the mass of which is 48.1 Da less than that of the methionine residue. Because methionine does not occur with high frequency in proteins, the CNBr-generated peptide fragments are few and large. These fragments are analyzed conveniently by ESI and MALDI mass spectrometry techniques or if needed, can be further subjected to a secondary

Table 8.1. Typical protein cleaving agents

Cleaving Agent	Specificity	Digestion Conditions [buffer; pH; temperature (°C)]
Chemical agents		
Cyanogen bromide	Met–X	70% TFA
N-Chlorosuccinimide	Trp–X	50% acetic acid
N-Bromosuccinimide	Trp–X	50% acetic acid
Highly specific proteases		
Trypsin	Arg–X, Lys–X	50 mM NH$_4$HCO$_3$; 8.5; 37
Endoproteinase Glu-C	Glu–X	50 mM NH$_4$HCO$_3$; 7.6; 37
Endoproteinase Arg-C	Arg–X	50 mM NH$_4$HCO$_3$; 8.0; 37
Endoproteinase Lys-C	Lys–X	50 mM NH$_4$HCO$_3$; 8.5; 37
Endoproteinase Asp-N	X–Asp	50 mM NH$_4$HCO$_3$; 7.6; 37
Nonspecific proteases		
Chymotrypsin	Phe–X, Tyr–X, Trp–X, Leu–X	50 mM NH$_4$HCO$_3$; 8.5; 37
Thermolysin	X–Phe, X–Leu, X–Ile, X–Met, X–Val, X–Ala	50 mM NH$_4$HCO$_3$; 8.5; 37
Pepsin	Phe–X, Tyr–X, Trp–X, Leu–X	0.01 M HCl; 2.0; 37

digestion with proteolytic enzymes to provide convenient-size smaller fragments. *N*-Chloro- and *N*-bromosuccinimide are other common chemical reagents that have found some utility to cleave a protein. These reagents also generate large peptide fragments.

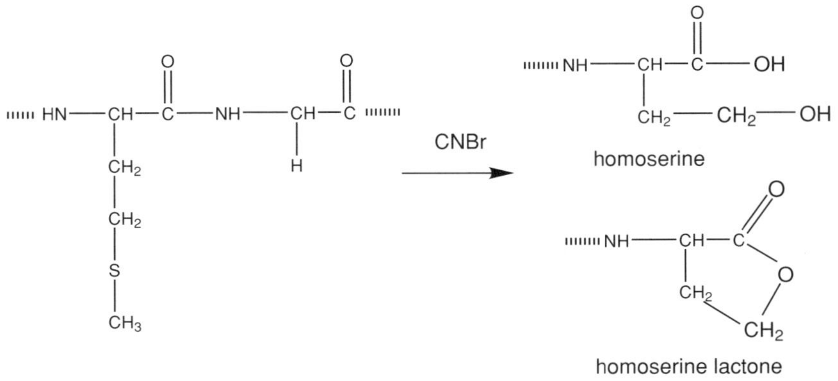

Scheme 8.1

Digestion with Proteolytic Agents. A variety of amino acid–specific enzymes are available that will yield reproducible peptide maps of a protein (see Table 8.1) [30,34,35]. The use of any particular enzyme chemistry depends on the primary structure of a protein and the information that is desired. Trypsin is generally a universal choice for most applications because of its excellent properties as a protease. Also, it is inexpensive, and produces peptides of ideal size for mass spectrometry analysis. Because trypsin cleaves an amide bond on the C-terminal side of lysine and arginine residues, the tryptic peptides that result contain a basic amino acid at the C terminus. This aspect is especially useful for the ESI process and for subsequent sequencing by tandem mass spectrometry. However, the Lys–Pro bond is not cleaved by trypsin. Also, when a protein contains adjacent lysine and/or arginine residues, single amino acid fragments will result. Such fragments may escape detection to create difficulty in mass balancing. To avoid this situation, larger and overlapping fragments are generated by reacting the protein with other endopro-teases (Table 8.1). As an example, endoproteinase Lys-C can generate peptide fragments by cleaving the protein at a position that is C-terminal to lysine. Similarly, endoproteinase Arg-C and endoproteinase Glu-C (also known as *Staphylococcus aureus* V8 protease) cleave bonds C-terminal to arginine and glutamic acid residues, respectively. Endoproteinase Asp-N is specific to the bond that is N terminal to aspartic acid. Another enzyme that cleaves N-terminal side of the amide bond is thermolysin. This enzyme has specificity for leucine, isoleucine, methionine, phenylalanine, and tryptophan.

A standard procedure can be used for the digestion of a protein [34,35]. However, special care must be taken when dealing with low amounts of protein. The conventional procedure of digestion does not work well in dilute solutions because the rate of protein cleavage is diffusion-limited. Also, the possibility of adsorption losses of proteins and peptides is greater in dilute solutions. In order to overcome these problems, the use of immobilized trypsin that is packed in a small-diameter PEEK (polyetheretherketone) column has been recommended [38]. Care should also be taken to use volatile buffers; otherwise the ionic components of the buffer will interfere in the mass spectrometry analysis. Ammonium carbonate and ammonium bicarbonate both can be used to adjust the pH of the digestion mixture between 7.0 and 8.5. The volatile salts are removed by a lyophilization step. If a nonvolatile buffer is used or any other nonvolatile components or salts are present, then a solid-phase extraction (SPE) step may be incorporated.

Vapor-Phase Acid Hydrolysis. Specific cleavage agents seldom work well with posttranslationally modified proteins because of their limited accessibility to the substrate molecules and because the detection of phosphorylated and glycosylated sites requires high sequence coverage. These problems have been addressed by the use of vapor-phase acid hydrolysis [36,37]. Although partial acid hydrolysis of proteins in solution has been known for quite some time, the procedure has limitation of low cleavage specificity. In contrast, cleavage specificity of acid hydrolysis can be enhanced if carried out in the vapor phase under controlled conditions. Roepstorff and co-workers observed three distinct types of cleavage reaction with the vapor-phase hydrolysis: (1) a specific internal cleavage at Asp, Ser, Thr, or Gly residues and (2) cleavages that lead to the formation of N- and (3) C-terminal sequence ladders [37]. The technique is equally applicable to lyophilized samples as well as those from in-gel-separated proteins.

8.6.3. Fractionation of Protein Digest by RP-HPLC

Because of its speed, simplicity, extremely high resolving power, and compatibility with mass spectrometry detection, RP-HPLC is a preferred technique by the practitioners of mass spectrometry for the fractionation of protein digests [39–41]. The success of the HPLC separation depends largely on the appropriate choice of column dimensions, column packing, mobile phase, and gradient. The column dimensions are governed by sample size. A 250 × 4.6-mm column is preferred to resolve large sample amounts (nmol range), whereas a capillary column provides optimum separation and sensitivity with much smaller size samples (fmol range). A 250 × 2.1-mm narrow-bore column is a compromise between the two situations. This column is suitable for the separation of 50 pmol-size samples. Columns are usually packed with C-18 materials of 3–5-μm particles and pore diameters of approximately 300 Å. A C-4 column may be used to separate larger peptide fragments such as those generated by the CNBr treatment. Only volatile buffers and ion-pairing reagents should be used for purification and separation of peptides. TFA, acetic acid, formic acid, triethanol amine : formic acid, ammonium acetate, and ammonium bicarbonate are most suitable additives to the HPLC mobile phase.

The use of a disposable SPE cartridge is a simple and convenient way to concentrate the sample and remove salts from a protein sample. A variety of packing materials are available, including C-18, C-4, ion exchange, and size-exclusion.

8.6.4. Analysis of Peptide Maps

In peptide mapping, proteins are differentiated on the basis of their distinct fragmentation pattern. The rational is that the proteins that yield identical peptide maps have identical amino acid sequences. Traditional methods to generate peptide maps involve fractionation of complex mixtures of peptides in a protein digest with either one-dimensional SDS-PAGE or RP-HPLC [30,42,43]. The role of CE has also been defined for this purpose [44]. The mass spectrometry peptide mapping protocol, in principle, is similar to these techniques, but it provides an added dimension of structure-specific data (i.e., the molecular mass and fragment ions). As a consequence, mass spectrometry techniques have the potential to analyze proteins of unknown structures. Peptide mapping by mass spectrometry has been termed *peptide mass fingerprinting* [45]. FABMS (or its variation, liquid–SIMS) [46], MALDIMS [47], LC-ESIMS [48,49], and CE-ESIMS [50] are currently replacing the traditional biochemical approaches. In the past, ^{252}Cf-PD ionization–MS was also used to analyze peptide maps [51,52]. Not all peptide fragments of a protein digest are detected by a single mass spectrometry technique. Therefore, it is advantageous to use more than one technique (e.g., MALDI plus LC-ESIMS) for the analysis of peptide maps.

FAB-MS for the Analysis of Peptide Maps. Since its inception in 1981, FABMS has gained a general acceptance as a technique for the analysis of peptides [5]. The discovery of FABMS provided an opportunity for the mass-specific analysis of peptide maps [46,53–55]. FABMS, in combination with peptide mapping, has been used to verify and correct gene sequences [46,53]. With this combination, the frame shift, deletion, and addition errors in gene sequences have been identified.

In practice, the protein digest, without fractionation, is mixed with a suitable liquid matrix and analyzed in the positive-ion mode to produce $[M + H]^+$ ions of the intact peptide segments. The beam-induced fragmentation is undesirable in FAB mapping of the digest, and can be minimized if α-thioglycerol or a $5:1$ (w:w) mixture of dithiothreitol and dithioerythritol (DTT/DTE) is used as the liquid matrix [56]. The intensity of the $[M + H]^+$ ion signal is influenced by the surface composition of the sample–matrix mixture. The matrix surface composition may be altered by adjusting the pH or by addition of surfactants. A small amount (1%) of trifluoroacetic acid (TFA) is usually mixed with matrix to enhance the $[M + H]^+$ ion current. The use of FABMS for molecular mass measurements is restricted to 5–30-amino-acid-long fragments. Fortunately, most tryptic fragments fall within this mass range. These fragments can be analyzed directly without any prior separation. Typically, at least 10–50 pmol of a protein is required. Although the direct analysis of a protein digest is simple and less time-consuming, some hydrophilic peptides may escape detection. To circumvent this problem, the protein digest may be fractionated by RP-HPLC, or RP-HPLC may be combined with continuous-flow FAB analysis [57].

FABMS is currently losing ground in favor of ESIMS and MALDIMS because of its limitations in the analysis of high-mass (> 3000-Da) peptide fragments. Also, it is relatively less sensitive and somewhat complicated by matrix and other interferences. Low-mass peptide fragments are often lost in the matrix noise. However, FABMS is simple to use, and is readily available in most modern mass spectrometry laboratories.

MALDIMS for the Analysis of Peptide Maps. MALDIMS is also uniquely qualified for the analysis of protein digests [47,58–61]. Karas and Hillenkamp first demonstrated it in 1988 to be an extremely useful technique for the analysis of biomolecules [10]. Similar to FABMS analysis, MALDI also allows the direct analysis of unfractionated protein digests, but it is more tolerant to low amounts of sample impurities. A proper selection of a matrix is critical in a mixture analysis by MALDI. Sinapinic acid, α-cyano-4-hydroxy cinnamic acid, and 2,5-dihydroxybenzoic acid (DHB) are commonly used matrices. In some cases, the use of a comatrix, such as *m*-methoxysalicylic acid and fucose, produces significant enhancement in the spectral quality over matrix alone [59]. In combination with SDS-PAGE, MALDIMS has emerged as a standard protocol for peptide mapping (see Section 8.7).

On-Probe Digestion. Despite MALDI being tolerant of impurities, the samples from biological sources must be purified prior to MALDIMS analysis to achieve the optimum results. In addition, MALDIMS analysis of the peptide fragments that are generated by proteolytic digestion of proteins in solution can lead to the interference of autolysis products and the loss of precious sample during the transfer of the digest to the probe. In order to circumvent these problems, techniques for *on-probe digestion* have been developed, including: bioreactive probes, chemically modified probes, and membrane supports.

In *bioreactive probes*, [62] the proteolytic enzyme is covalently attached to the activated probe, and the protein is digested directly on the probe by the surface-tethered enzyme. The digestion is stopped after a period of time by adding the matrix to the probe. In *chemically modified probes*, self-assembled monolayers are created on the surface of the probe with a chemical derivative that can preferentially retain peptides. In one such approach, the surface of the probe was modified with octadecyl mercaptan, which reversibly binds peptides via hydrophobic inetractions to allow removal of the salts and enrichment of analytes [63]. In the *membrane support* technique, the sample is deposited on a polymer membrane; the sample is selectively adsorbed and the impurities can be washed away. Several forms of polymer membranes that include nylon, nitrocellulose, polyethylene, polyvinylidene difluoride (PVDF), and polyurethane have been used successfully [64–66].

Electrospray Ionization-MS Analysis of Peptide Maps. Strategies for analyzing peptide maps have improved significantly with the emergence of ESI [48,49]. The direct analysis of unfractionated protein digests is feasible by this technique. The tryptic fragments produce doubly charged ions (except the C-terminal fragment). One innovation in the field of ESIMS is the introduction of nanoES, which has made a significant impact in the analysis of biomolecules [38,67–72]. Several useful features—low flow rates ($<20\,nL/min$), low sample consumption, low sample volume ($\sim1\,\mu L$), robustness, speed of analysis, and opportunity for signal averaging—make nanoES ideal for the direct analysis of protein digests. In their seminal paper on nanoES, Wilm and Mann demonstrated its use for the analysis of an unseparated tryptic digest of carbonic anhydrase [67]. In combination with a desalting cartridge, nanoES can also be applied for the direct analysis of peptide maps from in-gel digested proteins [20,67].

Liquid Chromatography–ESIMS for Analysis of Peptide Maps. A major reason for the success of mass spectrometry for analysis of proteins and peptides is its coupling with HPLC and CE (see Chapter 7). With the on-line combination of RP-HPLC and ESI-MS, the separation of the peptide fragments in intricate enzyme digests and determination of their molecular mass can be accomplished simultaneously [73,74]. In addition, peptides free from salts and other impurities are presented to the ESI source. Also, protein samples at sensitivity levels much lower than those of the current protein sequencing techniques can be analyzed. Because of the need to analyze proteins at fmol levels, the combination of ESIMS online with capillary- or microLC is recommended for the analysis of peptide maps. If these columns are connected to a tandem instrument (see discussion below), it is feasible to obtain in an on-the-fly manner the amino acid sequence of each fragment. This development significantly reduces the sample losses that invariably occur during multistep characterization procedures. Several applications of LC-ESIMS have been reported for analysis of peptide maps [75–80].

Capillary Electrophoresis/ESI-MS Analysis of Peptide Maps. Another technique that is compatible for analysis of peptide maps from low amounts of proteins is the on-line combination of CE and ESIMS [81–84]. CE provides a unique specificity in separation that complements RP-HPLC, and the separation efficiency reaches over one million theoretical plates. Although CE has been coupled to mass spectrometry via continuous-flow FAB, the CE-ESIMS combination is much more practical for peptide mapping. Most peptide mapping experiments with CE-ESIMS have been performed with 100-μm-i.d. capillaries; smaller capillaries exhibit increased detection sensitivity. Peptide maps are separated by using volatile buffers such as ammonium acetate (at a pH between 4.0 and 5.0) in the 1:1 (v:v) water:acetonitrile solvent system. The buffer concentration should be limited to <0.05 M.

Two major innovations have been incorporated to achieve ultrasensitive detection of peptides in protein digests. CE has been combined with micro- and nanoESMS detection systems [86,87]. Peptides at detection levels of <300 amol have been analyzed by this combination. In another innovation, an on-line SPE device is included for the concentration of analytes from large volumes into small volumes that are compatible with CE sample introduction [88,89]. SPE consists of a small length of RP-HPLC packing within the main CE column. Using this approach, up to a 1000-fold enrichment in concentration can be achieved. However, CE performance is somewhat compromised with the inclusion of the SPE packing in the main capillary [90]. An improvement over this system is to use a polymeric membrane impregnated with chromatographic stationary phase [91,92].

Tandem Mass Spectrometry Analysis of Peptide Maps. When a peptide map is generated from a protein of unknown sequence, or when the identity of a peptide in a protein digest is in question, it is essential to determine the sequence of individual peptides. A detailed account of how to sequence peptides is presented in the next chapter. Tandem mass spectrometry (MS/MS) is one of the primary tools to sequence peptides. MS/MS is performed either on peptides that were previously fractionated by RP-HPLC, or in combination with on-line RP-HPLC. Any of the MS/MS instruments discussed in Chapter 4 is suitable for this purpose. For off-line MS/MS measurements, ionization of the peptide fragments is accomplished via FAB, MALDI, or ESI. The last method has the advantage that it forms doubly charged ions (at least from tryptic fragments), which have a high proclivity for the peptide bond fragmentation [93–95]. The advantage of the off-line MS/MS approach is that the precursor ion of interest can be selected manually, and that the CID conditions for efficient fragmentation can be optimized. However, this approach has limitations of sample losses (and hence lower sensitivity) due to the multistep nature of this protocol.

Alternatively, a more efficient approach is to use an on-line microLC/ESI-MS/MS combination [96–98]. The characterization of proteins has been assisted considerably by the emergence of this technology. A triple-sector quadrupole and an ion-trap mass analyzer are both ideally suited for an MS/MS combination with LC. Such dedicated instruments are commercially available from several manufacturers.

The peak analysis time with microLC, however, is very short (usually < 12 s), with the consequence that precursor ion selection and CID parameters cannot be optimized efficiently in that short time. Lee and co-workers have provided an imaginative solution to this problem. They have used a variable-flow LC to overcome this limitation [97–99]. By reducing the pressure just as a peak elutes from the column, its elution time can be prolonged without compromising chromatographic resolution. This step provides an ample opportunity to optimize the precursor ion selection and CID parameters. Other advantages available with the off-line MS/MS combination, such as multistage MS, analysis of more than one charge state, and selection of precursor ion of each coeluting component, can also be realized.

The separation of protein digests via RP-HPLC (off-line or on-line) prior to MS/MS, however, is time-consuming and requires excessive amounts of sample. The direct analysis of protein digests by using nanoES is an attractive alternative [38,67,100,101]. This approach permits ample time to tune the collision energy for optimal sequence information from multiply charged peptide ions. As an example, the acquisition of precursor ion scans of m/z 147 and 175, y_1-type ions derived from lysine- and arginine-containing peptides, respectively, has allowed an easy identification of tryptic peptides in a protein digest [100].

Tandem mass spectrometry is also an effective means to identify proteins in a mixture [102]. After digesting the sample, the MS/MS spectra of at least one peptide from each constituent protein of the mixture are acquired. The identification of proteins involves a search of the protein database on the basis of the MS/MS data (see Section 8.8). This strategy is further facilitated by the development of LC/ESI-MS/MS [96–98], and by the possibility of high-throughput data acquisition through data-dependant automation of a tandem mass spectrometer [103,104].

8.7. PROTEIN IDENTIFICATION BY GEL ELECTROPHORESIS—MASS SPECTROMETRY

For quite some time, gel electrophoresis has played a pivotal role in protein purification and characterization [30,105,106]. Currently, its combination with mass spectrometry has assumed a prominence in protein analysis. Therefore, a separate section is devoted to this particular technique. It is widely used in the form of SDS-PAGE for the separation of complex mixtures of proteins, and the molecular mass screening of the gel-separated proteins, albeit with a limited accuracy. Because it is an anionic detergent, SDS forms an anionic complex with denatured proteins. The following key steps are used in a typical SDS-PAGE separation of proteins:

- The first step is to denature the protein by heating it in the SDS buffer.
- When a protein contains disulfide bridges, the protein is treated with a reducing agent, such as 2-mercaptoethanol or dithiothreitol, to reduce the disulfide linkages.

- The next step is the separation of the desired protein under the influence of an electric field. For proteins that contain disulfide bridges, this step is performed under reducing conditions by mixing a reducing agent with the buffer. In the presence of SDS buffer, all proteins carry equal charge density, and thus are separated on the basis of their size.
- The separated proteins are visualized by staining with Coomassie brilliant blue.
- The separated proteins are transferred to an inert support by a process known as *blotting*. Electrophoretically driven transfer [107], popularly known as Western blotting, is the most widely used method. Commonly used inert supports are PVDF and nitrocellulose. This process is required for cleanup and concentration of the protein before its further analysis.
- Finally, for the removal of contaminating salts and SDS, these membranes are washed with deionized water.

One-dimensional (1D) and two-dimensional (2D) PAGE techniques have both been optimized for protein separation. 1D PAGE is commonly used for parallel processing of a large number of samples, whereas 2D PAGE provides a high-resolution separation and thus is a more effective for ultracomplex mixtures. 2D PAGE is a two-step process, in which each dimension makes use of complementary properties of proteins for their separation. The first dimension is usually performed in the isoelectric focusing (IEF) mode, and the second in the conventional, size-based separation mode. The IEF mode separates proteins on the basis of their isoelectric point (pI values). A mixture containing thousands of proteins can be separated in a single run by 2D PAGE.

8.7.1. Mass Spectrometry Characterization of Gel-Separated Proteins

Biochemists have used specific immunologic detection methods to further characterize gel-separated proteins [30]. Several specific high-affinity recognition systems (antibody–antigen, lectin–glycoproteins, protein–nucleic acid, biotin–avidin, ligand blotting, etc.) have been developed. With the emergence of MALDI [10,11] and ESI [8,9], mass spectrometry has become competent in the analysis of gel-separated proteins and peptides [108]. MALDI has an edge over ESI in this respect because it can tolerate many of the buffers and additives (except SDS) used in PAGE separation. However, ESI has the advantage of an on-line LCMS system, whereby peptide fragments from the protein digest can be individually fractionated and identified. Both of these techniques provide a much higher mass measurement accuracy than that obtained by conventional SDS-PAGE analysis.

Molecular Mass Measurement of Proteins. Because of its high-mass capability, MALDIMS is an appropriate technique for the molecular mass measurement of the gel-separated proteins [109–119]. In addition, it is also feasible to link gel electrophoresis directly to MALDIMS. Several laboratories have optimized conditions for the direct analysis of electroblotted proteins on polymer membranes after their 1D gel separation [109–117]. The blot membranes are used as mere substrates for laser irradiation. The MALDI matrix is incorporated into unstained membranes. The dye-adduct protein ions are observed when Coomassie blue is used to stain the protein spots. An approximate position of the spot is inferred from a stained duplicate lane. Soaking the membrane in a saturated matrix solution allows the incorporation of the matrix. MALDI mass spectra are obtained by taping the cut membranes onto the metal sample probe.

The direct mass measurement of proteins from a gel without resorting to electroblotting has also been reported [118,119]. Although conventional polyacrylamide gels crumble in the vacuum environment of MALDI-TOFMS, specially prepared ultrathin gels with a lower percentage of acrylamide have been found suitable for these experiments [119]. IEF gels have no such cracking problem. The matrix is spotted on the protein spot, or gels are soaked in the matrix solution.

Some applications of ESI [109,120–122] and nanoES [67] have also been reported for the molecular mass measurement of gel-separated proteins. One of the problems with ESIMS is that SDS can cause interference in the analysis by forming adducts with the intact protein. Fridriksson et al. have shown that the irradiation of these adducts with a beam of infrared radiation eliminates SDS from the adducts, with the consequence that a much higher concentration of SDS can be tolerated in the final ESI solution [122].

Proteolytic Cleavage and Subsequent Mass Spectrometry Analysis.

Three methods have evolved for cleaving the in-gel separated proteins:

- The intact protein is eluted from the gel, and digested with a protease [123].
- The protein is digested within the gel, and the peptide fragments are eluted from the gel [124].
- The protein is electroblotted onto a polymer membrane, where cleavage and elution steps can be performed [125].

The objective of any one of these methods is to obtain as many fragments and to maximize their yield. No one approach is demonstrably superior for all situations. Because of the possibility of adsorption losses of proteins and peptides (which results in a lower sensitivity), the first method (i.e., elution of proteins from the gel and subsequent proteolysis) is nearly abandoned in favor of in situ digestion methods. The in-gel digestion method is less compatible with the mass spectrometry–based analysis of proteolytic digests because the presence of SDS and other contaminants may interfere in the detection of peptide fragments. These impurities must be removed before mass spectrometry analysis. The presence of SDS also has a debilitating effect on the RP-HPLC separation of protein digests. In-gel digestion, however, has the advantage that an extra blotting step is avoided. Some researchers have reported excellent success with analysis of in-gel digested proteins by MALDIMS [62,64,126,127], with 5–6 min per analysis time [128].

Thus, of the three methods, the immobilization of proteins onto the membrane appears to be a more practical approach for analysis of protein digests. One major advantage is that most of the contaminants and SDS can be washed away prior to digestion [17,129–132]. Because of its high protein binding capacity, PVDF is the membrane of choice. Also, it provides a cleaner chemical background for subsequent mass spectrometry analysis. Carboxymethylcellulose and Immobilon^{TM}-CD membranes also exhibit a high recovery of the blotted proteins [133]. A typical gel electrophoresis–mass spectrometry protocol for the characterization of proteins is illustrated in Figure 8.3. The following steps are used in this protocol:

- The 1D or 2D gel-separated protein is electroblotted onto PVDF membrane, and the membrane is stained with Coomassie brilliant blue or sulforhodamine B.
- After visualization, the Coomassie-stained spot is destained prior to proteolytic digestion. The destaining step can be avoided with the sulforhodamine-stained spots.
- Next, the spot is excised, and the protein is cleaved either chemically or enzymatically.
- The cleaved peptides are extracted.
- A portion of the extract is analyzed directly by MALDIMS to obtain molecular mass information. LC-ESIMS (or LC/ESI-MS/MS) may be used to analyze the remaining extract.

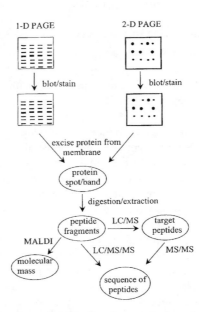

Figure 8.3. Mass spectrometry protocol for the characterization of gel-separated proteins.

- The LC fractions are collected (via split flow) and analyzed further for the amino acid sequence by an Edman sequencing or MS/MS procedure.

NanoESMS is another technique that has made a significant impact on the analysis of gel-separated proteins. Earlier, it was obligatory to pool a particular spot from multiple gels to provide enough material by mass spectrometry analysis. Now, the low sample requirement of nanoES allows a complete analysis of the protein with only a single spot.

8.8. PROTEIN IDENTIFICATION BY DATABASE SEARCHING

With the accumulation of an exhaustive list of oligonucleotide and protein databases, a new paradigm has emerged for the characterization of proteins. The rapid identification of proteins is now feasible by combining mass spectrometry data with database search [17–19]. The concept was originally developed by Henzel et al. [17], and later applied by a number of independent groups [18,19,134,135]. Several excellent reviews have appeared on this subject [136–138]. Various databases can be searched, including those for protein, DNA, and expressed sequence tags (EST) databases. This general approach is applicable to those proteins for which sequences are already known, and is based on the hypothesis that the primary structure of a protein is the direct product of genome transcription; therefore, its structure can be deduced from the knowledge of the organism's genome. Genomic sequences are becoming increasingly available. At present, genomes of several organisms, such as viruses (e.g., *Haemophilus influenzae*), prokaryotes (e.g., *Escherichia coli*), and eukaryotes (e.g., *Saccharomyces cerevisiae*), have been completely sequenced. The mapping of the human genome is also nearly complete. Biochemists have used this tool to search for a familial relationship or possible function of the target sequence. The partial or complete amino acid sequence of the target protein can be used to search a database to reveal a protein of identical or homologous sequence. The identification of proteins from database search protocols is much more rapid and requires quantities two orders of magnitude less than those required currently for other amino acid analysis methods. With an increase in the database entry, the dependence on de novo sequencing will be reduced.

Two general strategies, illustrated in Figure 8.4, have evolved. The first one makes use of peptide-mass fingerprints [17–19,130,131,134,135], which are generated, as described above, by the digestion of a small amount (pmol or less) of the protein with a suitable protease (usually trypsin). The molecular mass of each fragment is accurately determined with ESIMS or MALDIMS, and is matched against the theoretical peptide fragment mass values of each protein in the database. These theoretical mass values are created for the protease that was used specifically in the digestion step (e.g., trypsin, endoproteinase Lys-C) by applying the protease-specific fragmentation rules to all of the protein sequences in a database. A multitude of computer programs have been written to identify proteins on the basis of the measured mass values of the peptide fragments of a protein digest. In a typical

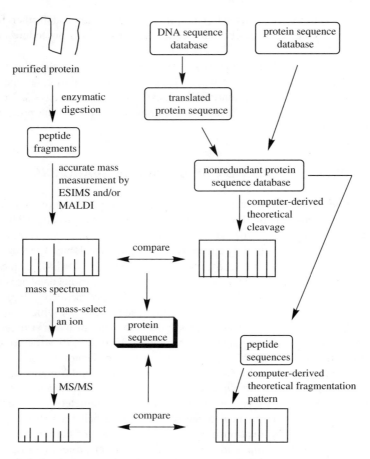

Figure 8.4. Database search and mass spectrometry for the characterization of proteins.

search algorithm, the cleavage database is first selected, and the criteria of the data search are defined [17,19]. The criteria used can be

- The species
- The mass-matching tolerance
- An approximate upper value of the molecular mass of the protein
- The pI value of the protein
- A minimum number of matches
- The number of cleavage sites missed by the trypsin digestion

From a list of the experimentally derived $[M + H]^+$ values of the peptides, the output of the search gives a ranked list of most likely candidates. The database sequence that produces the best score is assumed to be the sequence of the protein. Mismatches of the listed peptide candidates can reveal the possibility of mutation or posttranslational modifications. A successful search could be completed on the basis of the specified masses of just four to six peptide fragments [17,135].

Considerable efforts have been undertaken to improve the database search selectivity. The efficacy of the search decreases with an increase in the mass-matching tolerance. Therefore, to maximize the specificity of a database search, this parameter should be kept low (<3). Similarly, the accuracy of mass measurement also effects the ambiguity of the database search. Most search methods provide conclusive results with peptide masses accurate within ± 0.5 Da. The use of delayed extraction TOFMS [139] and FT-ICRMS [140] improves the mass accuracy. One can also use additional digests with enzymes of different selectivity [131,134]. The search can also be narrowed down if the number of cysteine [140] or some other amino acid [141] residues is known. A protein search engine called *ProFound* that employs a Bayesian algorithm has been described to identify proteins from databases with peptide mass mapping data [142].

The second approach (Figure 8.4, lower part) is based on a correlation of the MS/MS fragmentation patterns with the amino acid sequences in the protein databases [103,143–148]. The use of fragmentation patterns is far more discriminating than the approach based on peptide mass fingerprinting, and it allows the identification of proteins with just one MS/MS spectrum. In practice, the protein is cleaved as usual into smaller segments. A single peptide from this protein digest is sequenced by MS/MS, and the observed pattern of fragment ions is matched with the patterns of fragment ions that are calculated from the database sequences. Yates' research group has developed an algorithm, called *SEQUEST*, which first identifies amino acid sequences in the database with the measured mass of the selected peptide ion, and predicts the fragmentation pattern that is expected for each sequence [144]. These uninterpreted fragmentation patterns are matched with the experimentally derived MS/MS spectrum. The highest-scoring amino acid sequences are reported. This program was latter modified to include the high-energy collision-induced dissociation (CID) data [103]. For the identification of posttranslational modifications, the search method considers each putative modification site as modified and unmodified in one pass through the database. The same group has used a cross-correlation procedure to compare MS/MS spectra of peptides, and demonstrated its potential for library searching and subtractive analysis [147]. Mass information and fragmentation patterns are both used in this comparison.

Mann and Wilm have developed a peptide tag technique to search databases with the information from MS/MS spectra [143]. In this approach, a short stretch of sequence (called a "tag"), together with the molecular mass of the preceding and tailing region, is considered to be a unique signature of the peptide. Because of a higher level of specificity of this approach, a search on a single sequence tag is often sufficient to identify a protein unambiguously. The technique is also applicable to the ESI source-formed fragments and to posttranslationally modified peptides. A combination of in-source CID to generate sequence tags from an intact protein and FT-ICRMS to measure the molecular mass of the tags has also been used to identify proteins [149].

Apart from these two protocols, the database search can also be performed using chemical sequence tags [150]. Multiple small sequences from the N-terminal of peptides are generated by stepwise chemical degradation of the unseparated protein digest. Chemical cleavage involves thioacetylation of the N-terminal amino acid residue and its cleavage by acid (TFA) treatment. Such short sequence tags, along with the mass of the parent peptide, can be used to identify the protein in the database.

Several databases, such as SWISSPROT, PIR, EMBL, TREMBL, GENBANK, GENPEPT, OWL, NBRK, and dbEST, are maintained by independent groups, and can be accessed through the Internet. They contain comprehensive, up-to-the-minute lists of protein and nucleotide sequences. It is not an easy task to decide which database to use, because that database may not contain all known sequences. A practical solution is to assemble all protein databases and translated nucleotide databases (into protein sequences) into a composite nonredundant database from which duplicate entries are eliminated. Regularly updated CD-ROMs can also be obtained from the National Center for Biotechnology Information (Bethesda, MD).

8.9. DE NOVO PROTEIN SEQUENCING

The computer database search is applicable to proteins of known sequences or when their ESTs are available in the database. Several situations require the sequencing of unknown proteins. In such cases, a sufficient sequence information needs to be generated. If partial sequence information is available, it can be used to design oligonucleotide primers suitable for polymerase chain reaction (PCR) cloning of cDNA. From the sequence of the cDNA, the sequence of the target peptide can be confirmed. Such partial sequence information can be generated by sequencing peptide fragments of the enzymatic digest of a protein via any of the approaches discussed above (off-line or on-line LC/ESI-MS/MS, nanoES-MS/MS, MALDI, high-energy CID, etc.). NanoES-MS/MS of selected peptides that are extracted directly from in-gel digested proteins has been a successful strategy to generate a sufficient sequence information for subsequent construction of oligonucleotide primers [64,101]. Tagging the peptide with ^{18}O can facilitate de novo sequence interpretation of MS/MS spectra [151,152]. The partial sequence information can also be obtained via a database search.

8.10. PROTEOMICS

Proteomics is a new field of biomedical research, and deals with systematic characterization of gene products. The concept of the "proteome" has been defined as the entire protein complement expressed by a genome or by a cell or tissue type (i.e., *proteome* is *prote*in + gen*ome*) [153]. The subject of proteome analysis has gained significant importance because the data from genetic and genomic techniques by themselves are insufficient to describe biological processes, whereas the data from the proteome analysis can fulfill those needs. Mass spectrometry is playing an ever-increasing role in current proteome analysis [20,28,137,154]. This function is the consequence of the ability of mass spectrometry to rapidly identify gel-separated proteins, and its integration with protein databases. The following key steps are involved in a typical proteome analysis:

- Tissue/fluid acquisition
- Protein precipitation and redissolution
- 2D gel separation of proteins (see Section 8.8)
- The molecular mass determination of selected proteomes
- Enzymatic or chemical digestion of separated proteins (see Section 8.6.2)
- Measurement of the molecular mass of individual peptide fragments of protein digests (see Section 8.6.4)
- Tandem mass spectrometry of selected peptide fragments (see Section 8.6.4)
- Database search, using either mass profile or tandem mass spectrometry data (see Section 8.8)

This protocol has allowed the identification of proteins from the yeast proteome [20], human pituitary proteome [155], the *E. coli* proteins affected by sulfate starvation [156], and the major membrane and core proteins of the *vaccina virus* [157].

An alternative high-throughput technique for proteome analysis uses a combination of capillary isoelectric focusing (CIEF) and ESI/FT-ICRMS [158–160]. CIEF is a special mode of capillary electrophoresis, in which a pH gradient is set up throughout the capillary by filling it with a polyampholyte mixture. Proteins are separated by migration through this pH gradient until they have zero net charge. Protein bands are mobilized and detected on-line by ESIMS. This technique has several advantages over the 2D gel procedure, including speed, sensitivity, ease of automation, and accuracy in mass measurement.

8.11. AUTOMATED HIGH-THROUGHPUT ANALYSIS OF PROTEINS

In view of the vast volume of information being unearthed by genome and proteome projects, the rapid identification of protein sequences is of considerable interest [148,158–162]. The high-speed, high-sensitivity, and high-throughput protein identification is paramount to the biotechnology and pharmaceutical industries as well. Although an automated protein sequenator based on Edman degradation has been developed, the method is not ideal for high-throughput analysis because it is slow and lacks the sensitivity required for the analysis of low amounts of protein. Also, it is limited to proteins with approximately <30 amino acid residues. For a sensitive and high-throughput analysis of proteins, several of the techniques discussed above can be combined into an integrated approach. One system, shown in Figure 8.5, consists of an autosampler, microHPLC equipment, and an ESI-tandem mass spectrometer [161]. Proteins are separated by 2D gel electrophoresis, and each spot is digested with trypsin. The tryptic digests are loaded onto an autosampler of

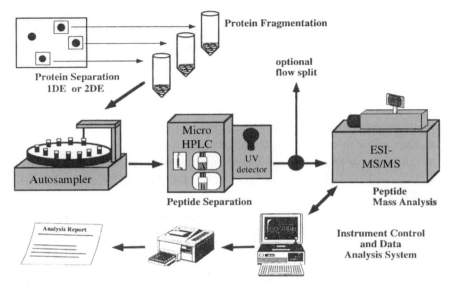

Figure 8.5. A high-throughput system for the characterization of proteins with LC/ESI-MS/MS. [Reprinted from Ref. 161 by permission of Marcel Dekker, New York (copyright 1998).]

the automated workstation, and each sample is analyzed unattended by LC/ESI-MS/MS. At the end of each LC-MS/MS run, the data are used to search protein sequence database. The search program is automatically launched in the background as a postacquisition application while the next HPLC run is in progress. Stahl et al. have developed an automated data control system for the real-time acquisition of a CID spectrum from an LCMS run [104]. The precursor ion mass selection is automatically performed from LCMS data using a preset S/N value. An automated LC-MS/MS, combined with a database search, is a powerful approach for the analysis of low fmol (<100) amounts of protein.

Mann and co-workers have developed a "layered" strategy for the large-scale protein identification [20]. A key feature of this approach is the use of automated MALDIMS analysis for high-throughput peptide mapping and nanoES-MS/MS for high-sensitivity and specific identification of the peptide fragments. In the first screen, the tryptic digests of the gel-separated proteins are analyzed by MALDIMS, and the measured masses are correlated with databases. This first screen can identify up to 90% proteins in a proteome. The remaining proteins are characterized by partially sequencing several peptides of the unseparated protein digest by nanoES-MS/MS and database searching with the peptide tag approach. This strategy has demonstrated that mass spectrometry by itself is a very effective method for the high-throughput sequencing of a complex mixture of proteins in a proteome. Using this approach, a large-scale automated sequencing of proteins from the yeast proteome was successfully accomplished [20].

In another automated workstation, the chemistry aspect of protein analysis (reduction–alkylation and digestion) is integrated with capillary LC-ESIMS [162].

With this system, all sample-handling steps prior to mass spectrometry analysis can be performed much more rapidly, and sample losses can be minimized. In this system, dilute protein solutions are trapped in a small-diameter column. After desalting (by washing the column), proteins are reduced and alkylated, and transferred to another column filled with immobilized trypsin, where the protein is digested and the peptide fragments of the digest are trapped onto a C-18 cartridge. The protein digest is analyzed by LC/ESI-MS/MS.

8.12. CHARACTERIZATION OF MUTATIONS IN PROTEINS

The characterization of mutations in proteins is the subject of considerable clinical diagnostic importance. Several genetic disorders are caused by point mutations in proteins. Electrophoresis and chromatography techniques have been used in the past for the detection of mutations in proteins. This task can now be accomplished by several mass spectrometry–based approaches. Matsuo and colleagues conducted pioneering studies in this field in 1981 using FDMS [163]. Later, several groups used FABMS [164–166], ESIMS [167,168], MALDIMS [169], and database search [148] to characterize aberrant proteins.

In principle, the protocol discussed above for confirmation of the sequence of a native protein is applicable here. A mutation in a protein can be detected by (1) mass measurement of the intact native and mutated proteins using ESIMS or MALDIMS techniques and (2) comparing the peptide maps from the two proteins. The rationale is that two proteins will have an identical mass and will yield identical peptide maps only when their amino acid sequences are identical. Conversely, dissimilar fragmentation patterns will result from different proteins. Comparison of the molecular mass of the peptide fragments that are generated from the native and mutated proteins will determine which region of the protein contains mutation. Further sequencing of the variant peptide fragments can unambiguously determine which amino acid(s) has (have) been replaced. The mass spectrometry–based approaches have been extensively applied for the characterization of abnormal hemoglobin variants (see Chapter 16 for specific examples) [148,164–170].

McLafferty and colleagues have developed a so-called top–down approach to verify the DNA-derived sequences and to detect mutation in proteins [171]. Instead of cleaving a protein extensively into smaller segments (as is done in the conventional bottom–up peptide mapping approach), a protein is cleaved into larger products that are divided into *complementary sets*. The advantage of this approach is that, from the measured and predicted masses, the disagreement can be localized to one or two fragments without analyzing all the other fragments.

8.13. DISULFIDE BOND LOCATION IN PROTEINS

Disulfide bonds are frequently encountered in proteins. This functionality is critical for biological activity of proteins because it plays an important role in the folding and refolding processes, and in maintaining the three-dimensional structure of proteins [172]. The formation of disulfide bonds is also obligatory for the expression of the activity of small neuropeptides such as oxytocin and vesopressin. Disulfide bonds are formed by the oxidation of cysteinyl thiols.

A well-established conventional protocol for location of disulfide bonds in proteins includes cleaving a protein between the half-cysteinyl residues, leaving the disulfide bonds intact, isolating the cysteine-containing peptides, and subjecting them to either amino acid analysis or sequence analysis [173,174]. A limitation of this method is that isolation and identification of cysteine-containing peptides is tedious and time-consuming. In an alternate approach, disulfide bonds are reduced sequentially to thiols, which are derivatized with radiolabeled iodoacetic acid [172,175]. The completely reduced protein is cleaved enzymatically, and the site of the disulfide bonds is ascertained from the identity of the radiolabeled peptides.

With the development of desorption and spray ionization techniques, mass spectrometry has emerged as a viable alternative for location of disulfide bonds in proteins. FAB was the first mass spectrometry technique applied to identify the disulfide-containing peptides [176]. Since then, a number of approaches have been developed to assign inter- and intradisulfide linkages [177–187]. A general mass spectrometry–based protocol is illustrated in Figure 8.6.

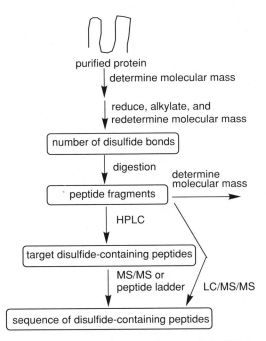

Figure 8.6. A mass spectrometry–based protocol for location of disulfide bonds in proteins.

8.13.1. Determination of the Number of Disulfide Bonds

The first step in the analysis of the disulfide-containing peptides is to determine the number of disulfide bonds [186]. The purified protein is reduced by treatment with either dithiothreitol or 2-mercaptoethanol, and the sulfhydryl groups that result are alkylated with iodoacetic acid (see Section 8.5.1). The molecular mass of the native protein (M_{nat}) and reduced–alkylated protein (M_{r+a}) is determined with MALDIMS or ESIMS. S-Carboxymethylation will increase the mass of each cysteine residue by 59 Da. From the change in the molecular mass, the number of cysteine residues (N_{Cys}) and hence the number of disulfide bonds (N_{S-S}) can be estimated (Eq. 8.3). The number of free sulfhydryl groups (N_{SH}) must be taken into account for these calculations, the presence of which is determined by following the same steps except that the reduction step is avoided.

$$N_{Cys} = \frac{M_{r+a} - M_{nat}}{59}; \qquad N_{SH} = \frac{M_{r+a} - M_{nat}}{59 - 1}; \qquad N_{S-S} = \frac{N_{Cys} - N_{SH}}{2} \qquad (8.3)$$

Isotopic labeling of the cysteine-containing peptides can also be used to determine the number of cysteine residues. In one such procedure, carboxymethylation of the cysteine residues in the protein is carried out with $^{13}C_2$-labeled bromoacetic acid, the protein is digested, and the peptide fragments are analyzed by high-resolution mass spectrometry. The isotopic pattern allows determination of the cysteine residues [187].

8.13.2. Generation of Disulfide-Containing Peptides

A common procedure for generarttion of disulfide-containing peptides is to cleave the protein with chemical and proteolytic agents. Ideally, the method chosen should produce peptides that contain only a single disulfide bond and prevent disulfide exchange during cleavage reaction. Specific and nonspecific bond-cleaving reagents have both found success in specific examples. The use of specific reagents (e.g., CNBr, trypsin, endoproteinase Glu-C) has the advantage that the C-terminal residues are known from the cleavage specificity of the reagent. However, often a specific reagent is unable to cleave a protein between half-cysteinyl residues. Under such circumstances, it is appropriate to use partial acid hydrolysis or nonspecific proteases such as pepsin. Furthermore, mild acidic conditions that are used with these reagents are ideal in preserving disulfide bonds [181]. Disulfide bonds are also stable during reaction with CNBr and endoproteinase Glu-C (highest activity at pH 4.0 and 7.6), but the pH at which trypsin has the highest activity, disulfide interchange is likely to occur. One disadvantage, however, of these nonspecific cleaving processes is that a complex mixture of the protein digest is obtained; this complexity makes the isolation and detection of the disulfide-containing peptides a difficult proposition. Table 8.2 lists various cleaving reagents and conditions appropriately modified (compare with Table 8.1) for disulfide bond-containing proteins.

Table 8.2. Typical cleaving conditions for disulfide-containing proteins

Cleaving Agent	Digestion Conditions [Buffer; pH; Temperature (°C)]	Time (h)
Cyanogen bromide	88% formic acid	24
Trypsin	50 mM $NH_4CH_3CO_2$; 6.5; 37	24
Trypsin	0.1% TFA	5
Chymotrypsin	Water; 7.0; 35	6
Thermolysin	50 mM NH_4HCO_3; 8.0; 37	3
Pepsin	5% formic acid; 37	6
Endoproteinase Glu–C	Water; 35	14
Endoproteinase Glu–C	0.1 M $NH_4CH_3CO_2$; 4.0; 37	20

8.13.3. Identification of Disulfide-Containing Peptides

The detection of disulfide-containing peptides in a protein digest is conveniently achieved by FABMS [177,178] and PDMS [179]. When specific cleaving agents are used to generate peptides, the position of the disulfide bonds in a protein can be assigned by simply measuring the molecular mass of the disulfide-bridged peptides. In order to identify these peptides, the protein digest is analyzed before and after the dithiothreitol reduction of disulfide bonds. The peaks due to the disulfide-containing peptides disappear, and new peaks related to the reduced peptides appear. Alternatively, disulfide bonds can also be reduced in situ during irradiation of the peptide/matrix (dithiothreitol) mixture by a high-energy atom beam. If the protein contains an intramolecular disulfide bond, then its molecular ion will be shifted by 2 Da after prolonged (>5-min) irradiation. When a peptide is composed of two chains that are joined by an intermolecular disulfide bond, the molecular ions of both of the constituent chains will be observed. This reduction approach is not applicable to peptides that contain a free thiol.

To extend the FABMS analysis for the detection of thiol-containing peptides, an on-probe oxidation approach has been developed [184]. The on-probe treatment of the thiol- and disulfide-containing peptides with performic acid converts each cysteine residue to cysteic acid, with a concomitant increase of 98 Da in the mass of the peptide that contains intramolecular disulfide bonds. The signal due to the intermolecular disulfide bond-containing peptides disappears. Instead, two new peaks at 49 Da higher than the mass of the constituent peptides are observed. One pitfall of this procedure is that methionine and tryptophan residues are also prone to oxidation by performic acid.

The FABMS analysis is limited to small-size peptides (<3000 Da). In addition, the poor surface activity of the disulfide-containing peptides may prevent some of them from yielding a measurable signal in FAB mass spectrum. MALDIMS and ESIMS may overcome these shortcomings [186,187]. As with the FAB analysis, the disulfide-containing peptides are detected by analyzing the protein digest before and after chemically reducing disulfide bonds. A MALDIMS method based on partial reduction has been described [188]. In this procedure, the nascent thiols that are formed after partial reduction are cyanylated, followed by their N-terminal cleavage in aqueous ammonia. The remaining disulfides are reduced, and the peptides are analyzed by MALDIMS.

In the MALDIMS method developed by Qin and Chait, the protein is cleaved enzymatically or chemically under the conditions that preserve disulfide bonds [189]. The multichain polypeptides that are linked by interchain disulfide bonds are allowed to fragment spontaneously in the MALDI/ion-trap (IT)MS method at the disulfide linkages to yield fragments that contain various combinations of the disulfide-linked peptide chains, from which it is highly feasible to establish the correct disulfide connectivity.

Although success has been achieved with FABMS, ESIMS, or MALDIMS for the detection of disulfide-containing peptides directly from protein digests, it is often desirable to isolate the peptide fragments by HPLC prior to the analysis by these mass spectrometry tools.

8.13.4. Sequence Verification of Disulfide-Containing Peptides

The unambiguous identification of the location of disulfide bonds demands that the amino acid sequence of the disulfide-containing peptides of the protein digest be determined. Any one of the techniques described in Chapter 9 for sequencing peptides, such as FABMS, MALDIMS, peptide ladders, MS/MS, and LC/ESI-MS/MS, can be used for this purpose.

8.14. ANALYSIS OF PHOSPHOPROTEINS

Protein phosphorylation is the most important and ubiquitous posttranslational event known to occur in proteins [190,191]. Many cellular functions are controlled by phosphorylation, which plays an essential role in the control of cell growth, metabolism, differentiation, and in the function of many proteins, hormones, neurotransmitters, and enzymes. Phosphorylation is mediated by certain substrate-specific phosphotransferase enzymes called protein kinases, which catalyze the transfer of the terminal phosphate moiety of a nucleoside triphosphate (ATP or GTP) to the nucleophilic hydroxyl group of serine, threonine, and tyrosine residues of proteins and peptides. A large number of protein kinases have been identified, and each contains the consensus sequence for recognition and protein phosphorylation [190]. One class of kinases is involved in phosphorylation of the hydroxyl group of serine and threonine [190–192]. Since the mid-1980s, tyrosine-specific kinases have also become the focus of attention because of their participation in transmitting signals that lead to cell growth, proliferation, and differentiation [190,192,193]. Phosphorylation is reversed by a group of enzymes collectively known as *protein phosphatases* [192].

8.14.1. Traditional Approaches for the Analysis of Phosphoproteins

Because of the unusual importance and current high level of interest in phosphorylation, a highly sensitive and specific experimental approach is needed to assay phosphorylated peptides and proteins. This knowledge provides insight into the mechanism of regulation of the kinases and phosphatases. The traditional approach for the detection of phosphorylation sites typically involves radiolabeling cells and tissues to a steady state by incubation with radioactive [32][P]-phosphate, followed by the isolation of the labeled protein by immunoprecipitation or acid extraction/precipitation, and separation with SDS-PAGE or a chromatographic technique [30,194,195]. The isolated radiolabeled protein is cleaved into smaller fragments, which, after HPLC fractionation, are subjected to Edman sequencing for the identification of individual phosphorylated amino acids. The detection of individual labeled amino acids is accomplished with thin-layer chromatography (TLC), HPLC, or electrophoresis.

This protocol, however, has several drawbacks. Besides being a potential radioactive hazard, the entire procedure is labor-intensive, is prone to sample losses, and often fails to identify the exact site of phosphorylation. In some cases, the complete incorporation of [32][P] does not occur. Also, multiplicity of phosphorylated species within the cell creates problems in achieving the steady state. The identification of phosphorylated amino acids by the Edman procedure is also problematic. Under the harsh conditions employed in the chemical degradation step of this protocol, the phosphate ester bonds to serine and threonine are less stable. Both undergo β elimination to form phenylthiohydantoin (PTH)–dithioerythritol by-products. In addition, low recoveries of PTH derivatives of phosphotyrosine have been encountered.

8.14.2. Mass Spectrometry Protocol for Analysis of Phosphoproteins

Mass spectrometry has established itself as an indispensable technique for characterizing phosphorylated proteins and peptides. FABMS [196–199], ESIMS [35,200–202], LCMS [203–206], and MALDIMS [189,207] all have found a varying degree of success. Mass spectrometric methods for the characterization of phosphorylated proteins and peptides have several potential advantages:

- These methods are inherently faster and require fewer samples.
- When the sequence of a protein is known, the presence of phosphopeptides in the digest may be established quickly from the measured molecular masses. The mass of a phosphorylated peptide is higher by 80 Da for each phospho unit from the corresponding nonphosphorylated analog.
- The location of the phosphate group in the sequence of a peptide can be unambiguously pinpointed by sequencing the peptide.
- Hazardous radioactive isotope labeling is not required.

A standard procedure for the analysis of phosphoproteins is illustrated in Figure 8.7. Following the site-specific cleavage of a purified phosphoprotein, the molecular mass of the peptide fragments is determined in the unfractionated digest by FAB, MALDIMS, or ESIMS. This information can rapidly verify phosphorylation of the protein. The exact site of phosphorylation is ascertained from the amino acid sequence of peptide fragments after their fractionation into individual components by RP-HPLC, either of-line or on-line with tandem mass spectrometry [35]. In some cases, the combination of conventional mass spectrometry and FAB could also provide the sequence information [208]. The selective detection of phosphopeptides in the digest mixtures is accomplished by HPLC (or CE) on-line with ESIMS and ESI-MS/MS [203–206,209,210]. The protocol can also benefit from the capability of iron(III)-immobilized metal ion affinity chromatography (IMAC) as a tool for the selective isolation of phosphopeptides [211] and proteins from biological matrices, and from the specificity of phosphatases in the hydrolysis of the phosphate ester bond [212].

Caution should be exercised in the FABMS analysis of phosphopeptides from a complex mixture. Other surface-active hydrophobic peptides may suppress FABMS signals of small phosphopeptides because of their hydrophilic character. In addition, the signal intensity may vary from one peptide to another. One solution to this problem is to acquire the spectrum over an extended period. A representative

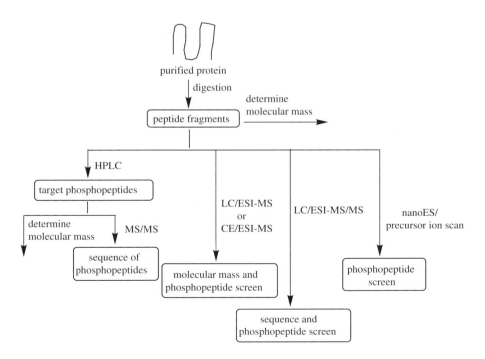

Figure 8.7. A scheme for the analysis of phosphoproteins.

spectrum is obtained by summing and averaging all the scans. Another alternative to avoid this situation is to generate larger peptide fragments by using a different degradation scheme (Table 8.1).

Although several researchers have reported the detection of phosphopeptides via the off-line combination of HPLC and mass spectrometry [213,214], on-line LC-ESIMS has emerged as the most successful approach. In this experiment, a conventional full-scan measurement allows the determination of the molecular mass of each eluting fragment, and selected-ion monitoring (SIM) of the phosphate group marker ions permits the selective detection of phosphopeptides in the digestion mixture [203–206,209,210]. CID of phosphopeptides in the transport region of the ESI source yields a prominent phosphate marker ion at m/z 79 (i.e., the PO_3^- ion) [204]. Monitoring the ion current at m/z 79 provides a means of the detection of phosphoserine-, phosphothreonine-, and phosphotyrosine-containing peptides. With current advances in computer software, it is feasible to acquire the full-scan as well as the SIM data in a single run. The advantage of this selective screening of phosphorylated peptides is that only those HPLC peaks that exhibit a positive signal of the marker ion are collected and analyzed further for their sequence determination.

Methods based on the combined LC/ESI-MS/MS methodology have also emerged for the selective detection of phosphopeptides in complex peptide mixtures [209,215,216]. In one example, the fragmentation reaction that leads to the loss of 98 Da is utilized [217]. All three types of phosphopeptides exhibit this loss (due to the expulsion of H_3PO_4 and/or $HPO_3 + H_2O$) when activated collisionally or impacted by a high-energy particle beam [207,217]. This reaction can be monitored in the positive-ion mode via a neutral loss scan for the unambiguous identification of phosphopeptides [148,202].

NanoES in combination with tandem mass spectrometry has also proved efficient and useful in phophopeptide mapping [100,217]. In this procedure, phosphopeptides are identified in the negative-ion mode via a precursor ion scan of m/z 79 [217]. This elegant approach has been used for the analysis of the phosphopeptide maps from 2D gel-separated proteins [100,217–219]. A combination of nanoES with a QIT instrument can also provide easy identification of the site of phosphorylation via MS/MS spectrum [220].

Another technique that has attained success in the analysis of phosphopeptides is the coupling of iron(III)-IMAC with ESIMS [200,211] and LC/nanoES-MS/MS [159,222,223]. Iron(III)-IMAC exploits the strong affinity of iron(III) for the phosphate group to selectively isolate phosphopeptides from a mixture [211,224]. The on-line IMAC/LC/nanoES-MS/MS system is illustrated in Figure 8.8, and includes a micro iron(III)-IMAC column coupled to a capillary LC and nanoES–tandem mass spectrometry system. A variation of this system is the replacement of LC with CE, in which the SPE section of the CE column is replaced with an iron(III)-IMAC packing [159]. Zhou et al. have achieved sequencing of phospho-peptides via consecutive enzymatic reactions of analytes that are bound to IMAC beads with subsequent direct analysis of products by MALDI–MS [225].

The specificity of phosphatases for removal of the phosphate group has also been utilized for the selective identification of phosphopeptides and proteins [212]. The

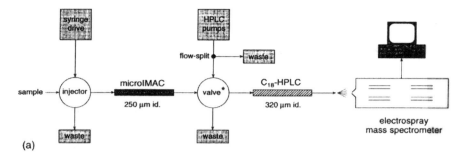

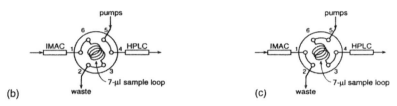

Figure 8.8. A schematic diagram of an IMAC-LCMS system. [Reprinted from Ref. 161 by permission of Marcel Dekker, New York (copyright 1998).]

molecular mass of phosphopeptides decreases by 80 Da for each phospho unit after phosphatase treatment. The reaction is monitored by MALDIMS [226]. Alternatively, immobilized phosphatase is packed in a small-diameter column and coupled to LC-ESIMS or CE-ESIMS [159,227]. The use of CE-ESIMS for determination of phosphopeptides via their dabsyl derivatives has also been described [228].

REFERENCES

1. C. L. Brooks, M. Karplus, B. M. Pettitt, *Proteins: A Theoretical Perspective of Dynamics, Structure, and Thermodynamics*, Wiley, New York (1988).

2. J. E. Shively, ed., *Methods of Protein Microcharacterization: A Practical Handbook*, Humana Press, Clifton, NJ (1986).

3. H. Towbin, T. Staehelin, and J. Gordon, *Proc. Natl. Acad. Sci.* (USA) **76**, 4350–4354 (1979).

4. A. Klapper, B. MacKay, and M. D. Rash, *BioTechniques* **12**, 650–654 (1992).

5. M. Barber, R. S. Bordoli, R. D. Sedgwick, and A. N. Tyler, *J. Chem. Soc. Chem. Commun.* 325–327 (1981).

6. W. Aberth, K. Straub, and A. L. Burlingame, *Anal. Chem.* **54**, 2029–2034 (1982).

7. B. Sundqvist and R. D. Macfarlane, *Mass Spectrom. Rev.* **2**, 421–460 (1985).

8. J. B. Fenn, M. Mann, C. K. Meng, S. F. Wong, and C. M. Whitehouse, *Science* **246**, 64–71 (1989).

9. R. D. Smith, J. A. Loo, R. R. Ogorzalek Loo, M. Busman, and H. R. Udseth, *Mass Spectrom. Rev.* **10**, 359–451 (1991).

10. M. Karas and F. Hillenkamp, *Anal. Chem.* **60**, 2299–2301 (1988).

11. K. Tanaka, H. Waki, H. Ido, S. Akita, and T. Yoshida, *Rapid Commun. Mass Spectrom.* **82**, 151–153 (1988).

12. E. C. Huang, T. Wachs, J. J. Conboy, and J. D. Henion, *Anal. Chem.* **62**, 713A–725A (1990).

13. T. Wachs, J. J. Conboy, F. Garcia, and J. D. Henion, *J. Chromatogr. Sci.* **29**, 357–366 (1991).

14. W. M. A. Niessen and A. P. Tinke, *J. Chromatogr. A* **703**, 37–57 (1995).

15. J. Cai and J. Henion, *J. Chromatogr. A* **703**, 667–692 (1995).

16. R. D. Smith, J. Wahl, and D. R. Goodlett, *Anal. Chem.* **65**, 574A–584A (1993).

17. W. J. Henzel, T. M. Billeci, J. T. Stults, and S. C. Wong, *Proc. Natl. Acad. Sci.* (USA) **90**, 5011–5015 (1993).

18. D. J. C. Pappin, P. Hojrup, and A. J. Bleasby, *Curr. Biol.* **3**, 327–332 (1993).

19. J. R. Yates, S. Speicher, P. R. Griffin, and T. Hunkapillar, *Anal. Biochem.* **214**, 397 (1993).

20. A. Shevchenko, O. N. Jensen, A. V. Podtelejnikov, F. Sagliocco, M. Wilm, O. Vorm, P. Mortensen, A. Shevchenko, H. Boucherie, and M. Mann, *Proc. Natl. Acad. Sci.* (USA) **93**, 14440–14445 (1996).

21. I. van Oostveen, A. Ducret, and R. Aebersold, *Anal. Biochem.* **247**, 310 (1997).

22. J. A. McCloskey, ed., *Methods in Enzymology*, Academic Press, San Diego, 1990, Vol. 193.

23. T. Matsuo, R. M. Caprioli, M. L. Gross, and Y. Seyama, eds., *Biological Mass Spectrometry, Present and Future*, Wiley, New York, 1994.

24. A. L. Burlingame and S. A. Carr, *Mass Spectrometry in the Biological Sciences*, Humana Press, Totowa, NJ, 1996.

25. B. S. Larsen and C. N. McEwen, eds., *Mass Spectrometry of Biological Materials*, Marcel Dekker, New York, 1998.

26. A. L. Burlingame, R. K. Boyd, and S. J. Gaskell, *Anal. Chem.* **66**, 634R–683R (1994).

27. A. L. Burlingame, R. K. Boyd, and S. J. Gaskell, *Anal. Chem.* **68**, 599R–651R (1996).

28. A. L. Burlingame, R. K. Boyd, and S. J. Gaskell, *Anal. Chem.* **70**, 647R–716R (1998).

29. W. S. Hancock, ed., *New Methods in Peptide Mapping for the Characterization of Proteins*, CRC Press, Boca Raton, FL, 1996.

30. I. M. Rosenberg, *Protein Analysis and Purification, Benchtop Techniques.* Birkhauser, Boston, 1996.

31. B. L. Karger and W. S. Hancock, eds., *Methods in Enzymology*, Academic Press, San Diego, 1996, Vol. 270.

32. M. Z. Atassi and E. Appella, eds., *Methods in Protein Structure Analysis*, Plenum Press, New York, 1995.

33. R. L. Moritz, J. S. Eddes, G. E. Reid, and R. J. Simpson, *Electrophoresis* **17**, 907–917 (1996).

34. T. D. Lee and J. E. Shively, in J. A. McCloskey, ed., *Methods in Enzymology*, Academic Press, San Diego, 1990, Vol. 193, pp. 361–374.

35. C. Dass, in B. S. Larsen and C. N. McEwen, eds., *Mass Spectrometry of Biological Materials*, Marcel Dekker, New York, 1998, pp. 247–280.

36. O. Vorm and P. Roepstorff, *Biomed. Mass Spectrom.* **23**, 734–740 (1994).

37. J. Gobom, E. Mirgordskaya, E. Nordhoff, P. Hojrup, and P. Roepstorff, *Anal. Chem.* **71**, 919–927 (1999).

38. R. K. Blackburn and R. J. Anderegg, *J. Am. Soc. Mass Spectrom.* **8**, 483–494 (1997).

39. T. Covey, E. C. Huang, and J. D. Hanion, *Anal. Chem.* **63**, 1193–1200 (1991).

40. K. L. Stone, J. I. Elliott, G. Peterson, W. McMurry, and K. R. Williams, in J. A. McCloskey, ed., *Methods in Enzymology*, Academic Press, San Diego, 1990, Vol. 193, pp. 389–412.

41. T. Covey, in J. R. Chapman, ed., *Protein and Peptide Analysis by Mass Spectrometry*, Humana Press, Totowa, NJ, 1996, pp.83–99.

42. W. S. Hancock, C. A. Bishop, R. L. Prestige, and M. T. W. Hearn, *Anal. Biochem.* **89**, 203–212 (1978).

43. B. A. Johnson, J. M. Shirokawa, W. S. Hancock, M. W. Spellman, L. J. Basa, and D. W. Aswad, *J. Biol. Chem.* **264**, 14262 (1989).

44. E. C. Rickard and J. K. Towns, in W. S. Hancock, ed., *New Methods in Peptide Mapping for the Characterization of Proteins*, CRC Press, Boca Raton, FL, 1996, pp. 97–117.

45. J. S. Cotrell, *Pept. Res.* **7**, 115–118 (1994).

46. H. R. Morris, M. Panico, and G. W. Taylor, *Biophys. Biochem. Res. Commun.* **117**, 299–305, 1983.

47. P. Lecchi and R. M. Caprioli, in W. S. Hancock, ed., *New Methods in Peptide Mapping for the Characterization of Proteins*, CRC Press, Boca Raton, FL, 1996, pp. 219–240.

48. D. Hess, T. C. Covey, R. Winz, R. W. Brownsey, and R. Aebersold, *Prot. Sci.* **2**, 1342–1351 (1993).

49. R. Moore, L. Licklider, D. Schumann, and T. D. Lee, *Anal. Chem.* **70**, 4879–4884 (1998).

50. J. Wahl, H. R. Udseth, and R. D. Smith, in W. S. Hancock, ed., *New Methods in Peptide Mapping for the Characterization of Proteins*, CRC Press, Boca Raton, FL, 1996, pp. 143–179.

51. A. Tsarbopoulos, G. W. Becker, J. L. Occolowitz, and I. Jardine, *Anal. Biochem.* **171**, 113–123 (1988).

52. P. F. Nielsen and P. Roepstorff, *Biomed. Environ. Mass Spectrom.* **18**, 131–137, (1989).

53. B. W. Gibson and K. Beimann, *Proc. Natl. Acad. Sci.* (USA) **81**, 1956–1960, 1984.

54. W. J. Henzel, J. H. Bourell, and J. T. Stults, *Anal. Biochem.* **187**, 228–233 (1990).

55. L. Cano, K. M. Swiderek, and J. E. Shively, in J. W. Crabb, ed., *Techniques in Protein Chemistry VI*, Academic Press, San Diego, 1995, pp. 21–30.

56. C. Dass, *J. Mass Spectrom.* **31**, 77–82 (1996).

57. R. M. Caprioli, *Continuous-Flow Fast Atom Bombardment Mass Spectrometry*, Wiley, New York, 1988.

58. M. Schar, K. O. Bornsen, and E. Gassman, *Rapid Commun. Mass Spectrom.* **5**, 319–326 (1991).

59. T. M. Billeci and J. T. Stults, *Anal. Chem.* **65**, 1709–1716 (1993).

60. A. Tsarbopoulos, M. Karas, K. Strupat, B. N. Pramanik, T. L. Nagabhushan, and F. Hillenkamp, *Anal. Chem.* **66**, 2062–2070 (1994).

61. G. Li, M. Waltham, N. L. Anderson, E. Unsworth, A. Treston, and J. N. Weinstein, *Electrophoresis* **18**, 391–402 (1997).

62. R. W. Nelson, *Mass Spectrom. Rev.* **16**, 353–376 (1997).

63. A. H. Brockman, B. S. Dodd, and R. Orlando, *Anal. Chem.* **69**, 4716–4720 (1997).

64. H. Zang and R. M. Caprioli, *J. Mass Spectrom.* **31**, 690–692 (1996).

65. T. A. Worrall, R. J. Cotter, and A. S. Woods, *Anal. Chem.* **70**, 750–756 (1998).

66. M. E. McComb, R. D. Oleschuk, A. Chow, W. Ens, K. G. Standing, H. Perreault, and M. Smith, *Anal. Chem.* **69**, 5142–5149 (1998).

67. M. Wilm and M. Mann, *Anal. Chem.* **68**, 1–8 (1996).

68. M. Wilm, A Shevchenko, T. Houthaene, S. Breit, L. Schweigerer, T. Fotsis, and M. Mann, *Nature* (London) **379**, 466–469 (1996).

69. R. Korner, M. Wilm, K. Monard, M. Schubert, and M. Mann, *J. Am. Soc. Mass Spectrom.* **7**, 150–156 (1996).

70. J. H. Chen, R. T. Timperman, D. Qin, and R. Aebersold, *Anal. Chem.* **71**, 4437–4444 (1999).

71. B. Küster and M. Mann, *Anal. Chem.* **71**, 1431–1440 (1999).

72. D. Figeys, G. L. Gorthals, B. Gallis, D. R. Goodlet, A Durect, and R. Aebersold, *Anal. Chem.* **71**, 2279–2287 (1999).

73. C. Dass, in F. Settle, ed., *Handbook of Instrumental Techniques for Analytical Chemistry*, Prentice-Hall PTR, Upper Saddle River, NJ, 1998, pp. 647–664.

74. C. Dass, *Current Org. Chem.* **3**, 193–209 (1999).

75. Y. Yang, K. D. Arbtan, J. Benan, and R. Orlando, *Rapid Commun. Mass Spectrom.* **12**, 571–579 (1998).

76. J. C. Le, J. Hui, M. Haniu, V. Katta, and M. F. Rohde, *J. Am. Soc. Mass Spectrom.* **8**, 703–712 (1997).

77. M. T. Davis and T. D. Lee, *J. Am. Soc. Mass Spectrom.* **9**, 194–201 (1998).

78. Y. Chen, X. Jin, D. Misek, R. Hinderer, S. M. Hanash, and D. M. Lubman, *Rapid Commun. Mass Spectrom.* **13**, 1907–1916 (1999).

79. M. Raida, P. Schulz-Knappe, G. Geine, and W.-G. Forssmann, *J. Am. Soc. Mass Spectrom.* **10**, 45–54 (1999).

80. K. Hirayama, R. Yuji, N. Yamada, K. Kato, Y. Arato, and I. Shimada, *Anal. Chem.* **70**, 2718–2725 (1998).

81. J. Wahl, D. C. Gale, and R. D. Smith, *J. Chromatgr. A* **659**, 217–222 (1994).

82. J. Wahl and R. D. Smith, *J. Capillary Electrophor.* **1**, 62–71 (1994).

83. J. Wahl, J. Hofstadtler, and R. D. Smith, *Anal. Chem.* **67**, 462 (1995).

84. K. C. Lewis, G. J. Opiteck, J. W. Jorgenson, and D. M. Sheely, *J. Am. Soc. Mass Spectrom.* **8**, 495–500 (1997).

85. W. Tong, A. Link, J. K. Eng, and J. R. Yates, III, *Anal. Chem.* **71**, 2270–2278 (1999).

86. D. Figeys, I. van Oostveen, A. Durect, and R. Aebersold, *Anal. Chem.* **68**, 1822–1828 (1996).

87. J. C. Severs, A. C. Harms, and R. D. Smith, *Rapid Commun. Mass Spectrom.* **10**, 1175–1178 (1996).

88. D. Figeys, A. Durect, and R. Aebersold, *J. Chromatgr. A* **763**, 295–306 (1997).

89. A. J. Tomlinson and S. Naylor, *J. Capillary Electrophor.* **2**, 225 (1995).

90. E. Rohde, A. J. Tomlinson, D. H. Johnson, and S. Naylor, *Electrophoresis* **19**, 2361 (1998).

91. Q. Yand, A. J. Tomlinson, and S. Naylor, *Anal. Chem.* **71**, 183A–189A (1999).

92. A. J. Tomlinson, M. L. Benson, S. Jameson, D. H. Johnson, and S. Naylor, *J. Am. Soc. Mass Spectrom.* **8**, 15–24 (1997).

93. J. A. Loo, C. G. Edmund, and R. D. Smith, *Anal. Chim. Acta*, **241**, 167–173 (1990).

94. A. L. Rockwood, M. Busman, and R. D. Smith, *Int. J. Mass Spectrom. Ion Proc.* **111**, 103–129 (1991).

95. M. W. Senko, J. P. Spier, and F. W. McLafferty, *Anal. Chem.* **66**, 2801–2808 (1994).

96. A. J. Link, E. Carmack, and J. R. Yates, III, *Int. J. Mass Spectrom. Ion Proc.* **160**, 303 (1997).

97. M. T. Davis, D. C. Stahl, S. A. Hefta, and T. D. Lee, *Anal. Chem.* **67**, 4549–4556 (1995).

98. M. T. Davis and T. D. Lee, *J. Am. Soc. Mass Spectrom.* **8**, 1059–1069 (1997).

99. M. T. Davis, D. C. Stahl, and T. D. Lee, *J. Am. Soc. Mass Spectrom.* **6**, 571–577 (1995).

100. M. Wilm, G. Neubauer, and M. Mann, *Anal. Chem.* **68**, 527–533 (1996).

101. A. Shevchenko, M. Wilm, O. Vorm, and M. Mann, *Anal. Chem.* **68**, 850–858 (1996).

102. A. L. McCormack, D. M. Schieltz, B. Goode, S. Yang, G. Barnes, and J. R. Yates, III, *Anal. Chem.* **69**, 767–776 (1997).

103. J. R. Yates, III, J. K. Eng, A. L. McCormack, and D. M. Schieltz, *Anal. Chem.* **67**, 1426–1436 (1995).

104. D. C. Stahl, K. M. Swiderek, M. T. Davis, and T. D. Lee, *J. Am. Soc. Mass Spectrom.* **7**, 532–540 (1996).

105. U. K. Laemmli, *Nature* (London) **227**, 680–685 (1970).

106. A. Klapper, B. MacKay, and M. D. Resh, *BioTechniques* **12**, 650–654 (1992).

107. S. D. Patterson, *Anal. Biochem.* **221**, 1–15 (1994).

108. S. D. Patterson, *Electrophoresis* **16**, 1104–1114 (1995).

109. S. D. Patterson and R. Aebersold, *Electrophoresis* **16**, 1791–1814 (1995).

110. J. C. Blais, P. Nagnan-Meillour, G. Bolbach, and J. C. Tabet, *Rapid Commun. Mass Spectrom.* **10**, 1–4 (1996).

111. C. Eckerson, K. Strupat, M. Karas, and F. Hillenkamp, *Electrophoresis* **13**, 664–665 (1992).

112. K. Strupat, M. Karas, F. Hillenkamp, C. Eckerson, and F. Lottspeich, *Anal. Chem.* **66**, 464–470 (1994).

113. M. Gordini, E. Bordini, C. Piubelli, and M. Hamdan, *Rapid Commun. Mass Spectrom.* **14**, 18–25 (2000).

114. M. M. Vestling and C. Fenselau, *Anal. Chem.* **66**, 471–477 (1994).

115. K. Strupat, F. Lottspeich, and C. Eckerson, *Electrophoresis* **17**, 954–961 (1996).

116. X. Liang, J. Bai., Y.-H. Liu, and D. M. Lubman, *Anal. Chem.* **68**, 1012–1018 (1996).

117. R. R. Ogorzalek Loo, T. I. Stevenson, C. Mitchell, J. A. Loo, and P. C. Andrews, *Anal. Chem.* **68**, 1910–1917 (1996).

118. R. R. Ogorzalek Loo, C. Mitchell, T. I. Stevenson, S. A. Martin, W. Hines, P. Juhasz, D. Patterson, J. Peltier, J. A. Loo, and P. C. Andrews, *Electrophoresis* **18**, 382–390 (1997).

119. R. R. Ogorzalek Loo, P. C. Andrews, and J. A. Loo, in B. S. Larsen and C. N. McEwen, eds., *Mass Spectrometry of Biological Materials*, Marcel Dekker, New York, 1998, pp. 325–343.

120. S. L. Cohen and B. T. Chait, *Anal. Biochem.* **247**, 257–267 (1997).

121. S. Haebel, C. Jensen, S. O. Andersen, and P. Roepstorff, *Protein Sci.* **4**, 394–404 (1995).

122. E. K. Fridriksson, B. Baird, and F. W. McLafferty, *J. Am. Soc. Mass Spectrom.* **10**, 453–455 (1999).

123. J. Matsudaira, *J. Biol. Chem.* **262**, 10035–10038 (1987).

124. J. Rosenfeld, J. Capdevielle, J. C. Guillemot, and P. Ferrera, *Anal. Biochem.* **203**, 173–179 (1992).

125. R. Aebersold, J. Leavitt, R.A. Saavedra, L. E. Hood, and S. B. H. Kent, *Proc. Natl. Acad. Sci.* (USA) **84**, 6970–6974 (1987).

126. B. Kuster, S. F. Wheeler, A. P. Hunter, R. A. Dwek, and D. J. Harvey, *Anal. Biochem.* **250**, 82–101 (1997).

127. N. M. Matsui, D. M. Smith, K. R. Clauser, J. Fichmann, J. E. Endrews, C. M. Sullivan, A. L. Burlingame, and L. B. Epstein, *Electrophoresis* **18**, 409–417 (1997).

128. A. Shevchenko, P. Mortensen, and M. Mann, *Biochem. Soc. Trans.* **24**, 893–896 (1996).

129. W. Zhang, A. Czenik, T. Yungwirth, R. Aebersold, and B. T. Chait, *Protein Sci.* **3**, 677–686 (1994).

130. D. Fabris, M. Vestling, M. M. Cordero, V. Doroshenko, R. J. Cotter, and C. Fenselau, *Rapid Commun. Mass Spectrom.* **9**, 1051–1055 (1995).

131. F. Gharahdaghi, M. Kirchner, J. Fernandez, and S. M. Mische, *Anal. Biochem.* **233**, 94–99 (1996).

132. C. P. Vogt, A. Willi, D. Hess, and P. E. Hunziker, *Electrophoresis* **17**, 892–898 (1996).

133. S. D. Patterson, D. Hess, T. Yungwirth, and R. Aebersold, *Anal. Biochem.* **202**, 193–203 (1992).

134. M. Mann, P. Hojrup, and P. Roepstorff, *Biol. Mass Spectrom.* **22**, 338–345 (1993).

135. P. James, M. Quadroni, E. Carafoli, and G. Gonnet, *Biochm. Biophys. Res. Commun.* **195**, 58 (1993).

136. I. Humphery-Smith, S. J. Cordwell, and W. Blackstock, *Electrophoresis* **18**, 1217–1242 (1997).

137. J. R. Yates, III, *J. Mass Spectrom.* **33**, 1–19 (1998).

138. D. Arnott, K. L. O'Connell, K. L. King, and J. T. Stults, *Anal. Biochem.* **258**, 1–18 (1998).

139. O. N. Jensen, A. V. Podtelejnikov, and M. Mann, *Rapid Commun. Mass Spectrom.* **10**, 1371–1378 (1996).

140. D. R. Goodlet, J. E. Bruce, G. A. Anderson, B. Rist, L. Pasa-Tolic, R. D. Smith, and R. Aebersold, *Anal. Chem.* **72**, 1172–1178 (2000).

141. T. D. Veenstra, J. E. Bruce, S. Martinovic, G. A. Anderson, L. Pasa-Tolic, and R. D. Smith, *J. Am. Soc. Mass Spectrom.* **11**, 78–82 (2000).

142. W. Zhang and B. T. Chait, *Anal. Chem.* **72**, 2482–2489 (2000).

143. M. Wilm and M. Mann, *Anal. Chem.* **66**, 4390–4399 (1994).

144. J. K. Eng, A. L. McCormack, and J. R. Yates, III, *J. Am. Soc. Mass Spectrom.* **5**, 976–989 (1994).

145. J. R. Yates, III, J. K. Eng, and A. L. McCormack, *Anal. Chem.* **67**, 3202–3210 (1995).

146. J. R. Yates, III, A. L. McCormack and J. K. Eng, *Anal. Chem.* **68**, 534A–540A (1996).

147. J. R. Yates, III, S. F. Morgan, C. L. Gatlin, P. R. Griffin, and J. K. Eng, *Anal. Chem.* **70**, 3557–3565 (1998).

148. C. L. Gatlin, J. K. Eng, S. T. Cross, J. C. Detter, and J. R. Yates, III, *Anal. Chem.* **72**, 757–763 (2000)

149. E. Mortz, P. B. O'Connor, P. Roepstorff, N. L. Kelleher, T. D. Wood, F. W. McLafferty, and M. Mann, *Proc. Natl. Acad. Sci.* (USA) **93**, 8264–8267 (1996).

150. S. Hoving, M. Munchbach, H. Schmid, L. Signor, A. Lehmann, W. Stavdenmann, M. Quadroni, and P. James, *Anal. Chem.* **72**, 1006–1014 (2000).

151. M. Schnolzer, P. Jedrzewski, and W. D. Lehmann, *Electrophoresis* **17**, 945–953 (1996).

152. A. Shevchenko, I. Chernushevich, W. Ens, K. Standing, B. Thompson, M. Wilm, and M. Mann, *Rapid Commun. Mass Spectrom.* **11**, 1015–1024 (1997).

153. M. R. Wilkins, C. Pasqualli, R. D. Appel, K. Ou, O. Golaz, J. C. Sanchez, J. X. Yan, A. A. Gooley, G. Hughes, I. Humphery-Smith, K. L. Williams, and D. F. Hochstrasser, *BioTechnology* **14**, 61–65 (1996).

154. P. Roepstorff, *Curr. Opin. Biotechnol.* **8**, 6–13 (1997).

155. S. Beranova-Giorgianni and D. M. Desiderio, *Rapid Commun. Mass Spectrom.* **14**, 161–167 (2000).

156. M. Quadroni, W. Staudenmann, M. Kertesz, and P. James, *Eur. J. Biochem.* **239**, 773–781 (1996).

157. O. N. Jensen, T. Houthaeve, A Shevchenko, S. Cudmore, T. Ashford, and M. Mann, *J. Virol.* **70**, 7485–7492 (1996).

158. L. Yang, C. S. Lee, S. A. Hofstadler, L. Posa-Tolic, and R. D. Smith, *Anal. Chem.* **70**, 3235–3241 (1998).

159. P. K. Jensen, L. Posa-Tolic, G. A. Anderson, J. A. Horner, M. S. Lipton, J. E. Bruce, and R. D. Smith, *Anal. Chem.* **71**, 2076–2084 (1999).

160. L. Posa-Tolic, P. K. Jensen, G. A. Anderson, M. S. Lipton, K. K. Peden, S. Martinovic, N. Tolic, J. E. Bruce, and R. D. Smith, *J. Am. Chem. Soc.* **121**, 7949–7950 (1999).

161. J. D. Watts, A Durect, D. Figeys, M. Gu, Y. Zhang, P. A. Haynes, R. Boyle, and R. Aebersold, in B. S. Larsen and C. N. McEwen, eds., *Mass Spectrometry of Biological Materials*, Marcel Dekker, New York, 1998, pp. 281–324.

162. D. B. Kassel, R. K. Blackburn, and B. Antonsson, in B. S. Larsen and C. N. McEwen, eds., *Mass Spectrometry of Biological Materials*, Marcel Dekker, New York, 1998, pp. 137–158.

163. Y. Wada, A. Hayashi, T. Fujita, T. Matsuo, I. Katakuse, and H. Matsuda, *Biochem. Biophys. Acta* **667**, 233–241 (1981).

164. Y. Wada, A. Hayashi, F. Masanori, I. Katakuse, T. Ichihara, H. Nakabushi, T. Matsuo, T. Sakurai, and H. Matsuda, *Biochem. Biophys. Acta* **749**, 244–248 (1983).

165. Y. Wada, T. Matsuo, and T. Sakurai, *Mass Spectrom. Rev.* **8**, 379–434 (1989).

166. T. D. Lee and S. Rahbar, in D. M. Desiderio, ed., *Mass Spectrometry of Peptides*, CRC Press, Boca Raton, FL, 1991, pp. 257–274.

167. H. E. Witkowska, F. Bitsch, and C. H. L. Shackleton, *Hemoglobin* **6**, 227–242 (1993).

168. H. E. Witkowska and C. H. L. Shackleton, *Anal. Chem.* **68**, 29A–33A (1996).

169. M. Hoshimoto, K. Ishimori, K. Imai, G. Miyazaki, H. Morimoto, Y. Wada, and I. Morishima, *Biochemistry* **32**, 13688–13695 (1993).

170. K. J. Light-Wahl, J. A. Loo, C. G. Edmonds, R. D. Smith, H. E. Witkowska, and C. H. L. Shackleton, *Biol. Mass Spectrom.* **22**, 112–120 (1993).

171. N. L. Kelleher, H. Y. Lin, G. A. Valaskovic, D. J. Aaserud, E. K. Fridriksson, and F. W. McLafferty, *J. Am. Chem. Soc.* **121**, 806–812 (1999).

172. T. E. Creighton, in F. Wold and K. Moldave, eds., *Methods in Enzymology*, Academic Press, San Diego, 1984, Vol. 107, pp. 305–329.

173. D. B. Wetlaufer, L. Parente, and L. Haeffner-Gormley, *Int. J. Pept. Prot. Res.* **26**, 83–91 (1985).

174. P. E. Staswick, M. A. Hermodson, and N. C. Nielsen, *J. Biol. Chem.* **259**, 13431–13435 (1984).

175. W. R. Gray, F. A. Luque, R. Galyean, E. Atherton, R. C. Sheppard, B. L. Stone, A. Reyes, J. Alford, M. McIntosh, B. M. Olivera, L. J. Cruz, and J. Rivier, *Biochemistry* **23**, 2796–2802 (1984).

176. H. R. Morris and P. Pucci, *Biochem. Biophys. Res. Commun.* **126**, 1122 (1985).

177. D. L. Smith and Y. Sun, in D. M. Desiderio, ed., *Mass Spectrometry of Peptides*, CRC Press, Boca Raton, FL, 1991, pp. 275–287.

178. D. L. Smith and Z. Zhou, in J. A. McCloskey, ed., *Methods in Enzymology*, Academic Press, San Diego, 1990, Vol. 193, pp. 374–389.

179. A. Tsarbopoulos, J. Varnerin, S. Cannon-Carlson, D. Wylie, B. Pramanik, J. Tang, and T. L. Nagabhushan, *J. Mass Spectrom.* **35**, 446–453 (2000).

180. H. Belva, C. Valois, and C. Lange, *Rapid Commun. Mass Spectrom.* **14**, 224–229 (2000).

181. Z. Zhou, and D. L. Smith, *J. Prot. Chem.* **9**, 523–532 (1990).

182. Y. Sun, D. L. Smith, and R. E. Shoup, *Anal. Biochem.* **197**, 69–76 (1991).

183. Y. Sun, P. C. Andrews, and D. L. Smith, *J. Prot. Chem.* **9**, 151–157 (1990).

184. Y. Sun and D. L. Smith, *Anal. Biochem.* **172**, 130–138 (1988).

185. R. Yazdanparast, P. C. Andrews, D. L. Smith, and J. E. Dixon, *Anal. Biochem.* **153**, 348–353 (1986).

186. Y. Sun, M. D. Bauer, T. W. Keough, and M. P. Lacey, in J. R. Chapman, ed., *Protein and Peptide Analysis by Mass Spectrometry*, Humana Press, Totowa, NJ, pp. 185–210.

187. J. M. Adamczyk, J. C. Gebler, and J. Wu, *Rapid Commun. Mass Spectrom.* **13**, 1813–1817 (1999).

188. S. D. Patterson and V. Katta, *Anal. Chem.* **66**, 3727–3732 (1994).

189. J. Qin and B. T. Chait, *Anal. Chem.* **69**, 4002–4009 (1997).

190. T. Hunter, in T. Hunter and B. M. Sefton, eds., *Methods in Enzymology*, Academic Press, San Diego, 1991, Vol. 200, pp. 3–37.

191. A. M. Edelman, D. K. Blumenthal, and E. G. Krebs. *Ann. Rev. Biochem.* **56**, 567–613 (1987).

192. B. E. Kemp, ed., *Peptides and Protein Phosphorylation*, CRC Press, Boca Raton, FL, 1990.

193. T. Hunter and J. A. Cooper, *Ann. Rev. Biochem.* **54**, 897–930 (1985).

194. L. Engstorm, P. Ekman, E. M. Humble, U. Ragnarsson, and O. Zetterquist, in F. Wold and K. Moldave, eds., *Methods in Enzymology*, Academic Press, San Diego, 1984, Vol. 107, pp. 130–154.

195. T. Hunter and B. M. Seften, eds., *Methods in Enzymology*, Academic Press, San Diego, 1991, Vol. 201.

196. C. Fenselau, D. N. Heller, M. S. Miller, and H. B White, III, *Anal. Biochem.* **150**, 309–314 (1985).

197. B. W. Gibson, A. M. Falick, A. L. Burlingame, L. Nadasdi, A. C. Nguyen, and G. L. Kenyon, *J. Am. Chem. Soc.* **109**, 5343–5348 (1987).

198. H. Michel, D. F. Hunt, J. Shabanowitz, and J. Bennett, *J. Biol. Chem.* **263**, 1123–1130 (1988).

199. B. W. Gibson and P. Cohen, in J. A. McCloskey, ed., *Methods in Enzymology*, Academic Press, San Diego, 1990, Vol. 193, pp. 480–501.

200. A. J. Rossomando, J. Wu, H. Michel, J. Shabanowitz, D. F. Hunt, M. J. Weber, and T. W. Sturgill, *Proc. Natl. Acad. Sci.* (USA) **89**, 5779–5783 (1992).

201. L. M. Nuwaysir and J. T. Stults, *J. Am. Soc. Mass Spectrom.* **4**, 662–669 (1993).

202. M. Busman, K. L. Schey, J. E. Oatis, Jr., and D. R. Knapp, *J. Am. Soc. Mass Spectrom.* **7**, 243–249 (1996).

203. T. Covey, B. Shushan, R. Bonner, W. Schröder, and F. Hucho, in H. Jörnvall, J. O. Höög, A. M. Gustavsson, eds., *Methods in Protein Sequence Analysis*, Birkhäuser Press, Basel, 1991, pp. 249–256.

204. M. J. Huddleston, R. S. Annan, M. F. Bean, and S. A. Carr, *J. Am. Soc. Mass Spectrom.* **4**, 710–717 (1993).

205. J. Ding, W. Burkhart, and D. B. Kassel, *Rapid Commun. Mass Spectrom.* **8**, 94–98 (1994).

206. X. Zhu and C. Dass, *J. Liq. Chromatgr. Rel. Technol.* **22**, 1635–1647 (1999).

207. R. S. Annan and S. A. Carr, *Anal. Chem.* **68**, 3413–3421 (1996).

208. C. Dass and P. Mahalakshmi, *Rapid Commun. Mass Spectrom.* **9**, 1148–1154 (1995).

209. C. R. Lombardo, T. G. Consler, and D. B. Kassel, *Biochemistry* **34**, 16456–16466 (1995).

210. E. Stimson, O. Truong, W. J. Richter, M. D. Waterfield, and A. L. Burlingame, *Int. J. Mass Spectrom. Ion Proc.* **169/170**, 231–240 (1997).

211. S. Li and C. Dass, *Anal. Biochem.* **270**, 9–14 (1999).

212. T. T. Yip and W. Hutchens, *FEBS Lett.* **308**, 149–153 (1992).

213. K. Palczewski, J. Buczylko, P. Van Hooser, S. A. Carr, M. Huddleston, and J. W. Crabb, *J. Biol. Chem.* **267**, 18991–18998 (1992).

214. K. A. Resing, R. S. Johnson, and K. A. Walsh, *Biochemistry* **34**, 9477–9487 (1995).

215. X. Zhang, C. Herring, P. Romano, J. Szczepanowska, H. Brzeska, A. Hinnebusch, and J. Qin, *Anal. Chem.* **70**, 2050–2059 (1998).

216. R. Verma, R. S. Annan, M. J. Huddleston, S. A. Carr, G. Reynard, and R. Deshaies, *Science* **278**, 455–460 (1997).

217. S. A. Carr, M. J. Huddleston, and R. S. Annan, *Anal. Biochem.* **239**, 180–192 (1996).

218. A. Weijland, J. C. Williams, G. S. A. Courtneidge, R. K. Wierenga, G. Sureti-Furga, *Proc. Natl. Acad. Sci.* (USA) **94**, 3590–3595 (1997).

219. G. Neubauer and M. Mann, *Anal. Chem.* **71**, 235–242 (1999).

220. S. Ogueta. R. Rogado, A. Moria, F. Moreno, J. M. Redondo, and J. Vázquez, *J. Mass Spectrom.* **35**, 493–503 (2000).

221. M. Affolter, J. D. Watts, D. L. Krebs, and R. Aebersold, *Anal. Biochem.* **223**, 74–81 (1994).

222. J. D. Watts, M. Affolter, D. Krebs, R. L. Wange, L. E. Samelson, and R. Aebersold, *J. Biol. Chem.* **269**, 29520–29529 (1994).

223. P.-C. Liao, J. Leykam, P. C. Andrews, D. A. Gage, and J. Allison, *Anal. Biochem.* **219**, 9–20 (1994).

224. S. Li and C. Dass, *Eur. Mass Spectrom.* **5**, 279–284 (1999).

225. W. Zhou, B. A. Merrick, M. G. Khaledi, and K. B. Tomer, *J. Am. Soc. Mass Spectrom.* **11**, 273–282 (2000).

226. L. N. Amankawa, K. Harder, F. Jirik, and R. Aebersold, *Protein Sci.* **4**, 113–125 (1995).

227. R. Aebersold, in A. Chrambach, M. J. Dunn, and B. J. Radola, eds., *Advances in Electrophoresis*, VCH-Verlag, Weinheim, Germany, 1991, Vol. 4, pp. 81–168.

228. H. E. Meyer, B. Eisermann, M. Heber, E. Hoffmann-Posorske, H. Korte, C. Weigt, A. Wegner, T. Hutton, A. Donella-Deana, and J. W. Perich, *J. FASEB* **7**, 776–782 (1993).

9

STRUCTURAL CHARACTERIZATION OF PEPTIDES

Peptides are chains of amino acid residues with a wide range of biological activities. They are important molecules in biochemistry, medicine, and physiology, and play significant roles in a multitude of biochemical and neurological processes. Many naturally occurring peptides are functionally active as hormones, neurotransmitters, analgesics, cytokines, and growth factors. Some peptides have shown clinical activity as inhibitors (of angiotensin-converting enzyme, renin, HIV protease, etc.) and hormones (e.g., oxytocin, leutinizing hormone-releasing hormone, vasopressin, calcitonin, and many more). Peptides are also useful as probes for research in enzymology, immunology, and molecular biology. They can also serve as the function of a diagnostic biomarker for various clinical disorders because their concentration levels in body tissues and fluids can be correlated with psychiatric and metabolic disorders.

The structural characterization of peptides is important in order to understand their biological functions and to profile metabolic changes in a particular peptide family that can occur as a result of stress or therapeutical treatment. As was seen in Chapter 8, the analysis of peptide fragments in a protein digest is a requirement for identification of proteins. For assessing the purity of newly synthesized peptides also demands their analysis. The purity of peptides is critical to their use as therapeutic agents.

9.1. TRADITIONAL APPROACHES FOR ANALYSIS OF PEPTIDES

Traditionally, many laboratories have used radioreceptor assay (RRA), radioimmu-noassay (RIA), enzyme-linked immunoassay (ELISA), and other bioassays for the identification of peptides [1]. Although some of these techniques are very sensitive, their molecular specificity is less than ideal. The unambiguous characterization of peptides requires the determination of their amino acid sequence with Edman degradation procedure [2]. Edman sequencing involves the sequential degradation of a peptide from the N terminus, and the identification of the released amino acids one at a time. The N-terminal amino acid of a peptide is first derivatized with phenylisothiocyanate (PITC), and is cleaved from the peptide by treating it with an acid to yield phenylthiohydantoin (PTH) derivative, which is then separated and identified by a suitable chromatographic analysis. This cycle is repeated with the remaining peptide. The technique has several limitations. Besides being time-consuming, it is not applicable to N-terminally blocked peptides. The carryover of the products from one cycle to the next is also a major concern, and is the cause of a decrease in sensitivity and an increase in interference with the progress of the Edman cycle. As a consequence, only a partial sequence is obtained for polypeptides. Also, as the C-terminus approaches, hydrophobic peptides may be washed out because of their increased solubility in the extraction solvent. Furthermore, harsh acidic conditions of the Edman cycle are not conducive to certain posttranslational modifications such as sulfation and phosphorylation. Although the whole protocol has been automated, the technique is not applicable to a large-scale analysis of proteins within a reasonable time.

9.2. OLDER MASS SPECTROMETRY METHODS

The analysis of peptides has posed a formidable challenge to the mass spectrometry community in the past. This problem was the result of the peptide's poor volatility, which stems from their zwitterionic, hydrogen-bonded character. Researchers relied on chemical derivatization of peptides to increase their volatility prior to analysis with electron ionization, which was the only ionization technique available during the early days of organic mass spectrometry. Several procedures were used to form the volatile derivatives of peptides: acylation followed by reduction and trimethyl-silylation to form acyltrimethylsilylpolyamino alcohols, N,O-acetylation and permethylation, acyl esters, and trimethylsilylation [3–7]. Large-sized peptides and proteins were subjected to enzymatic or partial acid hydrolysis to produce small-sized peptides, which after derivatization were separated and analyzed by gas chromatography (GC)MS. Liquid chromatography (LC)MS has also been used to sequence N-acetyl-N,O,S-permethylated peptides [8,9]. Chemical ionization (CI) [10] and desorption CI were also applied to analyze derivatized peptides [11,12]. Despite obvious limitations of these techniques, several important contributions emerged that include the discovery of the sequence of methionine and leucine enkephalins [13], the identification of thyroid-stimulating hormone (TRF) [14], and the sequencing of proteins [15,16].

The widespread use of mass spectrometry for the analysis of peptide was made feasible only after the advent of soft ionization techniques, namely, field desorption (FD) [17,18], fast-atom bombardment (FAB) [19,20], ^{252}Cf-plasma desorption ionization (PD) [21], electrospray ionization (ESI) [22,23], and matrix-assisted laser desorption/ionization (MALDI) [24]. Although efforts to produce mass spectra of small peptides were successful with FD [25,26] and PD techniques [27–30], it was the introduction of FAB in 1981 by Barber and colleagues that revolutionized the mass spectrometry analysis of peptides [19]. Concurrent with the emergence of desorption ionization methods, great strides were made to extend the mass range of the existing mass analyzers.

9.3. FRAGMENTATION CHARACTERISTICS OF PEPTIDES

Fragmentation characteristics of peptides are well documented [31–38]. Peptides are biopolymers that are formed by combining two or more amino acids. During the desorption/ionization or collision activation event, an excess amount of energy is deposited in the protonated ions of peptides to effect the cleavage of various bonds all along the peptide backbone. The site of protonation directs fragmentation to yield the amino acid sequence-specific fragment ions. Although protonation preferentially occurs at the N-terminal amino group or the amino group of basic residues (e.g., Arg and Lys), the charge can be randomly localized on the nitrogen atom of any one of the amide bonds. Thus, any of the peptide bonds can be broken. All three types of peptide backbone bonds—the alkyl–carbonyl bond (CHR−CO), the peptide–amide bond (CO−NH), and the amino–alkyl bond (NH−CHR) (Figure 9.1)—are susceptible to cleavage. Either N- or C-terminus fragments may retain the charge. Thus, in principle, six different sequence-specific ion series may be formed. A standard nomenclature has been proposed by Roepstorff and Fohlman [39] to designate the fragment ions. This proposal was later modified by Biemann [40]. According to these proposals, the N-terminus ions are represented by the symbols a_n, b_n, and c_n, and the corresponding C-terminus ions by x_n, y_n, and z_n as shown in the upper portion of Figure 9.1. The structures of all sequence-specific ions are shown in the lower portion of this figure. Further cleavage in the nth side chain of these newly formed ions produce secondary ions that are denoted by the symbols w_n, v_n, and d_n (Figure 9.2). The w ions are formed from $z + 1$ ions via cleavage of the β, γ bond, v ions from y ions via cleavage of the α, β bond (i.e., cleavage of the entire side chain), and d ions from $a + 1$ ions via cleavage of the β, γ bond [37,41]. Although some studies have claimed that the w_n, v_n, and d_n ions are formed only during the collision-induced dissociation (CID) step [42,43], research from the author's laboratory has shown that they are also present in a conventional mass spectrum [44]. The w and d ions are of special significance. Because they retain the β-carbon atom of the amino acid, they can be used as diagnostic ions to distinguish Leu from Ile [41,45]. The w ions of Leu are always 14 Da lower in mass than those of Ile. The

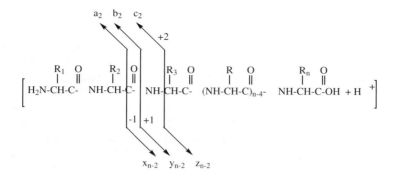

Figure 9.1. The nomenclature (upper portion) and structure (lower portion) of sequence-specific ions.

cleavage of two bonds in the peptide backbone produces immonium ions and internal fragments (see Figure 9.2). In the spectrum, the immonium ions are denoted by the one-letter code of the respective amino acid residue. The immonium ions may undergo further fragmentation as shown in Figure 9.3 [46]. The immonium ions and their fragments are diagnostic markers of the presence of certain amino acids.

Although, in principle, six types of primary sequence-specific ion series are possible, not all ion series are formed with equal facility. Certain features in a peptide have an overwhelming influence on the yield of sequence-specific ions. For example, the presence of a basic residue (Arg or Lys) near either terminus of a peptide favors the charge retention by that terminus fragment. Cleavage in the backbone of a peptide at Pro and Phe residues tends to produce intense y ions [34,47,48].

9.4. CHARACTERIZATION OF PEPTIDES BY FABMS

FAB and its variation, liquid secondary ionization mass spectrometry (liquid–SIMS), have emerged as invaluable tools for the analysis of peptides. In these techniques, the

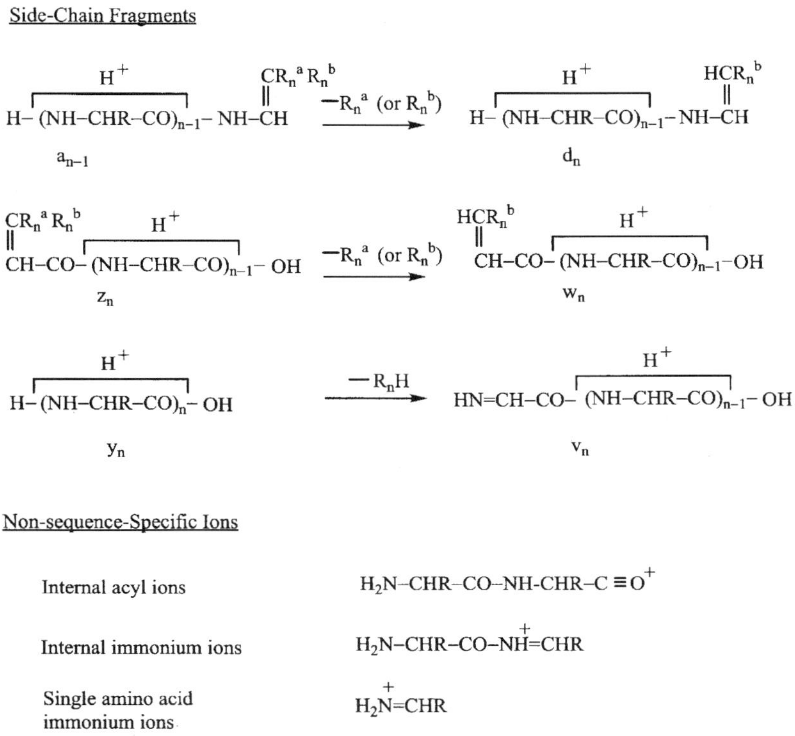

Side-Chain Fragments

Figure 9.2. The structure of side-chain fragments and non-sequence-specific ions.

peptide is mixed with a nonvolatile and polar liquid matrix, and that mixture is bombarded by a beam of high-energy Xe atoms or Cs^+ ions. Both ionization methods are conceptually similar except that liquid–SIMS provides an enhanced sensitivity and increased mass range, and therefore, is more widely accepted. However, both ionization methods are referred as FABMS in this chapter. The selection of a proper matrix is crucial to the success of the FABMS analysis. A common practice is to use glycerol, α-thioglycerol, or a 5:1 (w:w) mixture of dithiothreitol and dithioerythritol (DTT/DTE) as a liquid matrix.

The surface activity of the peptide sample, the solution pH, and the presence of salts are key issues in the FAB analysis of peptides [49]. In a mixture of hydrophilic and hydrophobic peptides, the latter have a tendency to occupy the surface of the matrix, and thus to overwhelm the mass spectrum. In contrast, hydrophilic peptides exhibit a poor response [50]. The matrix surface composition may be altered by adjusting the pH or by the addition of surfactants. A sample from a biological source should be purified to remove salts or any other metal cations because these impurities tend to form adducts with peptides (see Section 9.9). A publication from the author's laboratory has reported that the fragmentation of peptides can be

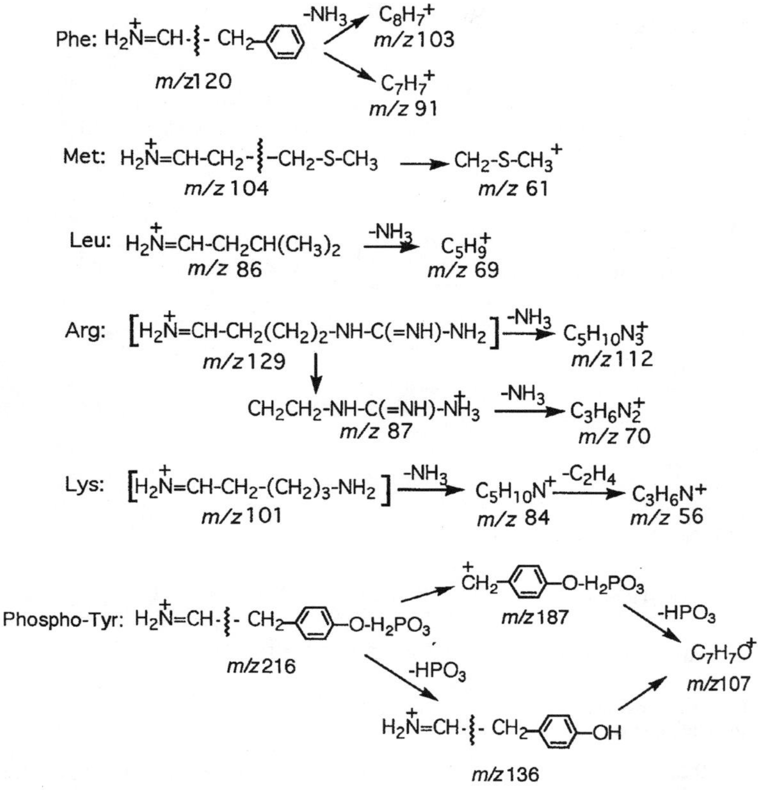

Figure 9.3. The fragmentation pathways of immonium ions. [Reproduced from Ref. 47 by permission of Marcel Dekker (copyright 1998).]

controlled with a proper choice of the matrix (51). α-Thioglycerol or the DTT/DTE mixture is used as a matrix for the molecular mass determination and tandem mass spectrometry (MS/MS) applications. In contrast, when conventional mass spectrometry is used to sequence peptides, glycerol is preferred as the matrix because it promotes fragmentation.

A majority of peptide analysis by FAB is performed in the positive-ion mode. A typical FAB spectrum of a peptide contains the protonated molecule ion ($[M + H]^+$) of the peptide and a few fragment ions; the former is usually the most abundant ion. The matrix background normally overwhelms the lower end of the spectrum. If metal ions are present, the cationic adducts with the peptide molecule will also be present. In addition, peptides also form adducts with glycerol; the most prominent adduct is seen at a mass that is 12 Da higher than the molecular mass of the peptide [52]. These adducts are absent when α-thioglycerol and DTT/DTE are used as matrices.

9.4.1. Sequence Determination by FAB and Conventional Mass Spectrometry

Although the FAB-produced molecular ions of peptides usually possess only a little excess energy over their ground state, often this energy is sufficient to generate a meaningful fragmentation profile in a number of cases. As mentioned above, the fragmentation of peptides can be increased by the use of glycerol as a liquid matrix [51]. The research from our laboratory [34,44,47,48,51] and from the laboratories of other researchers [53–56] has shown that 10–12-amino-acid-long peptides can be readily sequenced with this technique. With longer-chain peptides, fragmentation is sparse, and will provide only a partial sequence. Low-nanomole amounts of pure peptides are required for the sequence analysis.

The potential of FAB to sequence peptides is illustrated in Figure 9.4, using the example of a phosphopeptide, YpGGFMRGL [44]. This spectrum was obtained in the positive-ion mode for less than 5 nmol of the peptide that was dissolved in glycerol. Several sequence-specific ions are recognized in this spectrum. Foremost are the y- and $z + 1$-type C-terminal ions. A complete coverage of these two ion-series is observed. Most members of the x-type ion series are also present. The N-terminal sequence ions are very sparse. The dominance of the C-terminal ions occurs as a result of the presence of Arg in the vicinity of the C terminus. Because the

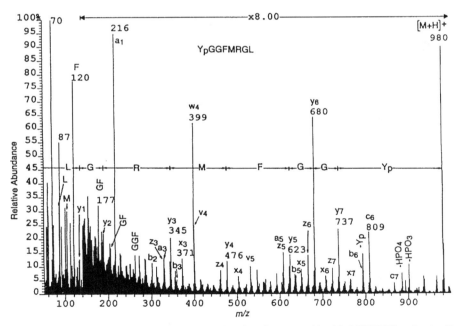

Figure 9.4. FAB mass spectrum of a phosphorylated octapeptide, YpGGFMRGL mixed with glycerol as a matrix. Ions a, b, c, x, y, and z are sequence-specific ions; L, M, and F are immonium ions; and GF and GGF are the internal fragments. [Reproduced from Ref. 44 by permission of John Wiley & Sons (copyright 1995).]

peptide analyzed in this example is of a known sequence, the sequence ions are readily discerned.

In some cases, the negative-ion FAB analysis may also be useful to obtain the sequence information. Because the structural features that stabilize a negative ion are usually different from those that stabilize a positive ion, the negative-ion analysis may provide complementary sequence-specific information [34,48]. If a sufficient amount of the peptide sample is available, then it is a good practice to obtain spectra in both ionization modes. The negative-ion FAB analysis has been successful in sequencing phosphorylated [44] and nonphosphorylated enkephalin peptides [34,48].

9.4.2. Chemical Derivatization of Peptides for FAB Analysis

Although a large number of underivatized peptides have been analyzed by FAB, some hydrophilic peptides, particularly those that contain a greater number of acidic groups, have posed a challenge. Several researchers have demonstrated improved results in terms of sensitivity as well as sequence coverage after derivatization of such peptides [56,57]. The hydrophobicity of peptides can be improved by esterification of the N terminus [50]. The analysis of pyridinium salts has been shown to improve the quality of FAB spectra [60]. These derivatives are prepared by reacting the N terminus of peptides with a pyrylium salt. The conversion of peptides to amides [61–63], iminothiolanes [64], ethyl triphenylphosphonium derivatives [65,66], quaternary ammonium salts [67,68], and other charged derivatives [69,70] is an additional means of peptide derivatizations.

9.5. CHARACTERIZATION OF PEPTIDES BY ESIMS

ESIMS has rapidly become an indispensable technique for the analysis of peptides [71–79]. Several unique features of this technique, such as the lack of fragmentation and the formation of multiply charged ions, have contributed to this success. In addition, it is much more sensitive than FAB, and is better suited to handling mixtures of peptides because of its tolerance to the sample impurities (except for excessive salts) and the absence of the matrix effects that are usually encountered in FAB. The background ion current limits the detection sensitivity. The main contribution to the background arises from water clusters, the magnitude of which can be reduced by operating the source at higher temperatures, drying gas flow, and sampling cone voltage [80]. Temperatures above 150°C, however, may induce fragmentation of peptides. The detection sensitivity of ESIMS has improved significantly (in the femtomole–attomole range) with the emergence of the nano-electrospray (nanoES) format of sample introduction [81].

One important application of ESI is to determine the molecular mass of peptides in a peptide digest. The simplicity of the spectrum depends on the number of basic residues in each fragment. The tryptic fragments always contain a basic residue at the C terminus; therefore, each fragment produces a single-charged and a double-

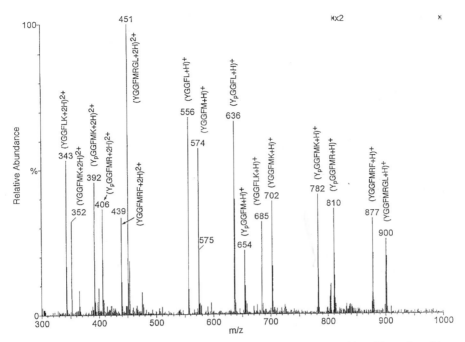

Figure 9.5. The positive-ion ESI mass spectrum of a mixture of 10 peptides. [Reproduced from Ref. 47 by permission of Marcel Dekker (copyright 1998).]

charged ion. An example of a mixture analysis is presented in Figure 9.5, which contains the spectrum of a mixture of 10 phosphorylated and nonphosphorylated peptides. Each peptide is clearly identified by the presence of singly and doubly charged ions.

Apart from the molecular mass determination, ESIMS can also be used to obtain the sequence ion information. It has been shown that large, multiple-charged peptide ions can be more efficiently fragmented compared to single-charged ions (82). Although the intact molecular ions of peptides are produced in the normal mode of the ESI operation, the fragmentation of those molecular ions can be induced in the ES ion source by adjusting the nozzle-skimmer voltage. This step, commonly referred to as *ion-source CID*, imparts an additional kinetic energy to the ions to induce their fragmentation in the transport region of the ion source. A single quadrupole machine can be used to mass-analyze the ion-source CID products. However, an unambiguous assignment of the sequence by this procedure requires a high-purity sample.

Some examples of the use of ESIMS for the analysis of peptides are given in Refs. 71–79, and others can be found in review articles published in *Analytical Chemistry* [83–85].

9.6. CHARACTERIZATION OF PEPTIDES BY MALDIMS

The current mass spectrometry literature is dominated by the applications of MALDIMS for the analysis of peptides and proteins [83–85]. The seminal work of Karas and Hillenkamp [24], who analyzed melittin and other peptides of molecular mass up to 3000 Da, has provided a major impetus to the analysis of biopolymers by mass spectrometry.

9.6.1. Sequencing of Peptides by MALDIMS

In the early days of MALDI, its primary role, in combination with TOFMS, was to determine the molecular mass of biomolecules. Currently, MALDI-TOFMS is finding an increased use as an analytic tool to sequence peptides. Apart from the abundant $[M + H]^+$ ions of peptides, fragmentation in the peptide backbone has been observed. Unlike simple organic molecules, which fragment promptly on electron ionization, the molecular ions of peptides that are formed by desorption/ionization processes fragment over an extended time period. Four different timescales of the fragmentation of the MALDI-generated ions have been characterized in a TOF instrument [86]:

- *Prompt dissociations*—occur within the time frame of the desorption event
- *Fast fragmentations*—occur in the source-accelerating region after the desorption event, but before the ion acceleration step
- *Fast metastable fragmentations*—have a decay time constant on the order of the acceleration event
- *Metastable fragmentations*—occur in the flight tube of a TOF mass spectrometer (field-free region)

Of these four types of fragmentation, only the source-accelerating region fast fragmentations (*ion-source decay*) and the field-free region (FFR) metastable fragmentation processes (*postsource decay*) have been utilized for sequence analysis. Ion activation is the result of low-energy collisions of sample ions with neutral matrix molecules that are present in the desorption plume [87,88], as well as due to the excess energy deposited to the ions during the ionization stage of the MALDI process [89]. Capture of electrons that are released through photoionization of the matrix can also be the cause of fragmentation [90]. The fragmentation of peptides is influenced by the plume density and collision energy, which are controlled by laser fluence and initial acceleration field, respectively. Another MALDIMS approach to sequencing a peptide is based on the generation of peptide ladders (discussed in Section 9.8).

Delayed-Extraction MALDI-TOFMS Analysis. The source-accelerating region fast dissociation creates a wealth of sequence-specific fragment ions [91]. MALDI is a pulsed event, whereas the ion extraction in the normal mode of TOFMS operation is performed in a continuous fashion. This arrangement does not allow enough time for ions to fragment in the ion source, and thus leads to the loss of a useful fragment ion information. If a delayed-extraction pulse is applied, fragments that are formed via an ion-source decay process can be accumulated between pulses. These fragments can be resolved and mass-analyzed with an ordinary linear TOFMS instrument for structural analysis [92,93]. This procedure also allows an enhancement in the mass resolution of TOFMS [94]. An entire mass spectrum can be obtained in one single scan—an obvious advantage over the postsource decay (PSD) approach (discussed below) for sequencing peptides. In addition, delayed extraction is applicable to peptides with much higher mass than PSD is.

Experimental evidence has shown that ion-source fragmentation involves a metastable decay that is different from the PSD process [95]. The ion-source fragmentation is observed at greater-than-threshold laser fluences, and is complete within 300 ns. Fragmentation is also influenced by the type of matrix used [95]; acidic matrices promote fragmentation. The ion-source decay leads to types of bond cleavage different from those observed in the PSD process. The ion-source decay preferentially forms c_n-type N-terminus ions along with y_n- and z_n-type C-terminus ions [95]. The presence of moderate amounts of salts in the sample has been shown to enhance the formation of c_n-type sequence ions [96]. An example of the successful application of this technique is presented in Figure 9.6 [91], which is the spectrum of the oxidized β-chain of bovine insulin. A partial sequence information from intact proteins has also been extracted by the fast ion-source decay process [97].

The Postsource Decay MALDI-TOFMS Technique. Several reports of *postsource decay* (PSD) to sequence peptides have appeared [87,88,98–106]. This technique makes use of metastable fragmentations in the first linear region of a reflectron–TOF mass spectrometer [98,99]. The principle behind the mass analysis of the FFR fragment ions was discussed in Section 4.3.4. Briefly, when a metastable ion decays in the flight tube, the resultant fragments retain the velocity of their precursor, but their kinetic energies are reduced in proportion to their mass. Thus, in a linear TOFMS, both precursor and product ions will reach the detector at the same time. However, the field-free region fragment ions can be dispersed in time in terms of their kinetic energies if a reflectron–TOFMS is used for mass analysis (see Figure 4.8). The PSD analysis on a reflectron–TOFMS has emerged as a standard procedure to sequence peptides with molecular mass up to 3000 Da. This combination, however, has one pitfall—it does not allow all fragment ions to be focused simultaneously; only a small segment of the spectrum can be recorded at each setting of the ion mirror potential. Therefore, to record a complete energy-focused spectrum, the voltage of the ion mirror is changed in small, discrete steps. Figure 9.7 is an example of PSD analysis of a peptide, GDHFAPAVTLYGK [98]. The fragmentation pattern is similar to that observed under high-energy CID conditions [99]. Similarity in the PSD mass spectra of some peptides with their FAB and low-energy CID mass spectra has also been noted [103]. A desirable feature of this technique is that the sequencing of peptides is feasible with as little as 10 fmol of the

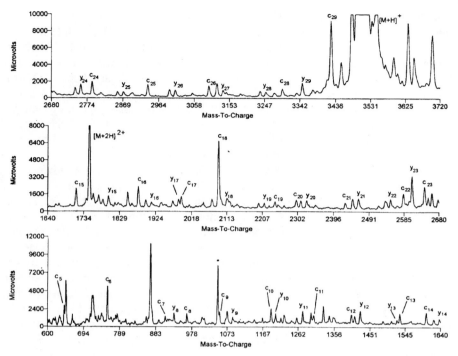

Figure 9.6. Delayed-extraction MALDI mass spectrum of the oxidized B chain of bovine insulin. [Reproduced from Ref. 90 by permission of American Chemical Society, Washington DC (copyright 1995).]

sample [103]. The PSD spectra can be simplified if the charge is fixed at one terminus [106]. This strategy was used by Shen and Allison, who prepared charged derivatives by attachment of the tris[2,4,6,-trimethoxyphenyl)phosphonium] acetyl group to the N-terminus of peptides. This modification mainly produces a_n-type ions. PSD data of protein digests can also be combined with a database search for identification of proteins [107].

9.7. SEQUENCE DETERMINATION BY TANDEM MASS SPECTROMETRY

Tandem mass spectrometry has made a tremendous impact on the structure elucidation of peptides and other biomolecules [35,40,41,108,109]. This technique is highly useful for those peptides for which conventional mass spectrometry does not produce sufficient sequence-specific fragmentation. Specifically, as the size of the peptide increases, the extent of fragmentation is reduced because less energy per bond is available to induce fragmentation. In addition, a conventional mass spectrum is often beset with chemical background noise from the matrix ions and sample impurities that are still present even after purification. Therefore, the unequivocal

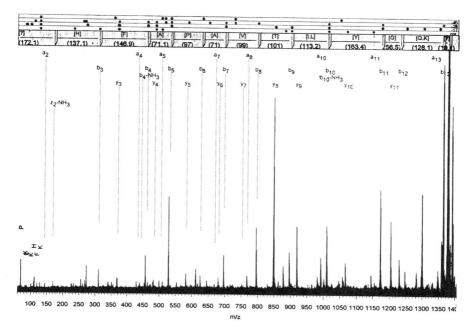

Figure 9.7. PSD mass spectrum of a peptide GDHFAPAVTLYGK. [Reproduced from Ref. 98 by permission of John Wiley & Sons (copyright 1997).]

assignment of the sequence ions may be at risk. With the use of tandem mass spectrometry, the high background of the matrix can be effectively filtered. Another unique feature of this technique is that it can tolerate heterogeneous mixtures. Therefore, the sequencing of peptide fragments in a protein digest can be accomplished without any prior separation. This situation holds true even when the peptide is a minor component. As discussed in Chapter 4, two stages of mass analysis are used in this procedure. The first stage (MS1) exclusively mass-selects the molecular ion of the desired peptide, which undergoes fragmentation in the intermediate region, and the second stage (MS2) mass-analyzes those fragment ions.

Ion activation in tandem mass spectrometry involves collisions of the fast-moving ions with a neutral gas to induce extensive fragmentation. Several researchers have achieved success with surface-induced dissociation (SID) of peptides [110–112]. The sequencing of peptides above mass 3500 Da is still a problem by tandem instruments, primarily because of the decreased molecular ion abundance and the reduced efficiency of CID of high-mass peptides.

FAB and ESI have both proved highly compatible as ionization methods prior to MS/MS analysis. With FAB, it is important to reduce the ion-source fragmentation and increase the molecular ion signal. As discussed above, the ion-source fragmentation can be reduced by an optimization of the matrix composition [51]. A further improvement in the $[M + H]^+$ ion signal can be achieved with acidification of the matrix–sample mixture; acetic acid and TFA are effective in this aspect. ESI has the advantage of producing multiply charged ions, which exhibit improved efficiency of fragmentation compared to singly charged ions [82]. All tryptic peptides fortunately (except for the C-terminal fragment) produce doubly charged ions, and therefore, are good candidates for CID studies with tandem mass spectrometry.

High- and low-energy CID processes have both been used to sequence peptides. The MS/MS spectra of peptides in the two collision regimes, however, are not identical. In high-energy CID, the fragmentation pathways reflect the relatively high center-to-mass collision energies and the short timescale of the decomposition process. Considerable dependence of low-energy CID on the internal energy of the precursor ion has been reported [113]. The fragmentation patterns that are observed under low-energy CID are qualitatively similar to the conventional mass spectra. Also, the products formed in a low-energy process are due principally to the charged-directed fragmentations [114,115]. The w_n, v_n, and d_n-type sequence-specific ions, which are frequently observed in the high-energy CID spectra, are absent in the low-energy CID spectra. Low-energy CID process is also limited to < 2000-Da-molecular-mass peptides.

Tandem mass spectrometry has achieved an unprecedented success in sequence determination of a large number of peptides. An example is presented in Figure 9.8.

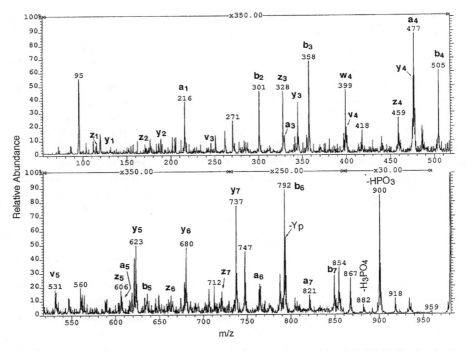

Figure 9.8. Tandem mass spectrum of a phosphorylated octapeptide, YpGGFMRGL mixed with α-thioglycerol as a matrix. The spectrum was acquired with linked-field scan at a constant B/E. [Reproduced from Ref. 47 by permission of Marcel Dekker (copyright 1998).]

Again, to compare this spectrum with the conventional mass spectrometry data, the MS/MS spectrum of the same octapeptide (YpGGFMRGL) is included [47]. This spectrum is a high-energy CID spectrum acquired with linked-field scan at constant B/E by using the front end (EBE section) of a hybrid tandem mass spectrometer of EBE-qQ geometry. Less than 5 nmol of the peptide was mixed with the matrix (α-thioglycerol).

In contrast to the positive-ion conventional mass spectrum (Figure 9.4), the MS/MS spectrum (Figure 9.8) of YpGGFMRGL contains a complete series of b-type ions. The y-type ion-series is also complete, with the exception that y_1 and y_2 ions are relatively weak. In addition, nearly all z-type ions and a few a-type ions are also formed.

9.8. SEQUENCE INFORMATION FROM A MASS SPECTRUM

The task of retrieving the sequence information from a mass spectrum is tricky and time-consuming, and involves finding all members of a certain sequence ion series. This task is made simpler if the sequence of the peptide or its homology to other peptides is known. The first important step to derive the sequence of a peptide from the conventional mass spectral data is to recognize the molecular ion of the peptide. This task is easily accomplished owing to the overwhelming abundance of the $[M + H]^+$ ions in the spectrum. Often, alkali-metal (Na^+ and K^+) adduct ions can aid in the assignment of the molecular ion. In the CID spectra, the molecular ion is the mass-selected precursor ion. The next step is to identify the presence of certain amino acids. The immonium ion peaks, which are present at the low-mass end of the spectrum, can provide this information. The masses of the amino acid residues along with the masses of their immonium ions are listed in Table 9.1. A hint of the presence of certain amino acids can also be gained from the loss of side-chain masses from the $[M + H]^+$ ion (also listed in Table 9.1). Next, it is important to recognize a specific sequence ion series. This task can be achieved by finding the mass difference between each fragment ion peak. Only the residue masses of 20 naturally occurring amino acids are considered. The first few ions at the high-mass end [50 Da < the mass of the $(M + H)^+$ ion] of the spectrum can always be ignored as these are due the loss of neutral molecules such as H_2O, NH_3, CO_2, and $HCOOH$. Consider the example shown in Figure 9.4 (or 9.8), which is the spectrum of YpGGFMRGL. The mass difference between the $[M + H]^+$ ion (m/z 980) and the ion at m/z 737 shows that phosphorylated Tyr (Yp) is present at the N terminus and that this ion is a member of the y-type ion series. If the identity of the terminal amino acids is doubtful and if sufficient peptide is available, one can resort to acetylation and methylation of the N- and C-terminal residues, respectively. CID of the alkali metal ion adducts of a peptide can also help to identify the C-terminal residue (see Section 9.9). The successive members of this ion series are recognized similarly by finding the residue masses from the difference in masses between the successive ions. This strategy is explained in Figure 9.9. The presence of satellite ions can be used as an aid in the unambiguous identification of a specific sequence ion series.

Table 9.1. Masses of amino acid residues, their immonium ions, and side chains

Amino acid	Symbol		Residue Mass (Da)	Immonium Ion (Mass (Da)	Side-Chain Mass (Da)
Alanine	Ala	A	71.04	44	15
Arginine	Arg	R	156.10	129	100
Asparagine	Asn	N	114.04	87	58
Aspartic acid	Asp	D	115.03	88	59
Cysteine	Cys	C	103.01	76	47
Glutamic acid	Glu	E	129.04	102	73
Glutamine	Gln	Q	128.06	101	72
Glycine	Gly	G	57.0	30	1
Histidine	His	H	137.06	110	81
Isoleucine	Ile	I	113.08	86	57
Leucine	Leu	L	113.08	86	57
Lysine	Lys	K	128.10	101	72
Methionine	Met	M	131.04	104	75
Phenylalanine	Phe	F	147.07	120	91
Proline	Pro	P	97.05	70	–
Serine	Ser	S	87.03	60	31
Threonine	Thr	T	101.05	74	45
Tryptophan	Trp	W	186.08	159	30
Tyrosine	Tyr	Y	163.06	136	107
Valine	Val	V	99.07	72	43

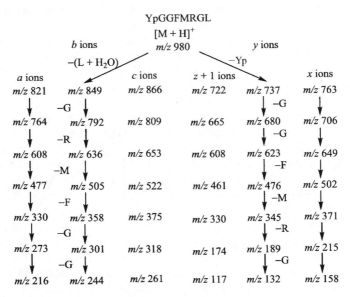

Figure 9.9. The strategy for finding sequence ion series from a mass spectrum.

For example, the identification of y-type ions is confirmed from the peaks at 26 Da above and 15 Da below, respectively, the m/z value of y-type ions. These peaks are due to x- and $z + 1$-type ions, respectively. The same strategy can be applied to recognize the N-terminal ions (e.g., b-type ions will be associated with a- and c-type ions). The determination of the entire sequence of a peptide requires that all members of at least one of the six sequence ion series be present in the spectrum. Often, two partial overlapping sequence ion series, one each from both ends, will also suffice. If one faces difficulty in the manual interpretation of the CID spectra, then the computer-aided approach can be used [116–117].

9.9. STUDY OF METAL ION ADDUCTS

The behavior of metal ion adducts under CID has been a subject of close scrutiny [118–130]. The objective had been to develop a method that can complement the existing sequencing methodology. The cationization with metals yields charged-localized ion species. This process offers an alternative and a simple means to direct fragmentation in peptides (the other is chemical derivatization). Several different metal ion adducts, such as those formed with alkali [118–123], alkaline-earth [124,125], and transition metal ions [126–130], have been studied. Many studies have focused on the elucidation of the mechanism of their fragmentation and on the structure of the fragment ions. The cationization occurs at the C terminus of peptides. Alkali-metal ion adducts mainly fragment by elimination of the C-terminal amino acid to yield $[b_{n-1} + OH + metal]^+$ ions, thus providing a simple means of identifing the C-terminal amino acid. The opinions about the mechanism of this reaction, however, are at variance, and beyond the scope of discussion here. The consensus, however, is that, on collision activation, the C-terminal O—X (where X is a cation) moiety is transferred to the carbonyl of the preceding (i.e., $n - 1$) residue to trigger the elimination of CO and imine, and produces a new peptide with one less amino acid [118–120,123].

In contrast to the limited sequence ion information that can be derived from the alkali-metal ion adducts, the Ag^+ adducts of a peptide yield increased sequence ion information [130]. ESI of a peptide solution that contains Ag^+ ions exhibits a stronger signal for the $[M + Ag]^+$ ion, which fragments under low-energy collision conditions to produce $[b_n + OH + Ag]^+$, $[b_n - H + Ag]^+$, $[a_n - H + Ag]^+$, and $[y_n + H + Ag]^+$ types of ion series. This information can help determine the sequence of peptides in many cases. Multiply charged Ag^+ ion adducts are also formed [129].

9.10. DE NOVO PEPTIDE SEQUENCING

Although rapid advances have been made to identify proteins via a database search with the molecular mass and tandem mass spectrometry data of the protein digest, the need for de novo protein sequencing will persist even after the genomes of

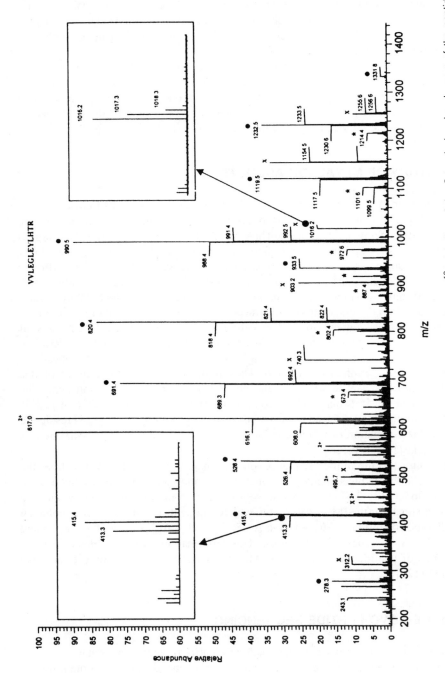

Figure 9.10. De novo sequencing of a MIHCK tryptic peptide VVEGLEYLHTR after ^{18}O labeling of the C-terminal carboxyl group of the peptide. [Reproduced from Ref. 132 by permission of John Wiley & Sons (copyright 1997).]

various species have been completely sequenced. To that end, tandem mass spectrometry will continue to play a major role. However, the recognition of sequence-specific ions in the MS/MS spectrum of an unknown peptide often becomes difficult. It has been observed that ^{18}O labeling of the C-terminal carboxyl group of a peptide can simplify this task. The fragmentation of this peptide produces y-type ions, and each is split into two ions that are separated by two mass units. This signature facilitates "readout" of the sequence. Shevchenko et al. have used this approach with quadrupole-TOFMS [131], and Qin et al. sequenced a mixture of peptides with LC-MS/MS on a QIT [132]. Figure 9.10 illustrates the usefulness of this approach in sequencing a MIHCK tryptic peptide VVEGLEYLHTR. Derivatization of a peptide is another means of simplifying the CID spectrum—a step that can help in de novo sequencing of peptides [133]. In another approach that simplifies de novo sequencing of peptides, MS^2 and MS^3 spectra are acquired. A 2-D plot of the MS^2 spectrum versus the intersection spectra (which is a spectrum that contains common ions in MS^2 and MS^3 spectra) provides enough sequence ion information [134].

9.11. ELECTRON-CAPTURE DISSOCIATION

The electron-capture dissociation (ECD) technique is relatively new [135–138]. In this process, the ESI-produced multiply protonated peptides and proteins, $[M + nH]^{n+}$, capture a low-energy (<0.2 eV) electron to produce the odd-electron ion $[M + nH]^{(n-1)+\cdot}$, which dissociates rapidly via an energetic H$\cdot$ transfer to the backbone carbonyl group to form c and $z\cdot$ sequence ions (Eq. 9.1). Minor amounts of $a\cdot$ and y ions are also produced (Eq. 9.2). ECD produces far more backbone cleavages than CID, and thus can be used to sequence larger-sized peptides. ECD reactions are readily performed in the cell of an FTMS instrument.

$$R\text{–}\overset{\overset{\displaystyle O}{\|}}{C}\text{–NH–CHR'–} \rightleftharpoons R\text{–}\overset{\overset{\displaystyle OH}{|}}{C}\text{–NH–CHR'–} \longrightarrow R\text{–}\overset{\overset{\displaystyle OH}{|}}{C}\text{=NH} + \ ^{\cdot}\text{CHR'–} \qquad (1)$$

$$\qquad\qquad\qquad\qquad\qquad\qquad\qquad\qquad\qquad\qquad\qquad c \qquad\quad z\cdot$$

$$R\text{–}\overset{\overset{\displaystyle O}{\|}}{C}\text{–NH–CHR'–} \longrightarrow R\text{–}\overset{\overset{\displaystyle O}{\|}}{C}\text{–NH}_2\text{–CHR'–} \longrightarrow \ ^{\cdot}R + C\equiv O + NH_2CHR'– \qquad (2)$$

$$\qquad\qquad\qquad\qquad\qquad\qquad\qquad a\cdot \qquad\qquad\qquad\qquad\qquad\qquad y$$

9.12. PEPTIDE LADDER SEQUENCING

In this approach, the N- or C-terminal amino acids are removed sequentially to generate a mixture of a nested set of fragments, each differing from the other by a single amino acid residue. The molecular mass of each truncated peptide is determined after each degradation cycle, and the sequence of the peptide is deduced from the molecular mass differences [43,139–142]. Chait et al. have described a modified ladder sequencing approach [143]. This procedure uses two steps. In the first step, a sequence-defining concatenated set of peptide fragment is chemically generated in a controlled fashion by rapid stepwise degradation in the presence of a small amount of a terminating agent, 5% phenylisocyanate (PIC) in phenylisothiocyanate (PITC). The phenylthiocarbamyl (PTC) derivatives of the peptides are cleaved by the TFA treatment, whereas phenylcarbamyl (PC) derivatives are stable under these conditions. A predetermined number of cycles are performed, without any intermediate separation and analysis, to generate a set of PC peptides. In the second step, the mass spectral data are obtained in a single operation. The mass differences between successive peaks in the spectrum relate to specific amino acid residues—the order of their occurrence determines the sequence of the peptide.

9.12.1. Generation of Peptide Ladders

Peptide ladders are generated in several ways. The most widely used method to remove the C-terminal amino acids is via treatment of the peptide with carboxypeptidase. Carboxypeptidase Y is one of the most commonly used carboxypeptidases, although it exhibits a preference for hydrophobic residues and the release of hydrophilic residues is slow. Similar difficulties are observed with other carboxypeptidases (A, B, P, MII, etc.). Therefore, a cocktail of more than one type of carboxypeptidase is required to cover the entire sequence. If methionine enkephalin (YGGFM) were to be sequenced by digestion with a carboxypeptidase, the mass spectrum would contain ions shown in Figure 9.11.

A vapor-phase acid hydrolysis procedure has also been exploited for removal of the C-terminal amino acids [144–148]. Earlier research used acetic acid, TFA, phosphoric acid, or hydrochloric acid [144]. Studies from two different research groups have used pentafluoropropionic acid [145,148]. In addition to the C-terminal sequence ladders, Gobum et al. also observed specific internal cleavages at Asp, Ser,

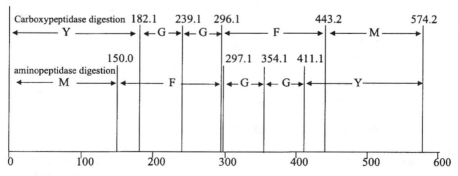

Figure 9.11. The N- and C-terminal sequence ladders generated by treatment of carboxypeptidase and aminopeptidase, respectively.

Thr, and Gly residues and the formation of the N-terminal sequence ladders [146]. Tsugita et al. have found perfluoropropionic acid anhydride to be more reliable than pentafluoropropionic acid for the generation of the C-terminal sequence ladders [146,147].

The N-terminal residues are removed by subjecting the peptide to controlled stepwise Edman degradation [143], or a treatment with aminopeptidases. Aminopeptidase M and leucine aminopeptidase can both be used for the removal of the N-terminal residues [149]. A set of molecular species that will be generated when YGGFM is treated with aminopeptidase M are shown in Figure 9.10. Hoving et al. have developed a different chemical degradation scheme for removal of N-terminal residues. This protocol can be carried out in solution with peptides immobilized on reverse phase material. A reagent (thiobenzoylthioglycolic acid, thioacetyl-thioethane, or thioacetylthioglycolic acid) is coupled to the N-terminal amino acid in the presence of a base (N-ethylmorpholin or N-methylpiperidine). The derivatized amino acid is cleaved by TFA treatment to produce peptide ladders [150].

REFERENCES

1. H. Van Vunakis and J. J. Langone, *Meth. Enzym.* **70** (1980); **73** and **74** (1981); **92** (1983).

2. P. Edman, *Acta Chim. Scand.* **4**, 238–293 (1950).

3. K. Biemann, F. Gapp, and J. Siebl, *J. Am. Chem. Soc.* **81**, 2274–2275 (1959).

4. B. C. Das, S. D. Gero, and E. Lederer, *Biochem. Biophy. Res. Commun.* **29**, 211 (1964).

5. P. A. Leclerq and D. M. Desiderio, *Anal. Lett.* **4**, 305–316 (1971).

6. K. Biemann, in G. C. Waller and O. C. Dermer, eds., *Biochemical Applications of Mass Spectrometry*, Wiley, New York, 1980, pp. 469–525.

7. K. Biemann and S. Martin, *Mass Spectrom. Rev.* **6**, 1–75 (1987).

8. T. J. Yu, H. A. Schwartz, R. W. Giese, B. L. Karger, and P. Vouros, *J. Chromatogr.* **218**, 519–533 (1981).

9. T. J. Yu, H. A. Schwartz, S. A. Cohen, P. Vouros, and B. L. Karger, *J. Chromatogr.* **301**, 425–440 (1984).

10. A. A. Kiryushkin, H. M. Fales, T. Axenrod, E. J. Gilbert, and G. W. A. Milne, *Org. Mass Spectrom.* **5**, 19–31, (1971).

11. V. N. Reinhold and S. A. Carr, *Anal. Chem.* **54**, 499–503 (1982).

12. S. A. Carr, K. Biemann, S. Shoji, D. C. Parmelee, and K. Titani, *Proc. Natl. Acad. Sci.* (USA) **79**, 6128–6131 (1982).

13. J. Hughes, T. W. Smith, H. W. Kosterlitz, L. Fothergill, B. A. Morgan, and H. R. Morris, *Nature* (Lond.) **258**, 577 (1975).

14. R. Burgus, T. F. Dunn, D. M. Desiderio, D. N. Ward, W. Vale, and R. Guillemin, *Nature* (Lond.) **269**, 1870 (1975).

15. B. W. Gibson and K. Biemann, *Proc. Natl. Acad. Sci.* (USA) **81**, 1956 (1984).

16. A. D. Auffret, T. J. Blake, and D. H. Williams, *Eur. J. Biochem.* **113**, 333 (1981).

17. H. D. Beckey, *Principles of Field Ionization and Field Desorption Mass Spectrometry*, Pergamon, Oxford, England, 1977.

18. L. Prokai, *Field Desorption Mass Spectrometry*, Marcel Dekker, New York, 1990.

19. M. Barber, R. S. Bordoli, R. D. Sedgwick, and A. N. Tyler, *J. Chem. Soc. Chem. Commun.* 325–327 (1981).

20. W. Aberth, K. Straub, and A. L. Burlingame, *Anal. Chem.* **54**, 2029–2034 (1982).

21. B. Sundqvist and R. D. Macfarlane, *Mass Spectrom. Rev.* **2**, 421–460 (1985).

22. J. B. Fenn, M. Mann, C. K. Meng, S. F. Wong, and C. M. Whitehouse, *Science* **246**, 64–71 (1989).

23. R. D. Smith, J. A. Loo, R. R. Ogorzalek Loo, M. Busman, and H. R. Udseth, *Mass Spectrom. Rev.* **10**, 359–451 (1991).

24. M. Karas and F. Hillenkamp, *Anal. Chem.* **60**, 2299–2301 (1988).

25. D. M. Desiderio and J. E. Sabatini, *Biomed. Mass Spectrom.* **8**, 565–568 (1981).

26. D. M. Desiderio, S. Yamada, and J. Z. Sabatini, and F. Tanzer, *Biomed. Mass Spectrom.* **8**, 10–12 (1981).

27. A. Tsarbopoulos, G. W. Becker, J. L. Occolowitz, and I. Jardine, *Anal. Biochem.* **171**, 113 (1988).

28. P. F. Nielsen and P. Roepstorff, *Biomed. Environ. Mass Spectrom.* **18**, 131–137 (1989).

29. P. C. Andrews, M. Alai, and R. J. Cotter, *Anal. Biochem.* **174**, 23, (1988).

30. G. J. Feistner, P. Hojrup, C. J. Evans, D. F. Barosfsky, K. F. Faull, and P. Roepstorff, *Proc. Natl. Acad. Sci.* (USA) **86**, 6013 (1989).

31. D. H. Williams, C. Bradley, G. Bojesen, S. Santikarn, and L. C. E. Taylor, *J. Am. Chem. Soc.* **103**, 5700–5704 (1981).

32. S. Seki, H. Kambara, and H. Naoki, *Org. Mass Spectrom.* **20**, 18–24 (1985).

33. K. M. Downard and K. Biemann, *J. Am. Soc. Mass Spectrom.* **5**, 966–975 (1994).

34. C. Dass and D. M. Desiderio, *Anal. Biochem.* **163**, 52–66 (1987).

35. K. Biemann and S. A. Martin, *Mass Spectrom. Rev.* **6**, 1–75 (1987).

36. T. Yalcin, I. G. Csizmadia, M. R. Peterson, and A. G. Harrison, *J. Am. Soc. Mass Spectrom.* **7**, 233–242 (1996).

37. R. S. Johnson, S. A. Martin, and K. Biemann, *Int. J. Mass Spectrom. Ion Proc.* **86**, 137 (1988).

38. D. R. Mueller, M. Eckersley, and W.-J. Richter, *Org. Mass Spectrom.* **23**, 217–222 (1988).

39. P. Roepstorff and J. Fohlman, *Biomed. Mass Spectrom.* **11**, 601 (1984).

40. K. Biemann, *Biomed. Environ. Mass. Spectrom.* **16**, 99–111 (1988).

41. R. S. Johnson, S. Martin, K. Biemann, J. T. Stults, and J. T. Watson, *Anal. Chem.* **59**, 2621–2625 (1987).

42. K. Biemann, in J. A. McCloskey, ed., *Methods in Enzymology*, Academic Press, San Diego, 1990, Vol. 193, pp. 455–479.

43. K. Biemann, in T. Matsuo, R. M. Caprioli, M. L. Gross, and Y. Seyama, eds., *Biological Mass Spectrometry, Present and Future*, Wiley, New York, 1994, pp. 275–297.

44. C. Dass and P. Mahalakshmi, *Rapid Commun. Mass Spectrom.* **9**, 1148–1154 (1995).

45. J. T. Stults and J. T. Watson, *Biomed. Environ. Mass Spectrom.* **14**, 583–586 (1987).

46. W. Heerma and W. Kulik, *Biomed. Environ. Mass Spectrom.* **16**, 155–159 (1988).

47. C. Dass, in B. S. Larsen and C. N. McEwen, eds., *Mass Spectrometry of Biological Materials*, Marcel Dekker, New York, 1998, pp. 247–280.

48. C. Dass, in D. M. Desiderio, ed., *Mass Spectrometry of Peptides*, CRC Press, Boca Raton, FL, 1991, pp. 327–345.

49. W. V. Ligon and S. B. Dorn, *Anal. Chem.* **58**, 1889–1892 (1986).

50. S. Naylor, A. F. Findeis, B. W. Gibson, and D. H. Williams, *J. Am. Chem. Soc.* **108**, 6359–6363 (1986).

51. C. Dass, *J. Mass Spectrom.* **31**, 77–82 (1996).

52. C. Dass and D. M. Desiderio, *Anal. Chem.* **60**, 2723–2729 (1989).

53. B. W. Gibson, Z. Yu, W. Aberth, A. L. Burlingame, and N. M. Bass, *J. Biol. Chem.* **263**, 4182–4185 (1988).

54. K. Eckart, H. Schwarz, M. Chorev, and C. Gilon, *Eur. J. Biochem.* **157**, 209–216 (1986).

55. H. R. Morrism M. Panico, A. Karplus, P. E. Lloyd, and B. Rinikar, *Nature* (Lond.) **300**, 643–645 (1982).

56. M. Barber, R. S. Bordoli, R. D. Sedgwick, and A. N. Tyler, *Nature* (Lond.) **293**, 270–275 (1981).

57. K. Rose, in D. M. Desiderio, ed., *Mass Spectrometry of Peptides*, CRC Press, Boca Raton, FL, 1991, pp. 315–325.

58. D. Renner and G. Spiteller, *Angew. Chem., Int. Ed. Engl.* **24**, 408–409 (1985).

59. A. M. Falick and D. A. Maltby, *Anal. Biochem.* **182**, 165–169 (1989).

60. J. P. Kiplinger, L. Contillo, W. L. Hendrich, and A. Grodski, *Rapid Commun. Mass Spectrom.* **6**, 747–752 (1992).

61. J. T. Stults, J. Lai, S. McCune, and R. Wetzel, *Anal. Chem.* **65**, 1703–1708 (1993).

62. S. Foti, D. Marletta, R. Saletti, and G. Petronne, *Rapid Commun. Mass Spectrom.* **5**, 536–539 (1991).

63. M. G. Bartlett and K. L. Busch, *Biol. Mass Spectrom.* **23**, 353–356 (1994).

64. D. S. Wagner, A. Salari, J. Fetter, D. A. Gage, J. Leykam, R. Hollingsworth, and J. T. Watson, *Biol. Mass Spectrom.* **20**, 419–425 (1991).

65. J. T. Watson, D. S. Wagner, Y.-S. Chang, S. Hanash, J. Strahler, and D. A. Gage, *Int. J. Mass Spectrom. Ion Proc.* **111**, 191–209, (1991).

66. J. E. Vath and K. Biemann, *Int. J. Mass Spectrom. Ion Proc.* **110**, 287–299, (1990).

67. D. A. Kidwell, M. M. Ross, and R. Colton, *J. Am. Chem. Soc.* **106**, 2219–2220 (1984).

68. J. Zaia and K. Biemann, *J. Am. Soc. Mass Spectrom.* **6**, 428–436 (1995).

69. J. Zaia, in J. R. Chapman, ed., *Protein and Peptide Analysis by Mass Spectrometry*, Humana Press, Totowa, NJ, 1996, pp. 29–41.

70. Z.-H. Huang, J. Wu, K. D. W. Roth, Y. Yang, D. A. Gage, and J. T. Watson, *Anal. Chem.* **69**, 137–144 (1997).

71. C. Dass, J. J. Kusmierz, D. M. Desiderio, S. A. Jarvis, and B. N. Green, *J. Am. Soc. Mass Spectrom.* **2**, 149–156 (1991).

72. L. Yan, J.-L. Tseng and D. M. Desiderio, *Life Sci.* **55**, 1937–1944 (1994).

73. E. Gowing, A. E. Roher, A. S. Woods, R. J. Cotter, M. Chaney, S. P. Little, and M. J. Ball, *J. Biol. Chem.* **269**, 10987–10990 (1994).

74. I. Nylander, K. Tan-No, A. Winter, and J. Silberring, *Life Sci.* **57**, 123–129 (1995).

75. S. S. Smart, T. J. Mason, P. S. Bennell, N. J. Maeij, and H. M. Geysen, *Int. J. Pept. Prot. Res.* **47**, 47–55 (1996).

ESI peptides 71-79 (handwritten margin note)

76. C. D. Marquez, S. T. Weintraub, and P. C. Smith, *J. Chromatogr. B* **694**, 21–30 (1997).
77. P. A. D'Agostino, J. R. Hancock, and L. R. Provost, *J. Chromatogr. A* **767**, 77–85 (1997).
78. N. Sadagopan and J. T. Watson, *J. Am. Soc. Mass Spectrom.* **11**, 107–118 (2000).
79. S. Li and C. Dass, *Anal. Biochem.* **270**, 9–14 (1999).
80. X. Zhu and C. Dass, *J. Liq. Chromatgr. Rel. Technol.* **22**, 1635–1647 (1999).
81. M. Wilm and M. Mann, *Anal. Chem.* **68**, 1–8 (1996).
82. J. A. Loo and R. D. Smith, *Anal. Chem.* **63**, 2488–2499 (1991).

Reviews: ESI for peptides (handwritten margin note)

83. A. L. Burlingame, R. K. Boyd, and S. J. Gaskell, *Anal. Chem.* **66**, 634R–683R (1994).
84. A. L. Burlingame, R. K. Boyd, and S. J. Gaskell, *Anal. Chem.* **68**, 599R–651R (1996).
85. A. L. Burlingame, R. K. Boyd, and S. J. Gaskell, *Anal. Chem.* **70**, 647R–716R (1998).
86. E. Nordhoff, F. Kirpekar, and P. Roepstorff, *Mass Spectrom. Rev.* **15**, 67–138 (1996).
87. R. Kaufmann, D. Kirsch, and B. Spengler, *Int. J. Mass Spectrom. Ion Proc.* **131**, 355–385, (1994).
88. B. Spengler, D. Kirsch, and R. Kaufmann, *Rapid Commun. Mass Spectrom.* **6**, 239 (1992).
89. M. Karas, U. Bahr, K. Strupat, F. Hillenkamp, A. Tsarbopoulos, and B. N. Pramanik, *Anal. Chem.* **67**, 675–679 (1995).
90. M. Karas, M. Glückmann, and J. Schäfer, *J. Mass Spectrom.* **35**, 1–12 (2000).
91. R. S. Brown and J. J. Lennon, *Anal. Chem.* **67**, 3990–3999 (1995).
92. M. L. Vestal, P. Juhasz, and S. A. Martin, *Rapid Commun. Mass Spectrom.* **9**, 1044–1050 (1995).
93. R. M. Whittal and L. Li, *Anal. Chem.* **67**, 1950–1954 (1995).
94. R. S. Brown and J. J. Lennon, *Anal. Chem.* **67**, 1998–2003 (1995).
95. R. S. Brown, B. L. Carr, and J. J. Lennon, *J. Am. Soc. Mass Spectrom.* **7**, 225–232 (1996).
96. V. Katta, D. T. Chow, and M. F. Rohde, *Anal. Chem.* 4410–4416 (1998).
97. J. J. Lennon and K. A. Walsh, *Protein Sci.* **6**, 2446–2453 (1997).
98. B. Spengler, *J. Mass Spectrom.* **32**, 1019–1036 (1997).
99. R. Kaufmann, B. Spengler, and F. Lutzenkirchen, *Rapid Commun. Mass Spectrom.* **7**, 902–910 (1993).
100. B. Spengler, D. Kirsch, R. Kaufmann, and E. Jaeger, *Rapid Commun. Mass Spectrom.* **6**, 105–108 (1992).
101. M. C. Huberty, J. E. Vath, W. Yu, and S. A. Martin, *Anal. Chem.* **65**, 2791–2800 (1993).
102. W. Yu, J. E. Vath, M. C. Huberty, and S. A. Martin, *Anal. Chem.* **65**, 3015–3023 (1993).
103. J. C. Rouse, W. Yu, and S. A. Martin, *J. Am. Soc. Mass Spectrom.* **6**, 822–835 (1995).
104. B. Spengler, in J. R. Chapman, ed., *Protein and Peptide Analysis by Mass Spectrometry*, Humana Press, Totowa, NJ, 1996, pp. 43–56.
105. B. Spengler, D. Kirsch, and R. Kaufmann, *Rapid Commun. Mass Spectrom.* **5**, 198–202 (1991).
106. T. L. Shen and J. Allison, *J. Am. Soc. Mass Spectrom.* **11**, 145–152 (2000).
107. P. R. Griffin, M. J. MacCoss, J. K. Eng, R. A. Blevines, J. S. Aaronson, and J. R. Yates, III, *Rapid Commun. Mass Spectrom.* **67**, 3202–3210 (1995).
108. D. F. Hunt, J. R. Yates, J. Shabanowitz, S. Winston, and C. R. Hauer, *Proc. Natl. Acad. Sci. (USA)* **83**, 6233–6237 (1986).

109. K. Biemann, *Prot. Sci.* **4**, 1920–1927 (1986).

110. V. H. Wysocki, J. Jones, A. R. Dongre, A. Somogyi, and A. L. McCormack, in T. Matsuo, R. M. Caprioli, M. L. Gross, and Y. Seyama, eds., *Biological Mass Spectrometry, Present and Future*, Wiley, New York, 1994, pp. 249–254.

111. A. L. McCormack, A. Somogyi, A. R. Dongre, and V. H. Wysocki, *Anal. Chem.* **65**, 2859–2872 (1993).

112. A. R. Dongre, A. Somogyi, and V. H. Wysocki, *J. Mass Spectrom.* **31**, 339–350 (1996).

113. G. W. Kilby and M. M. Sheil, *Org. Mass Spectrom.* **28**, 1417–1423 (1993).

114. Y.-S. Chang and J. T. Watson, *J. Am. Soc. Mass Spectrom.* **3**, 769–775 (1992).

115. J. T. Stults, J. Lai, S. McCune, and R. Wetzel, *Anal. Chem.* **65**, 1703–1708 (1993).

116. R. S. Johnson and K. Biemann, *Biomed. Environ. Mass Spectrom.* **18**, 945–957 (1989).

117. W. M. Hines, A. M. Falick, A. L. Burlingame, and B. W. Gibson, *J. Am. Soc. Mass Spectrom.* **3**, 326–336 (1992).

118. D. Renner and G. Spiteller, *Biomed. Environ. Mass Spectrom.* **15**, 75–77 (1988).

119. R. P. Grese, R. L. Cerny, and M. L. Gross, *J. Am. Chem. Soc.* **111**, 2835–2842 (1989).

120. R. P. Grese and M. L. Gross, *J. Am. Chem. Soc.* **112**, 5098–5104 (1990).

121. L. M. Teesch and J. Adams, *J. Am. Chem. Soc.* **112**, 4110–4120 (1990).

122. L. M. Teesch and J. Adams, *J. Am. Chem. Soc.* **113**, 812–820 (1991).

123. S.-W. Lee, H. S. Kim, and J. L. Beauchamp, *J. Am. Chem. Soc.* **120**, 3188–3195 (1998).

124. H. Zhao, A. Reiter, L. M. Teesch, and J. Adams, *J. Am. Chem. Soc.* **115**, 2854–2863 (1993).

125. P. Hu and M. L. Gross, *J. Am. Chem. Soc.* **114**, 9153–9160 (1992).

126. S. Shields, B. K. Bluhm, and D. H. Russell, *J. Am. Soc. Mass Spectrom.* **11**, 626–638 (2000).

127. C. L. Gatlin, F. Turecek, and T. Vaiser, *J. Am. Chem. Soc.* **117**, 3637–3638 (1995).

128. P. Hu, C. Sorensen, and M. L. Gross, *J. Am. Soc. Mass Spectrom.* **6**, 1079–1085 (1995).

129. L. Hongbo, K. W. M. Siu, R. Guevremont, and J. C. Y. Le Blanc, *J. Am. Soc. Mass Spectrom.* **8**, 781–792 (1997).

130. I. K. Chu, X. Guo, T.-C. Lau, and K. W. M. Siu, *Anal. Chem.* **71**, 2364–2372 (1999).

131. A Shevchenko, I. Chernushevich, W. Ens, K. Standing, B. Thompson, M. Wilm, and M. Mann, *Rapid Commun. Mass Spectrom.* **11**, 1015–1024 (1997).

132. J. Qin, C. J. Herring, and X. Zhang, *Rapid Commun. Mass Spectrom.* **12**, 209–216 (1998).

133. I. Lindh, L. Hjelmqvist, T. Bergman, J. Sjövall, and W. Griffiths, *J. Am. Soc. Mass Spectrom.* **11**, 673–686 (2000).

134. Z. Zhang and J. S. McElvain, *Anal. Chem.* **72** 2337–2350 (2000).

135. E. Mirgorodskya, P. Roepstroff, and R. A. Zubarev, *Anal. Chem.* **71**, 4431–4436 (1999).

136. R. A. Zubarev, N. K. Kelleher, and F. W. McLafferty, *J. Am. Chem. Soc.* **120**, 3265–3266 (1998).

137. R. A. Zubarev, N. A. Kruger, E. K. Fridriksson, M. A. Lewis, D. M. Horn, B. K. Carpenter, and F. W. McLafferty, *J. Am. Chem. Soc.* **121**, 2857–2862 (1999).

138. N. A. Kruger, R. A. Zubarev, D. M. Horn, and F. W. McLafferty, *Int. J. Mass Spectrom. Ion Proc.* **185–187**, 787–793 (1999).

139. M. Schaer, K.O. Boernsen, and E. Gassmann, *Rapid Commun. Mass Spectrom.* **5**, 319–326 (1991).

140. I. Katakuse, T. Matsuo, H. Matsuda, Y. Shimonishi, Y.-M. Hong and Y. Izumi, *Biomed. Mass Spectrom.* **9**, 64–68 (1982).

141. R. J. Cotter, *Time-of-Flight Mass Spectrometry: Instrumentation and Applications in Biological Research*, ACS, Washington, DC., 1997.

142. D. H. Patterson, G. E. Tarr, F. E. Ragnier, and S. A Martin, *Anal. Chem.* **67**, 3971–3978 (1995).

143. B. T. Chait, R. Wang, R. C. Beavis, and S. B. H. Kent, *Science* **262**, 89–92 (1993).

144. O. Vorm and P. Roepstorff, *Biomed. Mass Spectrom.* **23**, 734–740 (1994).

145. A. Tsugita, K. Takmoto, M. Kamo, and H. Iwadate, *Eur. J. Biochem.* **206**, 691–696 (1992).

146. A. Tsugita, K. Takmoto, and K. Satake, *Chem. Lett.* 235–238 (1992).

147. A. Tsugita, M. Kamo, K. Miyazalki, M. Takayama, T. Kawakami, R. Shen, and T. Nozawa, *Electrophoresis* **206**, 928–938 (1998).

148. J. Gobom, E. Mirgordskaya, E. Nordhoff, P. Hojrup, and P. Roepstorff, *Anal. Chem.* **71**, 919–927 (1999).

149. D. Jayawardene and C. Dass, *Peptides* **20**, 963–970 (1999).

150. S. Hoving, M. Münchboch, H. Schmid, L. Signar, A. Lehmann, W. Stavdenmann, M. Quadroni, and P. James, *Anal. Chem.* **72**, 1006–1014 (2000).

10

ANALYSIS OF GLYCOPROTEINS AND OLIGOSACCHARIDES

The attachment of a carbohydrate chain to a protein is one of the most ubiquitous co- and posttranslational modifications observed in eukaryotic systems. It is estimated that $\sim 90\%$ of all mammalian proteins exist in the glycosylated form at some point during their existence. The carbohydrate side chains of a glycoprotein play several critical roles in cell biology, and also have a profound influence in modulating the physiochemical (e.g., solubility and stability) and biological (e.g., immunologic and proteolytic stability) properties of proteins [1]. Carbohydrate side chains serve as recognition markers for cell–cell and cell–molecule interactions, and as receptors for viruses, bacteria, and parasites. In addition, sugar units may orient glycoproteins in membranes and help determine the destination of a glycoprotein. Glycosylation can also affect the stability and secondary structure of peptides and proteins, including rigidification of peptide chains from random coils to rod-like structures. O-Glycosylated regions of mucins and the extracellular regions of many membrane glycoproteins are examples of these extended conformations.

10.1. STRUCTURAL FEATURES OF GLYCANS AND GLYCOPROTEINS

Carbohydrates have the potential for great structural diversity. Unlike amino acids that combine in a linear unbranched fashion to form peptides, monosaccharides can be linked through a variety of hydroxyl groups on the sugar to form a range of glycosidic linkages. In addition, there exist numerous stereochemical centers, an anomeric center, and extensive branching in the carbohydrate chain. Furthermore, each glycosylation site has a heterogeneous population of oligosaccharides that are attached to the same structural class. The heterogeneous distribution of oligosaccharides on glycoproteins is the consequence of the action of trimming enzymes within the endoplasmic reticulum and Golgi complex. The action of these enzymes

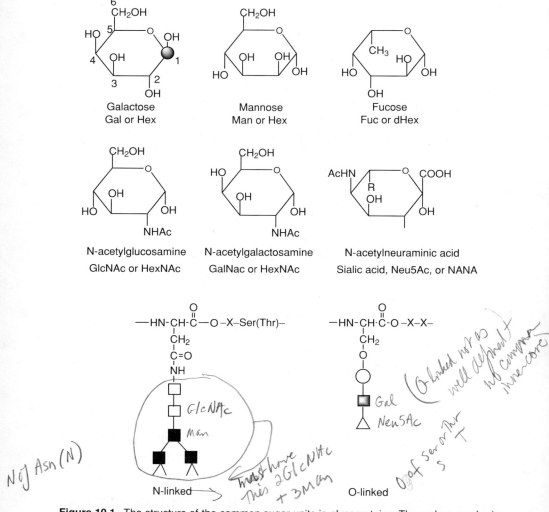

Figure 10.1. The structure of the common sugar units in glycoproteins. The carbon numbering is shown in the galactose structure. Carbon 1 is the anomeric center; carbons 2–5 are the stereochemical centers. In the β anomer, the OH group is in the up position, and in the α anomer, it is in the down position.

on a nascent glycoprotein generates an array of structures from a common precursor. Consequently, many more different oligosaccharides can be formed from four sugar units than oligopeptides from four amino acids. The most commonly occurring monosaccharides in the carbohydrate side chain of mammalian proteins are shown in Figure 10.1. These include D-mannose (Man or Hex), D-galactose (Gal or Hex), L-fucose (Fuc or dHex), N-acetylglucosamine (GlcNAc or HexNAc), N-acetylgalactosamine (GalNAc or HexNAc), and N-acetylneuraminic acid (sialic acid, Neu5Ac, or NANA). Glucose and xylose are less-encountered sugar units. The stereochemical and anomeric centers are also shown in Figure 10.1 with respect to galactose.

Carbohydrate side chains may be attached to proteins via an amide or acyl linkage. A typical glycoprotein may contain an N-glycosidic bond (N-linked structures) or a less common O-glycosidic bond (O-linked structures). These bonds are formed through the nitrogen atom of the primary amide of an asparagine residue or through the oxygen atom of serine and threonine side chains, respectively (Figure 10.1). The sugar unit most commonly attached to both types of carbohydrate may be found in the same protein usually attached to those amino acids that are located in turn or loop regions of the glycoprotein. An asparagine residue can accept a carbohydrate unit only if it is a part of the motif Asn–X–Ser (Thr), where "X" is any amino acid except proline. The carbohydrate units in N-linked oligosaccharides have a common inner-core structure that consists of two N-acetylglucosamine (GalNAc) and three mannose residues. The N-linked oligosaccharides are further classified into three common motifs: complex, high mannose, and hybrid; these structures are shown in Figure 10.2 [2]. The complex-type structures contain no mannose residues other than in the core pentasaccharide. The largest number of structural variations can be found in this subclass. The complex-type structures can have two to five antennas (or branches) that are formed by the addition of one or more lactosamines to the core structure. The antennas usually terminate in galactose or sialic acid. The high-mannose structures contain several additional mannose residues apart from those found in the inner core. The hybrid-type oligosaccharides are composed of both complex- and high-mannose-type branches. The O-linked oligosaccharides are not well defined. Although a consensus sequence for their attachment is not well understood, O-glycosylation occurs in sequence regions with a high density of Ser and Thr residues. Also, unlike the N-linked glycans, these oligosaccharides do not have a common inner core.

10.2. ANALYSIS OF GLYCOPROTEINS

Because the carbohydrate moieties of glycoproteins are not encoded in DNA sequences, their structural information can be obtained only by direct analytic measurements. The structure elucidation of a carbohydrate chain, however, is a challenging task and a major undertaking. This difficulty is the result of the structural complexity mentioned above, the heterogeneity of the glycoprotein structure, and small sample quantities that are usually encountered in biological systems. However, to determine whether a protein is glycosylated is straightforward.

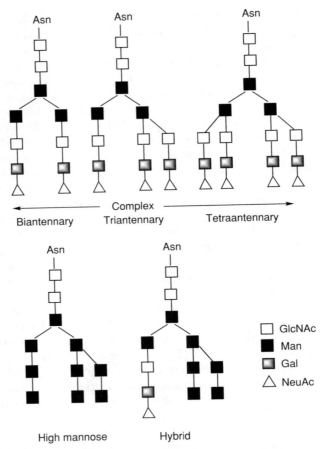

Figure 10.2. Representation of the common *N*-linked oligosaccharides. All possess a common core of two *N*-acetylglucosamine residues and three mannose residues.

During their separation on sodium dodecyl sulfate (SDS)–polyacrylamide gel electrophoresis (PAGE) gels, glycoproteins appear as broad, fuzzy bands. On treatment with glycosidases, these bands become sharper [3]. The detailed structural characterization of glycoproteins must include determination of the glycosylation sites and the type of glycosidic linkage, the position of the glycosidic linkage ($1 \rightarrow 2$, $1 \rightarrow 4$, $1 \rightarrow 6$, etc.), the anomeric configuration of each sugar, the sequence and branching of each monomers, and the identity of each sugar (Man, Gal, etc.). A general scheme for complete characterization of a glycoprotein involves three steps:

- Determination of the molecular mass of a glycoprotein
- Identification of the presence, type, and sites of glycosylation

- Structural characterization of the carbohydrate side chains after their release from the glycoprotein

The methods employed to identify the sites of glycosylation have been termed *glycosylation site mapping* or *carbohydrate mapping*, whereas the carbohydrate structure analysis methods are called *carbohydrate fingerprinting*.

A number of mass spectrometry approaches have been developed to analyze glycoproteins [4–7]. The early work in this field used electron ionization (EI) and chemical ionization (CI) of permethylated derivatives of glycopeptides prior to and following cleavage of the carbohydrate side chain [8]. A few attempts have been made to use field desorption (FD)MS [9] and [252]Cf-plasma desorption (PD)MS [10,11] to determine the extent of glycosylation and heterogeneity by comparing the tryptic maps of glycosylated and unglycosylated peptides. Fast-atom bombardment (FAB)MS [12], electrospray ionization (ESI)MS [13,14], and matrix-assisted laser

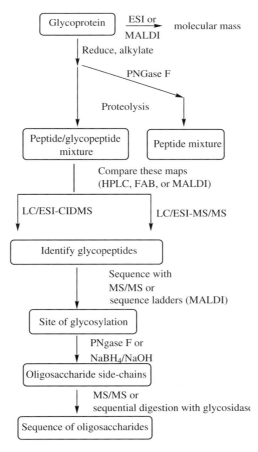

Figure 10.3. A general protocol for the identification of glycoproteins.

desorption/ionization (MALDI)MS [15] have all become popular for the analysis of glycoproteins. These three methodologies are described in detail in the following sections. A general strategy that can be used for complete analysis of glycoproteins is illustrated in Figure 10.3. The steps involved in this protocol are

- Molecular mass determination of the intact glycoprotein—ESI and MALDI are both applicable.
- Comparison of the peptide maps before and after the PNGase treatment—either HPLC, FAB, MALDI, or LC-ESIMS can be used.
- Selective detection of glycopeptides in the protein digest (before the PNGase treatment)—LC-ESIMS with in-source collision-induced dissociation (CID) is appropriate for this analysis.
- Determination of the site of glycosylation—either tandem MS or ladder sequencing with MALDI can be used.
- Structural characterization of the oligosaccharide side chain—tandem MS and sequential digestion with glycosidases are both useful.

10.3. DETERMINATION OF THE MOLECULAR MASS OF GLYCOPROTEINS

The first step in the characterization of a glycoprotein is to determine its molecular mass. Often, this information can provide a direct global assessment of glycosylation [16]. As discussed in Chapter 8, the two mass spectrometry approaches that can readily handle the molecular mass range of glycoproteins are ESI and MALDI. The former produces multiply charged ions of a protein in the 600–2000-m/z region of the spectrum, from which the molecular mass of the protein can be deduced. Although mixtures are difficult to handle by ESIMS, it is feasible to obtain the molecular mass information from the ESI spectrum of a heterogeneous sample of glycoproteins. Each glycoform and nonglycosylated protein will produce its own series of multiply charged ions. However, with the increase in the sites of glycosylation, the number of probable glycoforms also increases. Such a heterogeneous mixture of glycoproteins will yield a complex ESI spectrum, from which it may be difficult to retrieve a useful molecular mass information.

MALDIMS, on the other hand, produces a simpler mass spectrum. Each glycoform generally produces a singly charged molecular ion that will provide its molecular mass information. Therefore, for the molecular mass determination of glycoproteins, MALDIMS is preferred over ESIMS. However, the resolution and mass accuracy of the technique is poor compared to that obtained with ESIMS. In addition, highly heterogeneous glycoproteins may be difficult to analyze by MALDI because, with the increase in the heterogeneity of a glycoprotein, the MALDI signal becomes broad and less intense.

10.4. IDENTIFICATION OF THE PRESENCE, TYPE, AND SITES OF GLYCOSYLATION

In order to detect the presence of glycosylation, a *comparative peptide mapping*, in which the peptide maps are analyzed before and after the release of glycans, is a useful approach. The task becomes easier if prior knowledge of the primary structure of the glycoprotein is available from the cDNA sequence. In addition, this knowledge, along with the consensus sequence Asn–X–Ser(Thr), can also predict the possible sites of *N*-linked glycosylation, with the caveat that not all Asn residues that are part of this consensus sequence may exist in the glycosylated form. Peptide maps are generated as described in Chapter 8. The glycoprotein is first reduced and alkylated, and treated with a suitable protease. Although a number of proteolytic enzymes are available, treatment with trypsin is the most widely used procedure for generation of peptide maps from glycoproteins. To ensure complete cleavage, trypsin is added at two timepoints. Other enzymes listed in Table 8.1 are also used in some cases to cleave glycoproteins. Chemical cleavage methods should be avoided because the *N*- and *O*-glycosidic linkages are labile under the cleavage conditions. From the knowledge of the primary structure of a glycoprotein and the specificity of trypsin, the molecular mass of the peptides and glycopeptides that are generated in a tryptic digest can be easily predicted.

10.4.1. Release of Carbohydrate Side Chains from Glycopeptides

In order to release the *N*-linked carbohydrate chains, glycopeptides are treated with an endoglycosidase. Because of its broad specificity, the most ideal reagent is peptide *N*-glycosidase F (PNGase F; also known as peptide N^4-[*N*-acetyl-*β*-glucosaminyl]-asparagine), which cleaves the bond between Asn and GlcNAc and converts Asn to Asp. Other endoglycosidases that are suitable for this purpose are endo-*β*-*N*-acetyl-glucosaminidase H and F (endo H and endo F). However, these reagents are less broadly active. A chemical reagent, trifluoromethanesulfonic acid, has also been used for the deglycosylation of glycopeptides. The reaction with this reagent leaves one or two pendant GalNAc residues with Asn. *O*-Linked chains are released from the peptide backbone by reductive *β* elimination (i.e., by treatment with $NaBH_4/NaOH$).

The peptide maps are compared before and after release of the carbohydrate chain. The next step is to separate and identify the glycopeptide fractions. FABMS, ESIMS, and MALDIMS can be used for this purpose.

10.4.2. FABMS Analysis

The usual FABMS approach to determine the molecular mass of the peptide fragments of a protein digest cannot be relied on for location of the site of carbohydrate attachment because glycopeptides rarely exhibit a signal in a FAB spectrum. Several factors, such as the greater mass of glycopeptides, the heterogeneity of carbohydrate side chains, and their hydrophilicity, are responsible for their poor FABMS signal. Therefore, the FAB mass spectra of glycoprotein digests are compared before and after the PNGase F treatment [4,17,18]. The appearance of new peaks in the peptide map of the PNGase-treated fraction indicates the presence of glycosylated peptides in the untreated portion of the protein digest. From the m/z value of a new peptide, its location in the intact protein sequence can be ascertained. With the knowledge of the N-glycosylation motif, the task of determining the precise location of the glycosylation site becomes easier. Low-nanomole amounts of glycoproteins are required for this analysis. However, this quantity is seldom available with naturally occurring glycoproteins. FABMS has been used successfully to identify the N-linked glycosylation sites in tissue plasminogen activator (tPA), the pre-S2 region of the hepatitis B surface antigen, the spermine-binding protein from rat prostate, and the CD4 receptor glycoprotein [4,17].

10.4.3. ESIMS Analysis

ESIMS has proved to be a powerful tool for characterizing glycopeptides. The technique is usually combined with RP-HPLC. Because of their high detection sensitivity, it is a common practice to use microbore HPLC columns. Several ESIMS-based procedures have evolved for detection of the presence of glycopeptides in a protein digest [19–22]:

- Comparative protein mapping
- The use of contour plots
- Monitoring of carbohydrate-specific marker ions
- Precursor ion scan of specific marker ions

A simple approach is to compare the LCMS peptide maps before and after the release of the carbohydrate chain from the peptide backbone [22]. In another procedure, glycopeptides are located in a contour plot, which is a visual representation of all the ions that are generated in the entire LCMS separation [21,23]. During LCMS mapping, glycopeptides produce marker ions that appear in the plot of m/z versus scan number as a "cloud" in the form of a diagonal ladder with a negative slope. An example of the use of this procedure for detection of the presence of glycopeptides is shown in Figure 10.4; the upper trace contains the contour plot of a trypsin digest of the T103N mutant of r-tPA [21]. The corresponding plot of the control r-tPA is shown in the lower trace. Comparison of the two traces indicates that mutation (an extra glycosylation) occurs on a tryptic peptide T11.

A more specific procedure is to monitor the carbohydrate-specific marker ions in a protein digest [24–26]. This procedure is useful in the selective detection of N- and O-linked glycopeptides. These carbohydrate-specific ions are generated by CID of the molecular ions of glycopeptides, and include those oxonium ions that are a signature of Hex–HexNAc$^+$ (m/z 366), NeuAc$^+$ (m/z 274 and 292), HexNAc$^+$ (m/z 204), and Hex$^+$ (m/z 163). The m/z 204 is the most common marker ion because it is produced from most structural types of N- and O-linked carbohydrate side chains.

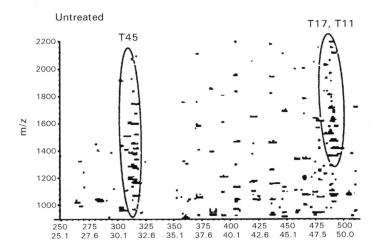

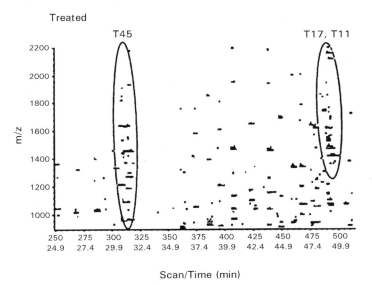

Scan/Time (min)

Figure 10.4. A contour plot of the trypsin digest of the T103 mutant (upper trace) and of the control r-tPA (lower trace). [Reproduced from Ref. 21 by permission of American Chemical Society, Washington, DC (copyright 1993).]

CID of the ESI-generated molecular ions of glycopeptides is performed in the high-pressure transport region of the ESI source by increasing the potential between the sampling orifice and skimmer (in-source CID). In practice, during an HPLC-ESIMS separation, the entire mass range is scanned at a low orifice potential to record the intact molecular ions of peptides and glycopeptides. The analysis is repeated at a high orifice potential, and the data are recorded in the selected-ion monitoring (SIM) mode to record only the suspected carbohydrate-specific oxonium ions. The first analysis provides the molecular mass of the tryptic fragments of the protein digest; the second analysis, the knowledge of the glycopeptides that are present. Carr et al. have developed a technique, called *stepped-orifice voltage scanning*, in which both analyses are performed in a single HPLC injection [24]. It involves monitoring the ion current first in the entire mass range of (e.g., 150–2000 Da). The orifice potential is stepped up to generate the carbohydrate-specific oxonium ions, but only during the time that the low-mass end (150–400 Da) of mass scan is being scanned. The rest of the mass scan is monitored at a normal low orifice potential to detect the molecular ions of the peptide fragments of the protein digest. These steps are repeated back and forth during the entire HPLC separation. The in-source CID is not a tandem mass spectrometry (MS/MS) method; thus, it is applicable to pure HPLC-eluting components only. For suspected coeluting components MS/MS monitoring of carbohydrate-specific ions can provide the analyte-specific data. In a typical procedure with a triple-sector quadrupole MS/MS instrument, the glyco peptide ions are fragmented in the rf-only quadrupole, and scanning the third quadrupole monitors the oxonium ions.

The precursor ion scan of specific marker ions (e.g., m/z 204) is another viable means of selective detection of glycopeptides [26]. For example, the MS2 of a tandem instrument can be set to monitor m/z 204, and the MS1 is scanned to detect those glycopeptides that fragment to yield the m/z 204 oxonium ion. This technique provides improved selectivity, but it is less sensitive than the stepped-orifice voltage-scanning procedure.

Another cleverly designed ESIMS-based approach has been reported [27]. It involves the *electron–capture dissociation* (ECD) [28,29] of the ESI-generated $[M + nH]^{n+}$ ions of polypeptides in a Fourier transform (FT)MS. ECD of multiply charged peptide ions rapidly cleaves the backbone $-NH-CHR-$ bond to form c- and $\dot{z}$-type sequence ions, where the c ions are more stable. This technique has been used to provide direct evidence of glycosylation in O-glycosylated peptides [27]. ECD is a truly "mild" process; the loss of carbohydrate, which is a common feature of the CID process, is not highly favored in ECD.

Once the glycopeptides have been detected by any of the ESIMS methods described above, the corresponding RP-HPLC fractions are collected and treated further with exoglycosidases to release specific sugar residues. The molecular mass is determined again, and the difference between the two masses provides the identity of the carbohydrate residue attached to the glycopeptide.

10.4.4. MALDIMS Analysis

Several different MALDI-based approaches have emerged for characterization of glycoproteins and glycopeptides [30–33]. A preliminary indication of the type and extent of glycosylation can be obtained by measuring the molecular mass of the intact glycoprotein, and treating it sequentially with endoglycosidases (e.g., with PNGase F and O-glycanase). The molecular mass is determined again by MALDI

TOFMS at the completion of each digest. On the basis of the known specificity of the enzyme reaction and the observed mass difference, the type and extent of glycosylation can be determined. This procedure, however, does not provide the necessary structural details for identification of the site of glycosylation. This information can be obtained by determining the molecular mass of the tryptic fragments. Combined with the knowledge of the known amino acid sequence of the protein, the measured molecular mass can identify the site of glycosylation. The comparison of peptide maps before and after the release of glycans will also provide a similar information.

The treatment of a glycopeptide with a carboxypeptidase or aminopeptidase to generate peptide ladders, followed by MALDIMS analysis of those ladders, is another viable approach to identification of the site of glycosylation [32]. This approach may not be effective with multiglycosylation sites because the enzymatic activity of these proteases is impaired at or near the site of glycosylation [32]. Another MALDI-based approach makes use of postsource decay (PSD) processes to identify the site of glycosylation [33].

10.4.5. Tandem Mass Spectrometry Analysis

For a detailed analysis of the glycosylation sites, the glycopeptide fractions are isolated by RP-HPLC, and treated with PNGase or $NaBH_4/NaOH$ to release *N*- or *O*-linked carbohydrate chains, respectively; peptides free from carbohydrate chains are analyzed by MS/MS to obtain the amino acid sequence of the glycopeptides [4,20,34–39]. The amino acid sequencing of intact glycopeptides is not a practical proposition. First, a FAB signal from glycopeptides is not always easy to obtain because of the attachment of heavy, hydrophilic carbohydrate side chains. Second, the fragmentation information is not very useful because fragmentation of glycopeptides usually occurs at more labile peptide–carbohydrate linkages. The signal suppression problem of FAB ionization can be avoided by using ESI to generate precursor ions of peptides [20,38–40].

10.5. STRUCTURAL CHARACTERIZATION OF THE CARBOHYDRATE SIDE CHAINS

The structural characterization of carbohydrate side chains that are attached to glycoproteins and glycolipids requires knowledge of the sugar sequence, stereochemistry, interglycosidic linkage, and anomeric configuration of each individual monosaccharide residues. The carbohydrate chain, which is alternatively called glycan, can be cleaved from glycopeptides by the action of PNGase F and *O*-glycanase, and from glycolipids by the action of an endoglycoceramidase or ceramide–glycanase. A range of mass spectrometry–based strategies has been used to analyze oligosaccharides and oligosaccharide conjugates to determine their molecular mass, sequence, and heterogeneity. However, native oligosaccharides are relatively difficult to ionize. Therefore, derivatization strategies have been developed, and include permethylation or peracetylation [41] and reductive amination with 2-aminopyridine [42] or 2-aminoacridone [43–46].

10.5.1. Linkage Analysis by GCMS

A GCMS method that was developed by Lindberg and colleagues has been a highly successful approach to determine the linkage position [47]. The *linkage analysis* or *methylation analysis*, as the procedure is commonly called, involves a complete alkylation of all free hydroxyl groups. The fully methylated oligosaccharides are hydrolyzed, reduced, and acetylated; the methylated alditol acetates that result are analyzed by GCMS in combination with either EI or CI. Identification of the location of the free hydroxyl group within a monomer will reveal the mode of attachment in the native sugar. Reductive cleavage [48] and periodate oxidation [49] procedures were later developed to accomplish the same objective. A multivariant data analysis procedure was later applied to the GCMS data to assign the linkage position and anomeric configuration [50].

These techniques, although of great utility, are time-consuming and labor-intensive, and require larger sample amounts. Attention has now shifted to several desorption/ionization techniques. In the past, PDMS was used to analyze complex carbohydrates [51]. More recent procedures, however, have relied on the use of FABMS [52–57], ESIMS [58,59], LC-ESIMS [60–62], and MALDIMS techniques [63].

10.5.2. FABMS Analysis

It is customary before FABMS analysis to derivatize oligosaccharides. Permethylation or peracetylation is the common approach [7]. The use of permethylated and peracetylated derivatives has the advantage that it leads to increased sensitivity, and an enhanced and well-defined fragmentation pattern. In addition, these derivatives are easily separated from peptides and salts by a simple extraction with an organic solvent (chloroform or dichloromethane). The FABMS analysis of the extracted derivatives provides compositions, molecular heterogeneity, and sequence of oligosaccharides. α-Thioglycerol is the matrix of choice. In some cases, it may be necessary to separate the mixture of oligosaccharides into individual fractions. By using high-performance ion–exchange chromatography on pellicular quaternary amine bonded resin, underivatized oligosaccharides are separated mainly on the basis of charge and size. The spectra of the purified fractions is usually structurally more informative because of the absence of mutual signal suppression, which is a common occurrence with mixture analysis. Poulter and Burlingame have developed a reductive amination strategy prior to FABMS analysis of glycans [64]. They have prepared a series of *n*-alkyl (methyl, ethyl, butyl, hexyl, octyl, decyl, and tetradecyl) *p*-aminobenzoic acid derivatives of oligosaccharides, and have found that a steady increase in the molecular ion abundances occurs with the increasing alkyl chain length [64]. The attachment of these chromophores also helps in their UV detection during the LC separation step.

Glycopeptides can also be analyzed directly by FABMS after permethylation or peracetylation of the carbohydrate portion to determine the monosaccharide sequence. Fragments are formed mainly by cleavage of the glycosidic bonds. Linkage information is obtained by usual methylation analysis. Alternatively, periodate oxidation of oligosaccharides, followed by reduction, permethylation or peracetylation, and the analysis by FABMS, can also be used to determine the binding position between monosaccharide residues.

10.5.3. Analysis by ESIMS and MALDIMS

Negative-ion ESI is also a valuable technique in the analysis of oligosaccharides. In particular, in-source CID can provide information on the linkage and anomeric configurations [58]. Derivatization improves the HPLC and CE separation and fragmentation [60,61,65,66]. Reductive amination is the most popular derivatizing procedure. This procedure produces an open-ring derivative from the reducing sugar. The formation of a closed-ring sugar derivative is advantageous because it provides much more linkage information [57]. This aspect was demonstrated by the ESI in-source CID of the *p*-aminobenzoic acid ethyl ester–labeled disaccharides and oligosaccharides [60].

ESI has been combined with precursor ion scan to provide a rapid identification of *N*-linked glycans in a pool of side chain carbohydrates that are derived from glycoproteins. In this procedure, the glycans are first derivatized with 2-aminoacri-done, and after a preliminary isolation by an RP-HPLC cartridge, they are analyzed by monitoring the precursors that generate certain sugar-specific marker ions. Figure 10.5 [66] is the precursor ion scan of 2-aminoacridone-labeled glycan pool of a glycoprotein from ovomucoid. The loss of oxonium ions, *N*-acetylhexosamine (m/z 204) and hexose-*N*-acetylhexosamine (m/z 306), were monitored. The *N*-linked glycans shown in Figure 10.6 were identified from these scans [66].

Profiling of constituent glycans can be performed with MALDI mass spectrometry technique also. In the procedure described by Harvey, the released glycans (by hydrazinolysis of glycoproteins) are N-acetylated and mixed with 2,5-dihydroxy-benzoic acid (DHB) MALDI matrix. The molecular mass of the glycans provides their identity [67].

10.5.4. Characterization by Molecular Mass and Database Search

Orlando and Yang have developed a procedure in which the primary structure of carbohydrate side chains can be determined from the molecular mass values and from a search of the Complex Carbohydrate Structure Database (CCSD) [5]. The suspected glycopeptide fraction is isolated by HPLC, and its molecular mass is determined by MALDIMS. The carbohydrate side chain is released by treatment with an exoglycosidase, and the molecular mass is determined again. The difference in the two mass values plus 19 Da (to account for the water loss plus the change of Asn to Asp) provides the molecular mass value of the carbohydrate side chain. This value is used as a search parameter in the search program CarbBank to provide

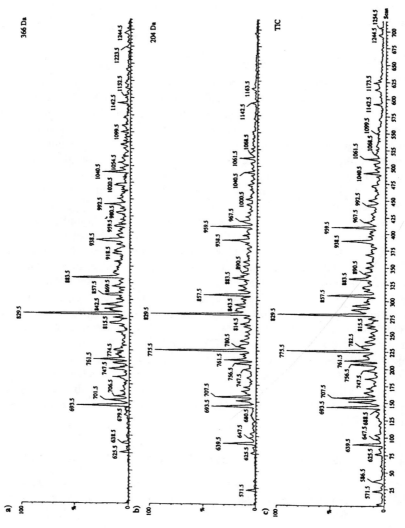

Figure 10.5. Precursor ion scan of 2-aminoacridone-labeled ovomucoid glycan pool. The loss of oxonium ions, N-acetylhexosamine (*m/z* 204) and hexose-N-acetylhexosamine (*m/z* 306), were monitored. [Reproduced from Ref. 66 by permission of John Wiley & Sons (copyright 1999).]

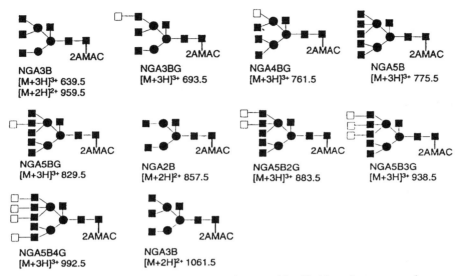

Figure 10.6. Structure of the oligosaccharides that were identified from the precursor ion scan of Figure 10.5. [Reproduced from Ref. 66 by permission of John Wiley & Sons (copyright 1999).]

structural information from the CCSD. The usefulness of this procedure was demonstrated by these researchers to determine the structure of the carbohydrate side chain that is linked to a tryptic peptide of bovine asialofetuin. This peptide has two glycoforms, and the molecular masses of its two carbohydrate side chains were determined to be 1642 Da (biantennary chain) and 2007 Da (triantennary chain). The CCSD search narrowed down the choices to 11 and 10 structures that have a molecular mass within $\pm 0.1\%$ of these experimental values, respectively.

10.5.5. Sequential Digestion Methods

Several degradation-based strategies have been developed to determine the stereo-chemical sequence, linkage, and anomeric configuration of an oligosaccharide [22,26,67–72]. Enzymatic degradation offers very specific bond cleavage. Primarily, it involves the sequential digestion of oligosaccharides with several exoglycosidases to release certain specific sugar units, and the measurement of the molecular mass of the oligosaccharide chain before and after each digestion step. The shift in mass, along with the knowledge of the specificity of the exoglycosidase, determines the stereochemical sequence, linkages, and anomeric centers. Orlando and Yang used sequential digestion with eight different exoglycosidases to establish that the triantennary carbohydrate side of a tryptic glycopeptide of bovine asialofetuin exists in two different forms. The successful use of this strategy has been reported in combination with ESIMS [22,26] and MALDIMS [30,67–71].

Mechref and Novotny have demonstrated an enzyme-based sequencing and linkage determination in which all enzymatic digestion steps are performed directly on the MALDI sample plate [71]. In this procedure, glycoproteins are first treated on the MALDI plate with PNGase F to release glycans, which are subjected to the action of three exoglycosidases (neuroaminidase, N-acetyl-β-glucosaminidase, and β-galactosidase). No other sample manipulation steps are used.

A chemical degradation procedure has also been used successfully to obtain the sequence and linkage information [72]. The underivatized oligosaccharides are treated with a strong base. Alkaline degradation, also called the "peeling reaction," cleaves a glycosidic bond at the reducing end through β elimination to produce a new reducing end. The prevailing oligosaccharides are analyzed by MALDI-FTMS. The linkage information is obtained from the cross-ring fragmentation from the new reducing end.

10.5.6. Collision-Induced Dissociation for Structural Analysis of Carbohydrates

As with other classes of biomolecules, CID can also provide an effective approach to obtain the sequence information either directly from glycopeptides and glycolipids or from carbohydrate side chains after their release from these glycoconjugates [73–81]. CID also provides interresidue linkage and branching details. The sequencing of glycans from intact glycopeptides and glycolipids is feasible because the glycosidic linkages are preferentially cleaved on CID. The target analyte is methylated to increase fragmentation. The molecular ions are generated by FAB or ESI; the latter is a preferred mode of ionization as a consequence of its increased sensitivity, less matrix interference, and choice of several multiply charged precursors. Furthermore, multiply charged ions have a higher probability of fragmentation. Low-energy and high-energy CID have both been used; the latter provides increased fragmentation [73]. Several different types of precursors, such as $[M + H]^+$, $[M - H]^-$, $[M + Li]^+$, $[M + Na]^+$, and $[M + 2Li - H]^+$, [52–54,59,73] have been chosen for the structural analysis. The $[M + H]^+$ ions provide less sensitivity, whereas the multiplicity (e.g., $[M + Li]^+$ or $[M + 2Li - H]^+$) of metal ion adducts may complicate the situation. Therefore, the attachment of a high-proton affinity site to the reducing terminal has been a successful strategy prior to CID of glycans.

Nomenclature of Fragment Ions. A scheme for the nomenclature of fragment ions has been proposed [81]. According to this scheme, shown in Figure 10.7, fragments that retain the charge on the nonreducing terminus (drawn on the left) are termed A, B, and C. The fragments that retain the charge on the reducing end (i.e., from the peptide or lipid end) are designated by the letters X, Y, and Z. The fragmentation within a carbohydrate ring is denoted by superscripts that precede the letter symbols. As an example, $^{2,5}A_3$ indicates the cleavage of bonds 2 and 5 in ring 3 and the charge retention by the nonreducing end. Branching is indicated by adding the subscripts α, β, and so on to the letter symbols.

CID of the Metal Ion Adducts. CID of the metal ion adducts have been used to distinguish linkage positions in oligosaccharides. These applications have included

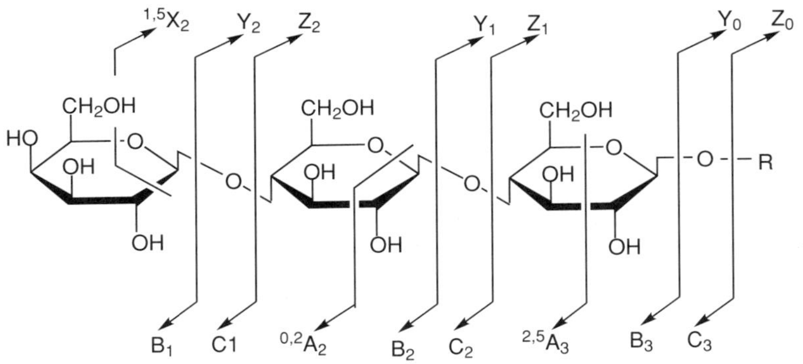

Figure 10.7. A scheme for the nomenclature of the fragment ions that are formed from oligosaccharides.

cationization with alkali [52,59,73,82], alkaline earth [59], and transition metals [84]. FAB- and ESI-produced cationized adducts have both been studied.

A key development in this field is the use of multistage MS/MS (MS^n) experiments that can be readily conducted in a quadrupole ion-trap (QIT) mass spectrometer. Asam and Glish studied the behavior of lithiated and sodiated adducts of several di-, tri-, and tetrasaccharides [84]. In a single-stage MS/MS experiment, some structural features in glycans, such as N-acetyllactosamine antennas, neuraminic acids, nonreducing terminal GlcNAc monosaccharides, usually suppress cross-ring and core saccharide cleavages. Once these structural features are removed, determination of branching patterns and intersaccharide linkages becomes an easier task. These experiments can be conducted in a QIT via higher-order MS/MS reactions. König and Leary have demonstrated that the linkage position can be determined in cobalt-coordinated pentasaccharides with this procedure [85]. The cobalt adducts of four isomeric lacto-N-fucopentaoses were generated with ESI, and the MS^2 and MS^3 spectra of the mass-selected C-type ions were acquired. The C-type ions participate in specific fragmentation reactions that allows differentiation among $1 \rightarrow 2$, $1 \rightarrow 3$, $1 \rightarrow 4$, and $1 \rightarrow 6$ linkage positions. In another study, nanoESI was combined with QIT to study multistage MS/MS reactions of oligosaccharides [86]. With just 1 µL of the sample, up to MS^7 experiments could be conducted. NanoESI affords the high detection sensitivity of metal ion adducts; the detection sensitivity in the fmol range is readily achieved.

CID of the Derivatized Oligosaccharides. NanoESI with low-energy CID on a quadrupole-TOF instrument was used by Mo et al. to structurally characterize 4-aminobenzoic acid 2-(diethylamino)ethyl ester (ABDEAE) and 2-aminopyridine derivatives of oligosaccharides that were released from several glycoproteins [87]. The ABDEAE derivatives provide an order of magnitude more sensitivity than the 2-

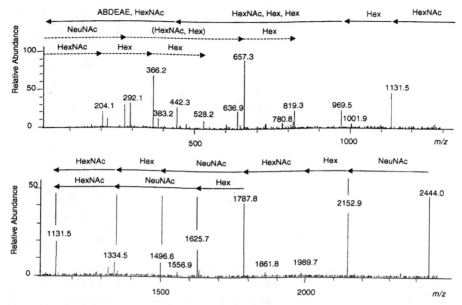

Figure 10.8. The low-energy CID spectrum of the 4-aminobenzoic acid 2-(diethylamino)ethyl ester derivative of NeuNAc$_2$Gal$_2$GlcNAc$_2$Man$_2$GlcNAc$_2$ from transferrin. [Reproduced from Ref. 87 by permission of American Chemical Society, Washington, DC (copyright 1999).]

aminopyridine derivatives. The multiply charged ions of these derivatives undergo fragmentation to yield the y and oxonium ions (b or internal series). The low-energy CID spectrum of the ABDEAE derivative of NeuNAc$_2$Gal$_2$GlcNAc$_2$Man$_3$GlcNAc$_2$ from transferrin is shown in Figure 10.8. In contrast to the PSD data, more structural information is obtained in this spectrum. Weiskopf et al. used multistage MS/MS strategy for the structural analysis of complex N-linked glycoprotein oligosaccharides [88]. In most instances, the MS^2 and MS^3 experiments provide the needed information on branching patterns and intersaccharides linkage. For oligosaccharides with a large number of labile saccharide units, higher-order MS^n analysis becomes a necessity. An example is the analysis of a GlcNAc$_8$Man$_3$ sugar from chicken ovalbumin. The MS^2 spectrum of the doubly sodiated adduct of this molecule is shown in Figure 10.9a. In this spectrum, the fragmentation of the mannose core is suppressed in favor of the facile loss of HexNAc from the oligomer. However, the sequential loss of one GlcNAc residue in each individual stage of MS^n produces an ion species GlcNAc$_2$Man$_3$(OH)$_6$, the MS^8 of which reveals the branching patterns of the GlcNAc residues that were previously attached to the oligomer (Figure 10.9b). This research group had used the same approach to the analysis of permethylated derivatives of maltoheptaose earlier [89]. Other researchers have also utilized the multistage MS^n facility of QITMS to characterize oligosaccharides [90]. FTMS has

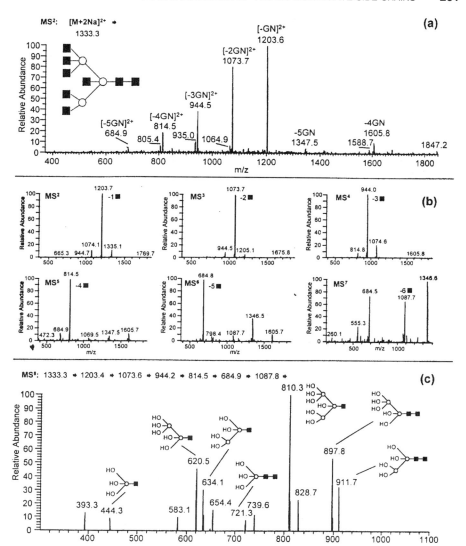

Figure 10.9. (*a*) The MS^2 spectrum of the doubly sodiated adduct of GlcNAc$_8$Man$_3$ and (*b*) the MS^8 of the same sugar reveals the branching patterns of the GlcNAc. [Reproduced from Ref. 88 by permission of American Chemical Society, Washington, DC (copyright 1998).]

also been used to conduct multistage tandem mass spectrometry experiments to sequence permethylated oligosaccharides [91]. A similar conclusion—multistage experiments provide more structural information than does the single stage MS/MS—was reached in this instrumental configuration.

Leary's group has developed a saccharide topology analysis tool (STAT), which is a web-based computational program that can quickly extract sequence information from the MS^n data [92]. This tool requires information such as precursor ion mass, probable monosaccharide residues, charge carrier, and product ion mass.

10.5.7. Post Source Decay for Structural Analysis of Carbohydrates

Post source decay has also emerged as a viable technique for characterization of carbohydrate side chains in glycoproteins [67,93]. A practical example is the analysis of released glycans from ovalbumin and related glycoproteins [67]. Some useful structural information can be derived from internal glycoside cleavages and cross-ring fragments that originate from the reducing terminal GlcNAc.

REFERENCES

1. A. Varki, *Glycobiology*, **3**, 97-130 (1993).
2. R. Kornfeld and S. Kornfeld, *Ann. Rev. Biochem.* **54**, 631 (1985).
3. P. H. Petra, P. R. Griffins, J. R. Yates, III, K. Moore, and W. Zhang, *Protein Sci.* **1**, 902 (1992).
4. S. A. Carr, J. R. Barr, G. D. Roberts, K. R. Anumula, and P. B. Taylor, in J. A. McCloskey, ed., *Methods in Enzymology*, Academic Press, San Diego, 1990, Vol. 193, pp. 501–518.
5. R. Orlando and Y. Young, in B. S. Larsen and C. N. McEwen, eds., *Mass Spectrometry of Biological Materials*, Marcel Dekker, New York, 1998, pp. 215–245.
6. V. N. Reinhold, B. B. Reinhold, and S. Chan, in T. Matsuo, R. M. Caprioli, M. L. Gross, and Y. Seyama, eds., *Biological Mass Spectrometry, Present and Future*, Wiley, New York, 1994, pp. 403–435.
7. A. Dell, in J. A. McCloskey, ed., *Methods in Enzymology*, Academic Press, San Diego, 1990, Vol. 193, pp. 647–660.
8. H. R. Morris, M. R. Thompson, D. T. Osiga, A. I. Ahmed, S. M. Chan, J. R. Vandenheede, and R. E. Feeney, *J. Biol. Chem.* **253**, 5155 (1978).
9. M. Linscheid, J. D'Angona, A. L. Burlingame, A. Dell, and C. E. Ballou, *Proc. Natl. Acad. Sci.* (USA) **78**, 1471 (1981).
10. R. R. Townsend, M. Alai, and C. Fenselau, *Anal. Biochem.* **171**, 180–191 (1988).
11. R. Cotter, L. Chen, and R. Wang, in D. M. Desiderio, ed., *Mass Spectrometry of Peptides*, CRC Press, Boca Raton, FL, 1991, pp. 17–40.
12. M. Barber, R. S. Bordoli, R. D. Sedgwick, and A. N. Tyler. *J. Chem. Soc. Chem. Commun.* 325–327 (1981).
13. J. B. Fenn, M. Mann, C. K. Meng, S. F. Wong, and C. M. Whitehouse, *Science* **246**, 64–71 (1989).
14. R. D. Smith, J. A. Loo, R. R. Ogorzalek Loo, M. Busman, and H. R. Udseth, *Mass Spectrom. Rev.* **10**, 359–451 (1991).
15. M. Karas and F. Hillenkamp, *Anal. Chem.* **60**, 2299–2301 (1988).
16. R. S. Rush, P. L. Derby, D. M. Smith, C. Merry, G. Rogers, M. F. Rohde, and V. Katta, *Anal. Chem.* **67**, 142–1452 (1995).
17. S. A. Carr and G. D. Roberts, *Anal. Biochem.* **157**, 396 (1986).

18. G. J. Rademaker and J. Thomas-Oates, in J. R. Chapman, ed., *Protein and Peptide Analysis by Mass Spectrometry*, Humana Press, Totowa, NJ, 1996, pp. 231–241.

19. C. A. Settineri and A. L. Burlingame, in J. R. Chapman, ed., *Protein and Peptide Analysis by Mass Spectrometry*, Humana Press, Totowa, NJ, 1996, pp. 255–278.

20. A. W. Guzzetta and W. S. Hancock, in W. S. Hancock, ed., *New Methods in Peptide Mapping for the Characterization of Proteins*, CRC Press, Boca Raton, FL, 1996, pp. 181–217.

21. A. W. Guzzetta, L. J. Basa, W. S. Hancock, B. A. Keyt, and W. F. Bennett, *Anal. Chem.* **65**, 2953–2962 (1993).

22. K. F. Medzihradszky, D. A. Maltby, S. C. Hall, C. A. Settineri, and A. L. Burlingame, *J. Am. Soc. Mass Spectrom.* **5**, 350–358 (1994).

23. V. Ling, A. W. Guzzetta, E. Canova-Davis, J. T. Stults, T. R. Covey, B. I. Shushan, and W. S. Hancock, *Anal. Chem.* **63**, 2909–2915 (1991).

24. S. A. Carr, M. J. Huddleston, and M. F. Bean, *Prot. Sci.* **2**, 183–196 (1993).

25. M. J. Huddleston, M. F. Bean, and S. A. Carr, *Anal. Chem.* **65**, 877–884 (1993).

26. P. A. Scindler, C. A. Settineri, X. Collet, C. J. Fielding, and A. L. Burlingame, *Prot. Sci.* **4**, 791–803 (1995).

27. E. Mirgorodskya, P. Roepstroff, and R. A. Zubarev, *Anal. Chem.* **71**, 4431–4436 (1999).

28. R. A. Zubarev, R. A. Kelleher, and F. W. McLafferty, *J. Am. Chem. Soc.* **120**, 3265–3266 (1998).

29. R. A. Zubarev, N. A. Kruger, E. K. Fridriksson, M. A. Lewis, D. M. Horn, B. K. Carpenter, and F. W. McLafferty, *J. Am. Chem. Soc.* **121**, 2857–2862 (1999).

30. D. J. Harvey, D. R. Wing, B. Küster, and I. B. H. Wilson, *J. Am. Soc. Mass Spectrom.* **11**, 564–571 (2000).

31. R. Cramer, W. J. Richter, E. Stimson, and A. L. Burlingame, *Anal. Chem.* **70**, 4939–4944 (1998).

32. D. H. Patterson, G. E. Tarr, F. E. Regnier, and S. A. Martin, *Anal. Chem.* **67**, 3971–3978 (1995).

33. S. Goletz, B. Thiede, F. G. Hanisch, M. Schultz, J. Peter-Katalinic, J. Muller, O. Seitz, and U. Karsten, *Glycobiology* **7**, 881–896 (1997).

34. K. F. Medzihradszky, B. L. Gillece-Castro, R. R. Townsend, A. L. Burlingame, and M. R. Hardy, *J. Am. Soc. Mass Spectrom.* **7**, 319–328 (1996).

35. M. J. Kieliszewski, M. O'Neil, J. Leykam, and R. Orlando, *J. Biol. Chem.* **270**, 2541–2549 (1995).

36. F.-G. Hanisch, B. N. Green, R. H. Bateman, and J. Peter-Katalinic, *J. Mass Spectrom.* **33**, 358–362 (1998).

37. G. J. Rademaker, S. A. Pergantis, L. Bloktip, J. I. Langridge, A. Kleen, and J. Thomas-Oates, *Anal. Biochem.* **257**, 149–160 (1998).

38. J. J. Canboy and J. D. Henion, *J. Am. Soc. Mass Spectrom.* **3**, 804–814 (1991).

39. K. Hirayama, R. Yuji, N. Yamada, K. Kato, Y. Arato, and I. Shimada, *Anal. Chem.* **70**, 2718–2725 (1998).

40. D. Lewis, A. W. Guzzetta, W. S. Hancock, and M. Costello, *Anal. Chem.* **66**, 585–595 (1994).

41. N. Viseux, E. de Hoffmann, and B. Domon, *Anal. Chem.* **69**, 3193–3198 (1997).

42. S. Honda, A. Makino, S. Suzuki, and K. Kakehi, *Anal. Biochem.* **191**, 228–234 (1990).

43. P. Camilleri, G. B. Harland, and G. N. Okafo, *Anal. Biochem.* **230**, 115–122 (1995).

44. G. N. Okafo, J. Langridge, S. North, A. Organ, A. West, M. Morris, and P. Camilleri, *Anal. Chem.* **69**, 4985–4993 (1997).

45. G. N. Okafo, L. M. Burrow, W. Neville, A. Truneh, R. A. G. Smith, and P. Camilleri, *Anal. Chem.* **68**, 4424–4430 (1996).

46. H. C. Birrell, J. Charlwood, I. Lynch, S. North, and P. Camilleri, *Anal. Chem.* **71**, 102–108 (1999).

47. C. G. Hellerqvist, in J. A. McCloskey, ed., *Methods in Enzymology*, Academic Press, San Diego, 1990, Vol. 193, pp 554–573.

48. G. R. Gray, in J. A. McCloskey, ed., *Methods in Enzymology*, Academic Press, San Diego, 1990, Vol. 193, pp 573–587.

49. A. Angel and B. Nilsson, in J. A. McCloskey, ed., *Methods in Enzymology*, Academic Press, San Diego, 1990, Vol. 193, pp. 587–607.

50. I. Fangmark, A. Jansson, and B. Nilsson, *Anal. Chem.* **71**, 1105–1110 (1999).

51. I. Jardine, G. Scanlan, M. McNeil, and P. J. Brennan, *Anal. Chem.* **61**, 416–422 (1989).

52. G. E. Hofmeister, Z. Zhou, and J. A. Leary, *J. Am. Chem. Soc.* **113**, 5964–5970 (1991).

53. R. A. Laine, E. Yoon, T. J. Mahier, S. Abbas, B. de Lappe, R. K. Jain, and K. L. Matta, *Biol. Mass Spectrom.* **20**, 505–514 (1991).

54. B. Domon, D. R. Müller, and W. J. Richter, *Biomed. Environ. Mass Spectrom.* **19**, 390–392 (1990).

55. B. Domon, D. R. Müller, and W. J. Richter, *Org. Mass Spectrom.* **29**, 713–718 (1994).

56. J. W. Dallinga and W. Heerma, *Biol. Mass Spectrom.* **20**, 215 (1991).

57. D. T. Li and G. R. Her, *Anal. Biochem.* **211**, 250–257 (1993).

58. B. Mulroney, J. C. Traeger, and B. A. Stone, *J. Mass Spectrom.* **30**, 127 (1995).

59. A. Fura and J. A. Leary, *Anal. Chem.* **65**, 2805–2811 (1993).

60. D. T. Li and G. R. Her, *J. Mass Spectrom.* **33**, 644–652 (1998);

61. D. T. Li, J. F. Sheen, and G. R. Her, *J. Am. Soc. Mass Spectrom.* **11**, 292–300 (2000).

62. M. Kohler and J. A. Leary, *Anal. Chem.* **67**, 3501–3508 (1995).

63. D. J. Harvey, *J. Chromatogr. A* **720**, 429–446 (1996).

64. L. Poulter and A. L. Burlingame, in J. A. McCloskey, ed., *Methods in Enzymology*, Academic Press, San Diego, 1990, Vol. 193, pp. 661–689.

65. J. Charlwood, J. Langridge, D. Tolson, H. Birrell, and P. Camilleri, *Rapid Commun. Mass Spectrom.* **13**, 107–112 (1999).

66. J. Charlwood, J. Langridge, and P. Camilleri, *Rapid Commun. Mass Spectrom.* **13**, 1522–1530 (1999).

67. D. J. Harvey, *J. Am. Soc. Mass Spectrom.* **11**, 572–577 (2000).

68. D. J. Harvey, *Am. Lab.* **26**, 22–28 (1994).

69. Y. Yang and R. Orlando, *Anal. Chem.* **68**, 570–572 (1996).

70. B. Küster, T. J. P. Naven, and D. J. Harvey, *J. Mass Spectrom.* **31**, 1131–1140 (1996).

71. Y. Mechref and M. V. Novotny, *Anal. Chem.* **70**, 455–463 (1998).

72. M. T. Cancilla, S. G. Penn, and C. B. Lebrilla, *Anal. Chem.* **70**, 663–672 (1998).

73. J. Lemoine, B. Fournet, D. Despeyroux, K. R. Jennings, R. Rosenberg, and E. de Hoffman, *J. Am. Soc. Mass Spectrom.* **4**, 197–203 (1993).

74. N. Viseux, E. de Hoffman, and B. Domon, *Anal. Chem.* **69**, 3193–3198 (1997).

75. V. N. Reinhold, B. B. Reinhold, and C. E. Costello, *Anal. Chem.* **67**, 1772–1784 (1995).

76. P. H. Lipniunas, R. R. Reid Townsend, A. L. Burlingame, and O. Hindsgaul, *J. Am. Soc. Mass Spectrom.* **7**, 182–188 (1996).

77. D. J. Harvey, R. H. Bateman, and B. N. Green, *J. Mass Spectrom.* **32**, 168–187 (1996).

78. B. Küster, T. J. P. Naven, and D. J. Harvey, *Rapid Commun. Mass Spectrom.* **10**, 1645–1651 (1996).

79. M. T. Cancilla, S. G. Penn, and C. B. Lebrilla, *Anal. Chem.* **68**, 2331–2339 (1996).

80. L. P. Brüll, V. Kovácik, J. E. Thomas-Oates, W. Heerma, and J. Haverkamp, *Rapid Commun. Mass Spectrom.* **12**, 1520-1532 (1998).

81. C. E. Costello and J. E. Vath, in J. A. McCloskey, ed., *Methods in Enzymology*, Academic Press, San Diego, 1990, Vol. 193, pp. 738–768.

82. A. A. Staemfli, Z. Zhou, and J. A. Leary, *J. Org. Chem.* **57**, 3590–3594 (1992).

83. A. Fura and J. A. Leary, *Anal. Chem.* **65**, 2805–2811 (1996).

84. M. R. Asam and G. L. Glish, *J. Am. Soc. Mass Spectrom.* **8**, 987–995 (1997).

85. S. König and J. A. Leary, *J. Am. Soc. Mass Spectrom.* **9**, 1125–1134 (1998).

86. U. Bahr, A. Pfenninger, and M. Karas, *Anal. Chem.* **69**, 4530–4535 (1997).

87. W. Mo, H. Sakamoto, A. Nishikawa, N. Kagi, J. I. Langridge, Y. Shimonishi, and T. Takao, *Anal. Chem.* **71**, 4100–4106 (1999).

88. A. S. Weiskopf, P. Vouros, and D. J. Harvey, *Anal. Chem.* **70**, 4441–4447 (1998).

89. A. S. Weiskopf, P. Vouros, and D. J. Harvey, *Rapid Commun. Mass Spectrom.* **11**, 1493–1504 (1997).

90. D. M. Sheeley and V. N. Reinhold, *Anal. Chem.* **70**, 3053–3059 (1998).

91. T. Solouki, B. B. Reinhold, C. E. C. M. O'Malley, S. Guan, and A. G. Marshall, *Anal. Chem.* **70**, 857–864 (1998).

92. S. P. Gaucher, J. Morrow, and J. A. Leary, *Anal. Chem.* 2331–2336 (2000).

93. D. J. Harvey, T. J. P. Naven, B. Küster, R. H. Bateman, M. R. Green, and G. Critchley, *Rapid Commun. Mass Spectrom.* **9**, 1556–1561 (1995).

11

FOLDING–UNFOLDING AND HIGHER-ORDER STRUCTURES OF PROTEINS

The four levels of protein structures—primary, secondary, tertiary, and quaternary— were discussed in Chapter 8. In their native state, proteins orientate themselves into relatively stable, tightly folded, compact structures that contain surface pockets and interior cavities that are the sites for the specific binding and chemical modification of biological ligands. Three-dimensional structures of folded proteins are the result of a combination of several noncovalent interactions such as short-range repulsive forces, electrostatic forces, van der Waals interactions, hydrophobic interactions, and hydrogen bonding. The native state of a protein is of a significantly lower energy than other allowed compact states. The changes in a localized as well as in a globular structure are frequently used in nature to regulate the function of enzymes and receptors [1,2]. Often, these changes are very subtle, as is the case of the oxidized and reduced forms of cytochrome c [3,4]. Other proteins undergo large conformational changes. In order to fully understand the function of a protein, it is essential to identify these changes in terms of atomic locations, folding–unfolding dynamics, and thermodynamics.

In solution, proteins can be denatured and unfolded by increasing the temperature and/or pH, and by subjecting them to high concentrations of detergents, organic solvents, and certain chaotropic compounds such as urea and guanidinium chloride. In addition, the disulfide bridge–containing proteins are denatured by the action of the disulfide bond-breaking reagents (e.g., dithiothreitol). The temperature at which thermal unfolding occurs varies from protein to protein, and is a measure of the strength of hydrogen bonding and hydrophobic interactions. The effect of pH on denaturation is viewed as a global effect that is triggered by ionizing/charging moieties on the side chains. Coulombic repulsions among these charge centers destabilize the folded structure, leading to swelling and ultimately unfolding of the protein molecule.

11.1. CONVENTIONAL METHODS FOR STUDY OF CONFORMATIONAL CHANGES IN PROTEINS

Biochemists traditionally have employed a variety of biophysical techniques for study of structural changes in proteins. These methods include infrared (IR) [2,5,6] and ultraviolet (UV) absorption spectroscopy [2,5,7,8], tryptophan fluorescence [9], circular dichroism (CD) [2,5,10], X-ray crystallography [2,5], nuclear magnetic resonance (NMR) spectroscopy [2,5,11], viscometry [2,5], and neutron diffraction [12]. IR [6] and UV [7,8] spectroscopy techniques, in conjuction with hydrogen/deuterium (H/D) exchange, have been used to characterize fluctuations between different protein conformations. Although these techniques can quantify the incorporated deuterium levels of a whole protein, they fail to assess the deuterium incorporation in the specific regions of a protein. One- and two-dimensional NMR, combined with H/D exchange, have found a great success in the detection of changes in protein structures. The former, however, is restricted to small, highly soluble proteins [13]. Because two-dimensional NMR is a high-resolution technique, it is applicable to proteins with increased number of peptide linkages. However, the large-sized proteins are still outside the realm of NMR analysis. The technique has been successfully used in a number of situations that include studies of protein–ligand binding [14,15], folding–unfolding dynamics [11,16,17], mutants [18,19], and functional variants [20]. The response in CD is sensitive to the secondary structure of a protein. Therefore, the relative proportions of α-helical, β-sheet, and random-coil structures are easily assessed with this technique [21]. CD has gained popularity as a convenient technique for studying variations in conformational states in proteins that are induced as a result of changes in temperature, pH, salt, ligand binding, and quaternary structure. Although CD and fluorescence methods are both simple, they provide only an overview of the protein structure.

Mass spectrometry has also been incorporated in the protocol of the determination of the conformational changes in proteins [22,23]. This development was made possible with the advent of electrospray ionization (ESI) as a means of transforming dissolved proteins and peptide molecules into gas-phase ions [24,25]. A rationale for the use of ESIMS for probing conformational structures of proteins is that during the ESI process the solution-phase structure of the protein is largely preserved. Therefore, the ESI mass spectra of a protein can be considered a reflection of its aqueous solution chemistry [26–29]. The multiple-charging feature of ESI is also a valuable asset because it allows the study of much larger proteins. In conjunction with H/D exchange, ESIMS has gained prominence as a forefront technique for studies concerned with the conformational structures of protein.

Smith et al. have discussed the merits of mass spectrometry over NMR in protein conformation studies [22,23]. Mass spectrometry offers several distinct advantages in terms of sensitivity, preserving protein stability, and extended mass range, and provides information complementary to NMR. The ESIMS approach is at least three orders of magnitude more sensitive than NMR. Also, the solubility and purity of proteins are of less concern in the mass spectrometry approach. NMR is limited to proteins with a molecular mass $< 30,000$ Da, whereas mass spectrometry can be used to study much higher molecular mass and complex multimeric, multidomain proteins (1,000,000 Da). Another advantage of mass spectrometry over NMR is the timescale. With the combined ESIMS H/D exchange procedure, the exchange rates of the most rapidly exchanging amide hydrogens can be determined. The NMR approach, on the other hand, is superior with respect to resolution, and also the sample is not destroyed.

11.2. CHARGE-STATE DISTRIBUTION AS A MEASURE OF CONFORMATION OF PROTEINS

The first attempt to use mass spectrometry to probe conformational states of proteins was to measure differences in the charge-state distribution of a protein in its ESI mass spectrum [30]. It has been argued that the charge-state distribution is a reflection of the differences in the folding of the protein molecule. The basis of this argument is that multiple charging of proteins in the positive-ion ESI spectra occurs as a result of protonation of the available basic sites in its structure, and availability of these sites is determined by the protein's conformational state. The protein in a folded (native) conformation will have fewer basic sites that are accessible to the solution for protonation, whereas these basic sites are fully exposed in the unfolded (denatured) protein. As stated above, the native state of a protein can be disrupted by a change in the solution pH and temperature, the presence of denaturing agents, and the cleavage of the disulfide linkages. Therefore, the ESI mass spectra of proteins, when observed under different pH and temperature conditions, will exhibit a different charge-state distribution. Chowdhary et al. were the first to apply this concept. They demonstrated that the changes in the ESI mass spectrum of cytochrome c can be correlated to fluctuations in its conformation that are induced by pH of the solution [30]. The ESI mass spectral profiles of ubiquitin and hen egg lysozyme (HEL) differ when those proteins are denatured in the presence of an excess of an organic solvent [31]. Several studies have shown that the changes with temperature in the charge distribution in the ESI mass spectra of proteins can be attributed to denaturation of proteins in solution [32–35]. Figure 11.1 shows the effect of denaturation on reduction of the disulfide linkages on the ESI charge distribution profile for HEL [36]. The spectrum (Figure 11.1a) of native HEL exhibits the most abundant charge state of 9+, and a maximum charge state of 11+. In contrast, the ion profile of the denatured (i.e., DTT-reduced, 4-pyridylated) HEL

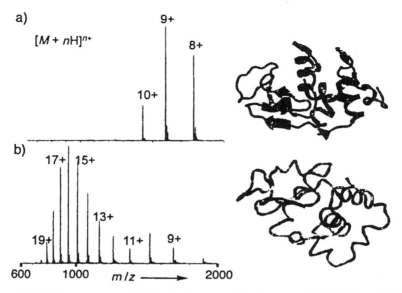

Figure 11.1. ESI mass spectra of (*a*) native hen egg-white lysozyme (14,307 Da) and (*b*) DTT-reduced, 4-pyridylethylated lysozyme (15,154). [Reproduced from Ref. 36 by permission of VCH Verlagagesellschaft, Weinheim, Germany (copyright 1996).]

shifts to higher charge states. Several other proteins have exhibited similar shifts to higher charge states on reduction of disulfide bridges [25,27,37,38].

11.3. REACTIVITY OF SOME FUNCTIONAL GROUPS AS A MEASURE OF CONFORMATION OF PROTEINS

The conformation of a protein also has an influence on the reactivity of some of its functional groups. As an example, the NH_2 groups on some of the Lys side chains in β-lactoglobulin are unreactive in its native conformation toward 2,4-dinitrofluoro-benzene, but the reactivity is restored once the protein is unfolded. This concept has been utilized to probe the conformational states of proteins. By comparing the ESI mass spectra of proteins after their selective chemical modifications, the differences in their tertiary structure can be determined [36,39]. The selective derivatization of Arg residues in HEL by 1,2-cyclohexanedione shows that the molecular ion profile in the ESI mass spectrum is identical to that of the native protein until four specific Arg residues are derivatized (Figure 11.2*a*) [36,39]. By increasing the time of reaction, a higher extent of modification is obtained. The ESI spectrum of this protein derivative shows additional higher charge-state ions (Figure 11.2*b*), suggesting a gradual unfolding in HEL.

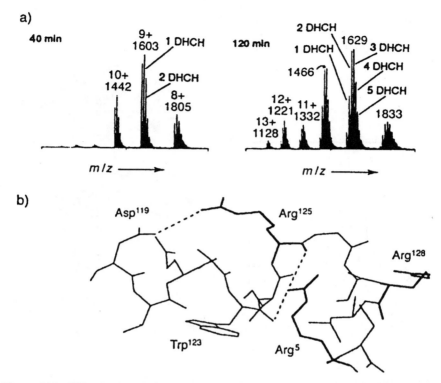

Figure 11.2. ESI mass spectra of (*a*) DHCH–Arg-modified hen egg-white lysozyme and (*b*) the tertiary structure of lysozyme around the Arg5 and Arg125 residues. [Reproduced from Ref. 36 by permission of VCH Verlagagesellschaft, Weinheim, Germany (copyright 1996).]

11.4. HYDROGEN EXCHANGE AS A PROBE OF CONFORMATIONAL CHANGES IN PROTEINS

Although differences in the charge profile of molecular ions are related to the conformational changes in a protein's tertiary structure, they cannot be a reliable measure of the protein's conformation because of the fact that the ESI ion profile may change with changes in certain experimental variables in the ESI ion source, such as gas flows, voltage settings, and unavoidable altered solution conditions (slight variations in pH, the presence of counterions, surface tension, rate of desolvation and declustering, protein concentration, etc.). A more precise method for gauging the conformational state is *isotopic hydrogen exchange*. This term refers to the replacement of the labile hydrogens in proteins with the hydrogens of different isotopic composition (deuterium or tritium) in solvent water. In combination with NMR, this approach has gained prominence as a standard method for study of the higher-order structures, structural changes, and structural dynamics of proteins [11,40,41]. Katta and Chait were the first to combine H/D exchange with mass spectrometry to probe

the conformational changes in proteins [42,43]. They demonstrated that on denaturation of bovine ubiquitin, not only is the ion profile shifted to a higher charge state,
but the incorporation of deuterium is also higher in the unfolded structure (see Figure
11.3) [43]. The basic premise of this approach is that more hydrogens are available
for H/D exchange in the unfolded conformer than in the native structure, where they
are not exposed to the solvent. Since then, several researchers have used this protocol
to probe changes in protein structures [23, 42–47].

In Katta and Chait's experiment [43], the H/D exchange was performed for the
native and unfolded bovine ubiquitin in different solvent environments. Wagner and
Anderegg have argued that such measurements must be performed in the same
chemical environment to ensure that the exchange rates are not affected due to

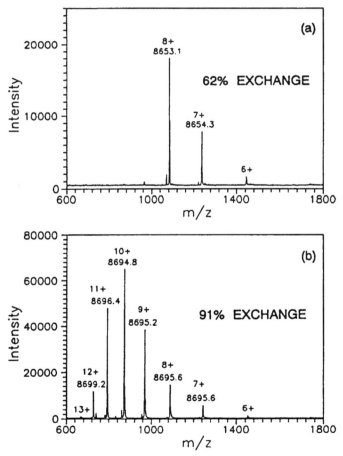

Figure 11.3. ESI mass spectra of bovine ubiquitin (a) 20 min after dissolving the protein in 1%
CH_3COOD in D_2O, (b) 23 min after dissolving the protein in CH_3OD:1% CH_3COOD in D_2O.
[Reproduced from Ref. 43 by permission of the American Chemical Society, Washington, D.C.
(copyright 1993).]

changes in a chemical environment [45]. In their experiment, the ESI spectrum of bovine cytochrome c at pH 6.0 mainly produces a charge-state distribution in the range of 7+ to 11+ (Figure 11.4a). The spectrum at pH 2.7 exhibits two distinct charge-state distributions; one is identical to that obtained at pH 6.0, and the other ranges between 11+ and 18+ (Figure 11.4b). This observation is a clear indication that two conformers are present in this solution. The lower charge-state distribution represents a tightly folded structure, and the higher charge-state distribution indicates the relaxed conformation. The spectrum at pH 2.5 is dominated by a

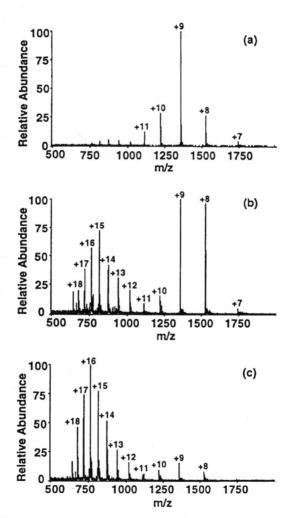

Figure 11.4. ESI mass spectra of bovine cytochrome c in H_2O (a) with no acetic acid, pH 6.0; (b) with 2% acetic acid, pH 2.7 (bimodal distribution is shown); (c) with 4% acetic acid, pH 2.5. [Reproduced from Ref. 45 by permission of the American Chemical Society, Washington, D.C. (copyright 1994).]

higher charge-state component, and the lower charge-state distribution is less abundant (Figure 11.4*c*).

11.5. AMIDE HYDROGEN EXCHANGE RATE AS A PROBE OF THE PROTEIN STRUCTURES

Determination of the amide hydrogen exchange rate in proteins has become a prominent technique for study of their conformational structures. This approach is based on the premise that the exchange rate of amide hydrogens is sensitive to the conformational states of proteins; the faster exchange is indicative of a more open structure, and the slower exchange is related to a tightly folded, compact state [48,49]. In the native conformer, only the surface hydrogens exchange at a faster rate, whereas the inner-core hydrogens exchange much more slowly, and often have no detectable exchange even after several days. The solvent inaccessibility and hydrogen bonding in the compact structure of proteins are the factors that have been implicated for slow exchange rates. In addition to a protein's structure, the isotopic exchange rate is also affected by the experimental variables such as pH and temperature. An additional factor that has a bearing on the exchange rate is the neighboring (e.g., the inductive and steric) effect of adjacent amino acid residues. Thus, considerable structural information can be gleaned by determining the hydrogen exchange rates of different amide linkages in proteins [48,49]

Hydrogen exchange in proteins is catalyzed by H^+ and OH^- ions. The rate constant for hydrogen exchange (k_{ex}) is the sum of the rate constants for acid- (k_H) and base-catalyzed (k_{OH}) reactions, as indicated in the following equation:

$$k_{ex} = k_H[H^+] + k_{OH}[OH^-] \tag{11.1}$$

In the case of polyalanine, the values of k_H and k_{OH} have been found to be 41.7 and $1.12 \times 10^{10}\,M^{-1}\,min^{-1}$, respectively, at 20°C [50]. The shape of the curve that illustrates the effect of pH on the exchange rate is given in Figure 11.5, which indicates that the slowest rate occurs close to pH 3. At more basic pH values, the exchange rate is affected by OH^- ion activity, and at more acid pH, by H^+ ion activity. The exchange rate changes by a factor of 10 for each unit change of pH. The opening of the secondary and tertiary structures of a protein may increase the amide hydrogen exchange rate by as much as 10^8. This large change in the exchange rate forms the basis of the use of amide hydrogen exchange as a sensitive probe to study the conformational changes in proteins. The side chain labile hydrogens (e.g., on Ser, Thr, Tyr, Glu, Arg, Lys, Asp, Gln, and Asn), as well as on the N and C terminii, exchange at a very fast rate on the experimental timescale; therefore, these hydrogens are not useful for conformational studies.

11.5.1. Conformational Dynamics of Proteins

In solution, native proteins are continuously fluctuating around the average conformation, which is represented by a rigid three-dimensional crystal structure. This

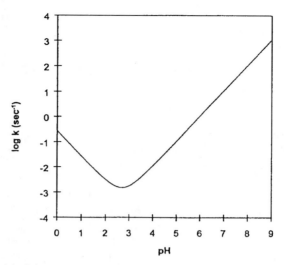

Figure 11.5. A plot of the rate constant for isotopic exchange of hydrogen located on peptide amide linkages in polyalanine as a function of pH. Results were calculated using Eq. 11.1. [Reproduced from Ref. 22 by permission of John Wiley & Sons, Ltd. (copyright 1997).]

dynamic change is the reason why even inaccessible hydrogens undergo isotope exchange. Various kinetic models have been developed to describe the amide hydrogen exchange in proteins [49,51–54]. The experimental data can be correlated to one of the two rate-limiting processes; one involves the fluctuations of the folded (F) state (Eq. 11.2), and the second corresponds to the global unfolding (Eq. 11.3). Equation 11.2 describes the folded-state mechanism, in which hydrogen exchange takes place directly from the folded protein. This mechanism is the representation of the internal motions of the native protein, and will occur only for those amide hydrogens that are accessible to the deuterated aqueous medium and OD⁻ catalyst. The amide hydrogens that are involved in hydrogen bonding in the folded structure will not participate in the exchange process. The experimentally observed exchange rate constant for a peptide amide linkage by this process is given by Eq. 11.4, where β is the probability of the contact of D_2O and OD^- with the site of hydrogen exchange, and k_2 is the exchange rate of amide hydrogens for the completely unfolded (U) protein. The value of β is equal to 1 for those peptides in which all amide hydrogens are exposed to solvents, and also to the completely unfolded proteins.

$$F(H) \xrightarrow{k_f} F(D) \tag{11.2}$$

$$F(H) \underset{k_1}{\overset{k_1}{\rightleftarrows}} U(H) \underset{k_2}{\overset{k_2}{\rightleftarrows}} U(D) \overset{k_1}{\underset{}{\rightleftarrows}} F(D) \tag{11.3}$$

$$k_{ex} = \beta k_2 \tag{11.4}$$

$$k_{ex} = \frac{k_2 k_1}{k_1 + k_2} \tag{11.5}$$

The global unfolding process is described by Eq. 11.3. It involves a reversible folding and unfolding of small regions of the protein, as well as of the entire protein. Isotopic exchange takes place only after the protein has undergone a complete unfolding. The rate constant (k_{ex}) of the overall process is given by Eq. 11.5, where k_1, and k_{-1} are the rates of the protein unfolding and refolding processes, respectively, and k_2 as described above.

At a neutral pH and in the absence of denaturants, the interconversion between the folded and unfolded structures is much faster than the isotopic exchange rate (i.e., $k_{-1} \gg k_2$). Under these conditions, the opening and refolding of a protein will occur many times before any isotopic exchange takes place. This exchange is referred to as an "EX2 mechanism" or an *uncorrelated exchange* [53], and the exchange rate is given by Eq. 11.6 (where K is the equilibrium constant that describes the unfolding process). Under denaturing conditions, the rate constant for the conversion of the unfolded to the folded state is much slower compared to the isotopic exchange rate (i.e., $k_{-1} \ll k_2$). Under these conditions, the exchange mechanism is referred to as an "EX1" or a *correlated exchange*, and the measured value of k_{ex} may be directly used to determine the rate constant of the unfolding process (Eq. 11.7). The correlated and uncorrelated processes can be distinguished from the characteristic mass spectrometry data:

$$k_{ex} = \frac{k_2 k_1}{k_{-1}} = K k_2 \qquad (11.6)$$

$$k_{ex} = k_1 \qquad (11.7)$$

The amide hydrogen exchange parameters of primary concern are the experimental exchange rate constant, k_{ex} and the protection factor, P, which is the ratio of the experimental (k_{ex}) and the calculated (k_2) values of the rate constants (i.e., k_2/k_{ex}). The value of the folding–unfolding equilibrium constant, K, can be estimated from the k_{ex} and k_2 values by using Eq. 11.6, and K, in turn, can provide an estimate of ΔG for the protein unfolding process.

11.5.2. Mass Spectrometry Measurements of Amide Hydrogen Exchange Rates

Several different mass spectrometry approaches are used to determine the amide hydrogen exchange rates. Continuous labeling and pulsed labeling are two common approaches; both can be used to study global, as well as localized structural changes.

Hydrogen Exchange for Detection of Global Changes in Proteins. In order to detect global changes, hydrogen exchange is performed by ESIMS on an intact protein. This experiment will provide the exchange rate averaged over all amide hydrogens in a protein. The unfolding (k_1) and folding (k_{-1}) exchange kinetics can both be studied. In a typical ESIMS procedure for the measurement of the rate constant for the folding process, the protein is dissolved in D_2O buffered to the desired pD [55]. The solution is heated to a higher temperature ($\geq 75°C$) to ensure that the protein is completely unfolded. For proteins that cannot withstand the extreme of higher temperatures, the unfolding is achieved alternatively by lowering the pH to 2.0, or by adding a suitable denaturing agent. Because of the unfolding of the protein, all amide hydrogens can be deuterated. The fully deuterated protein is lyophilized, and redissolved in D_2O. The exchange is initiated by diluting (100-fold) the concentrated D_2O solution of the protein with H_2O that has been adjusted to the desired pH. The samples are withdrawn at several timepoints, and are analyzed by ESIMS.

In order to study the kinetics of the unfolding process, the lyophilized folded protein is dissolved in D_2O at a pH where folded and unfolded states both coexist. The samples are withdrawn at different time intervals, and are analyzed by ESIMS. A particular charge state of the folded protein is selected, and deuterium incorporation is estimated from these measurements by the increase in mass of this peak. The intensity of the deuterated peak is the measure of the amount of the unfolded state. The change in abundance of the native (F) and deuterated peaks (U) is monitored at different time periods. The plot of either a simple U term or the expression $F/(F + U)$ versus the time of the reaction can provide the estimate of the rate constant (k_1) for the unfolding process.

Hydrogen exchange kinetics can be measured with a different approach. In this case, the protein is first dissolved in a H_2O solution that was adjusted to an appropriate pH, and the labeling is achieved by diluting the protein with an excess of a D_2O solution of the same pH. This procedure is suitable for proteins that are difficult to deuterate, but it requires that all traces of H_2O be removed from the ES interface, and only dry nitrogen be used during the ESI measurements.

Hydrogen exchange in the *pulsed-labeling* technique is studied in a quench-flow apparatus [55,56]. This approach is well suited to determination of the isotopic exchange rates of rapidly exchanging amid hydrogens. The mass spectrometry-based pulsed-labeling approach has been modeled on the pattern of NMR studies of a similar nature [57–59]. As with the continuous-labeling procedure, the protein is first dissolved in D_2O in which a strong denaturant is present. The unfolded protein is injected into a quench–flow system similar to that shown in Figure 11.6, and the folding is initiated by mixing (at mixing a tee marked T1) the injected protein solution with a large excess of H_2O solution that is buffered to the pH of 5–6. The length of the folding time (milliseconds to seconds) in the quench–flow apparatus can be controlled by varying the size of the folding tube. After this folding period, an exchange pulse is applied by mixing (at T2) the protein solution with a basic pH (> 7.0) H_2O solution. Because the rate of hydrogen exchange is very fast above pH 7, the exposed hydrogens will be exchanged instantaneously. The sites where deuteriums are retained represent the folded structure. The next step is to quench the labeling. As mentioned earlier in this chapter, a decrease in each pH unit lowers the exchange rate by a factor of 10, and a decrease in the exchange temperature from 20°C to 0°C further lowers the exchange rate 10-fold. Therefore, adjusting the pH of the solution to 2.5 and temperature to 0°C will effectively quench the labeling. A solution of the quench buffer is introduced at a mixing tee marked T3. The products

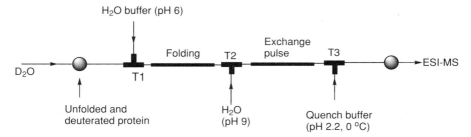

Figure 11.6. A schematic diagram of a quench–flow apparatus that can be used in the pulsed-labeling technique.

of this experiment are introduced directly into the ESI source. The reverse experiment in which the protein is dissolved in H_2O solution of a denaturant, and subjected to a short exchange pulse with an excess of D_2O solution, can also be performed in a similar manner.

Determination of Hydrogen Exchange in Short Segments of Proteins

Although hydrogen exchange measurements by ESIMS on an intact protein allow the global changes to be assessed readily, resolution of the method needed to identify the localized structural changes during hydrogen exchange is limited. In order to overcome this limitation, the *fragmentation–mass spectrometry* method has been developed, in which hydrogen exchange is combined with proteolytic fragmentation of the protein. In this approach, first developed by Smith and his associates, the

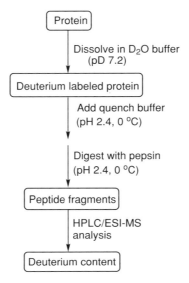

Figure 11.7. A flow diagram of a typical procedure used to determine deuterium levels at peptide amide linkages in short segments of intact proteins following H/D exchange.

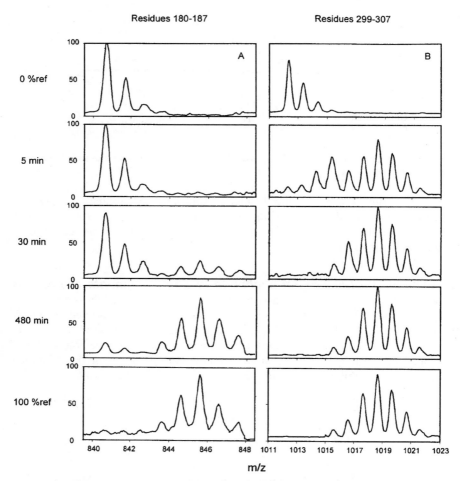

Figure 11.8. ESI mass spectra of peptide fragments that contain (*A*) residues 180–187 and (*B*) 299–307 of rabbit muscle aldolase labeled continuously in 3 M urea/D$_2$O for 5, 30, and 480 min. The top and bottom boxes are the spectra of reference aldolase that contains no deuterium and is fully deuterated, respectively. [Reproduced from Ref. 64 by permission of Elsevier Science Inc. (copyright 1999 by the American Society for Mass Spectrometry).]

deuterated protein is cleaved into small segments by pepsin digestion, followed by the mass spectrometry analysis of the protein digest [44,60,61]. This general approach also has parallel with the NMR studies of a similar nature [62,63].

A typical continuous-labeling procedure used in the protein fragmentation–mass spectrometry approach is illustrated in Figure 11.7. In this protocol, the in-exchange is performed for a defined time interval by incubating the folded protein in a D$_2$O solution at an appropriate pD. At the end of the defined time interval, the exchange process is quenched by adjusting the pD to 2.5 and the temperature to 0°C. Next, the protein is digested with pepsin, which is chosen because it has maximum proteolytic activity at an acidic pH. The deuterium levels of all peptide fragments of the digest are measured with HPLC on-line with ESIMS under conditions that minimize the hydrogen exchange at the peptide amide linkages [64–67]. The extent of deuterium incorporation is determined by comparing the ESI mass spectra with reference spectra of the same peptide that contains no deuterium and 100% deuterium. In the past, an on-line combination of HPLC and continuous-flow FAB was also implemented for this purpose [61,68–70].

Figure 11.8 shows the ESI mass spectral results of the unfolding of two peptide fragments of rabbit muscle aldolase, residues 180–187 and 299–307, following 5-, 30-, and 480-min labeling of the intact protein [64]. These two residues exhibit different types of amide hydrogen exchange behavior; both show a bimodal isotope pattern, but at different times of labeling. Unfolding of the 299–307 fragment is much faster than unfolding of the 180–187 fragment; that difference indicates the large values of β or unfolding equilibrium constant K for that fragment. The pattern for the 180–187 fragment shows that the exchange rate k_2 is much greater than k_{-1}. The pattern of the unfolding kinetics for the 180–187 segment is shown in Figure 11.9, which indicates that this aldolase segment unfolds in 3 M urea with a rate constant of $0.0061\,min^{-1}$. Figure 11.9 also shows that the region of aldolase that contains residues 284–298 unfolds at a much faster rate ($k_1 = 0.078\,min^{-1}$).

The pulsed-labeling approach can also be used to study the isotope exchange reaction in short segments of a protein [64,65,71,72]. A quench–flow apparatus is

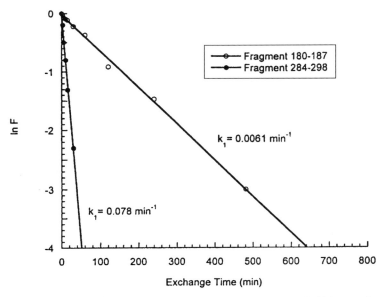

Figure 11.9. First-order kinetic plots of the natural log of the fraction of aldolase molecules that remained folded in segments containing residues 180–187 and 284–298 when aldolase was labeled continuously in urea/D_2O. [Reprinted from Ref. 64 by from Elsevier Science Inc. (copyright 1999 by the American Society for Mass Spectrometry).]

used for these experiments. The apparatus shown in Figure 11.7 can be modified to suit these experiments. First, a pepsin column is added to perform an on-line peptide digestion. Second, the proteolytic fragments are separated and analyzed by the HPLC-ESIMS combination. A schematic diagram of this modified apparatus is shown in Figure 11.10.

11.6. TIME-RESOLVED ELECTROSPRAY IONIZATION MASS SPECTROMETRY FOR STUDY OF CONFORMATIONAL CHANGES

Another convenient method of monitoring conformational changes and kinetics of folding/unfolding reactions in proteins is time-resolved ESIMS [73–76]. In this technique, the ESI mass spectrum of the protein is recorded at different time intervals after initiation of the reaction in a continuous-flow apparatus. Changes in the protein conformation can be detected by measuring differences in charge-state distribution. (Un)folding is initiated by exposing the protein solution to a buffer solution of required pH. The two solutions are driven through two separate narrow fused-silica capillary tubes by a syringe pump and mixed in a low dead volume-mixing tee. Varying the length of the reaction capillary that is connected between the mixing tee and the ESI source controls the reaction time. In many cases, this technique allows detection of transient intermediates of folding/unfolding reactions. This technique was applied successfully to study the folding of cytochrome c [73] and holo-myoglobin [74], and the acid-induced unfolding of holo-myoglobin [75, 76]. The kinetics of holo-myoglobin is a two step process [75]; in the first step, a short-lived intermediate that exists largely in the unfolded state, but still retains the heme group, is formed. In the second step, the heme–protein complex dissociates to unfolded apomyoglobin.

11.7. MATRIX-ASSISTED LASER DESORPTION/IONIZATION FOR STUDY OF CONFORMATIONAL CHANGES

In contrast to ESIMS, only a handful of studies has been reported that use MALDIMS to probe the conformational changes in protein and peptides [77–79].

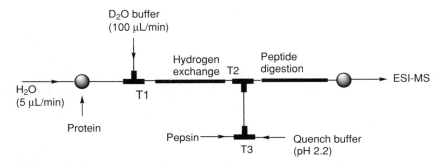

Figure 11.10. A schematic diagram of a quench–flow apparatus used in pulsed-labeling technique for the determination of H/D exchange in short segments of intact proteins.

Mandell et al. have used the combined fragmentation–mass spectrometry–H/D exchange approach with MALDIMS to determine the kinetics of H/D exchange in proteins [77]. As usual, the isotopic exchange of the amide hydrogens is performed by incubating the protein for variable times in D_2O at a pD of 7.2. The exchange is quenched at pH 2.5 and $0°C$. Next, the labeled protein is digested with trypsin, and the protein digest and the matrix are loaded onto the chilled MALDI target for mass analysis. α-Cyano-4-hydroxycinnamic acid at a pH of 2.5 is used as the matrix. This procedure was applied to measure the H/D exchange rate in cyclic AMP-dependent protein kinase. The advantages of using MALDIMS over ESIMS are the elimination of an HPLC separation step, the ability to analyze smaller sample amounts, and faster analysis times. Russell and colleagues have used MALDIMS and H/D exchange to probe the conformational changes in a few medium-sized peptides such as bradykinin, α-melanocyte-stimulating hormone, and melittin [78,79]. It was found that these peptides acquire a more compact structure in organic solvents.

Often, proteins come into contact with solid surfaces during isolation or experimental studies. The conformation of proteins can be altered when they are adsorbed on solid surfaces. As a consequence, the biological functioning of proteins may be affected. Therefore, it is important to know the stability of proteins that are physically adsorbed onto solid substrates. MALDI-TOFMS, in combination with

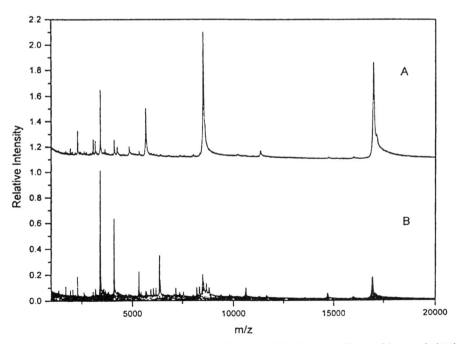

Figure 11.11. MALDI-TOF mass spectra of the immobilized trypsin digest of horse skeletal muscle myoglobin in the (*A*) absence and (*B*) presence of 25 mM *n*-octyl pyranoglucoside denaturant. [Reproduced from Ref. 84 with permission from John Wiley & Sons, Ltd. (copyright 1997).]

H/D exchange, has been successfully used to determine the conformational state of adsorbed proteins [80].

11.8. PEPTIDE MASS MAPPING FOR STUDY OF CONFORMATIONAL CHANGES

The conformational changes in proteins can also be investigated by using the peptide–mass mapping procedure alone (i.e., without any prior H/D exchange) [81–84]. In this simple procedure, the two conformational states of the protein under study are digested with trypsin, and the peptide maps are compared. The two conformational states will produce different peptide maps, which are conveniently analyzed by MALDIMS. On-probe digestion with immobilized trypsin facilitates MALDIMS analysis of the peptide maps [83]. Figure 11.11 compares the on-probe-generated peptide maps of horse skeletal muscle myoglobin in the absence and presence of a denaturant, *n*-octyl pyranoglucoside (NOG) [84]. Those data show that in the absence of denaturant, myoglobin is highly resistant to tryptic activity, whereas in the presence of NOG, extensive tryptic cleavage of the protein is observed.

11.9. ION MOBILITY MEASUREMENTS FOR STUDY OF CONFORMATIONAL CHANGES

Ion mobility spectrometry (IMS), in conjunction with mass spectrometry measurements, has been useful for determining the conformation and folding–unfolding kinetics of proteins [85,86]. In IMS, ions are separated on the basis of differences in cross-sectional area (see Section 3.6). Because different conformations of a protein have different shapes, they will exhibit different drift times in the IMS drift tube. The folded conformers have a small collision cross section and thus will have higher mobilities than will the larger open structures. The mobility of a protein also depends on that protein's charge state. By interfacing another quadrupole between an ESI source and the drift tube, a particular charge state can be mass-selected to examine the folding–unfolding process of proteins. In a typical experimental protocol, a protein is ionized with ESI, and a short pulse of the protein ions is injected into the drift tube containing helium as a buffer gas. Ions that exit the drift tube are analyzed with a second quadrupole, which provides the drift time distributions, from which information regarding the number of isomers can be obtained. IMS–mass spectrometry has been used to study the conformation and folding–unfolding behavior of cytochrome *c* [87–89], apomyoglobin [89], bovine pancreatic trypsin inhibitor [90], lysozyme [91], and ubiquitin [92,93].

REFERENCES

1. G. E. Schulz and R. H. Schirmer, *Principles of Protein Structure*, Springer-Verlag, New York, 1978.

2. T. E. Creighton, *Protein Structures and Molecular Properties*, Freeman, New York, 1984.

3. T. Takano and R. E. Dickerson, *Proc. Natl. Acad. Sci.* (USA) **77**, 6371 (1980).

4. T. Takano and R. E. Dickerson, *J. Mol. Biol.* **153**, 95 (1981).

5. T. E. Creighton, *Biochemeistry J.* **270**, 1–16 (1990).

6. H. B. Osbome and E. Nabedryk-Viala, *Meth. Enzymol.* **88**, 676–680 (1982).

7. J. J. Englander, D. B. Calhoun, and S. W. Englander, *Anal. Biochem.* **92**, 517–524 (1979).

8. J. A. Thomson, B. A. Shirley, G. R. Grimsley, and C. N. Pace, *J. Biol. Chem.* **264**, 11614–11620 (1988).

9. M. Eftink, *Biophys. J.* **66**, 481–501 (1994).

10. L. Swint and A. D. Robertson, *Prot. Sci.* **2**, 2037–2049 (1993).

11. Y. Bai, T. R. Sosnick, L. Mayne, and S. W. Englander, *Science* **269**, 192–197 (1995).

12. A. A. Kossiakoff, *Nature* (Lond.) **296**, 713–721 (1982).

13. M. F. Jeng, S. W. Englander, G. A. Elove, A. J. Wand, and H. Roder, *Biochemistry* **29**, 10433–10437 (1990).

14. Y. Paterson, S. W. Englander, and H. Roder, *Science* **249**, 755 (1990).

15. D. C. Benjamin, D. C. Williams, R. J. Poljak, and J. S. Rule, *J. Mol. Biol.* **257**, 866 (1996).

16. S. D. Hooke, S. E. Radford, and C. M. Dobson, *Biochemistry* **33**, 5867 (1994).

17. B. E. Jones and C. R. Matthews, *Prot. Sci.* **4**, 167 (1995).

18. J. Clarke, A. M. Hounslow, and A. R. Fersht, *J Mol. Biol.* **253**, 505 (1995).

19. S. Rothemund, M. Beyermann, E. Krause, G. Krause, M. Bienert, R. S. Hodges, B. D. Sykes, and F. D. Sonnichsen, *Biochemistry* **34**, 12954 (1995).

20. M. F. Jeng and H. J. Dyson, *Biochemistry* **34**, 611 (1995).

21. W. C. Johnson, *Annu. Rev. Biophys. Chem.* **17**, 145 (1988).

22. D. L. Smith, Y. Deng, and Z. Zhang, *J. Mass Spectrom.* **32**, 135–146 (1997).

23. D. L. Smith and Z. Zhang, *Mass Spectrom. Rev.* **13**, 411–429 (1994).

24. J. B. Fenn, M. Mann, C. K. Meng, S. F. Wong, and C. M. Whitehouse, *Science* **246**, 64–71 (1989).

25. R. D. Smith, J. A. Loo, R. R. Ogorzalek Loo, M. Busman, and H. R. Udseth, *Mass Spectrom. Rev.* **10**, 359–451 (1991).

26. R. Guevremont, K. W. M. Siu, J. C. Y. Le Blanc, and S. S. Berman, *J. Am. Soc. Mass Spectrom.* **3**, 216–224 (1992).

27. J. A. Loo, C. G. Edmonds, C. J. Barinaga, H. R. Udseth, and R. D. Smith, *Anal. Chem.* **62**, 693–698 (1990).

28. R. D. Smith, J. A. Loo, C. G. Edmonds, C. J. Barinaga, and H. R. Udseth, *Anal. Chem.* **62**, 882–899 (1990).

29. J. A. Loo, H. R. Udseth, and R. D. Smith, *Biol. Environ. Mass Spectrom.* **17**, 411–414 (1988).

30. S. K. Chowdhury, V. Katta, and B. T. Chait, *J. Am. Chem. Soc.* **112**, 9012–9013 (1990).

31. J. A. Loo, R. R. Ogorzalek Loo, H. R. Udseth, C. G. Edmonds, and R. D. Smith, *Rapid Commun. Mass Spectrom.* **5**, 101–105 (1991).

32. J. C. Y. Le Blanc, D. Beuchemin, K. W. M. Siu, and R. Guevremont, *Org. Mass Spectrom.* **26**, 831 (1991).

33. M. Hamdan and O. Curcuruto, *Rapid Commun. Mass Spectrom.* **8**, 144–148 (1994).

34. U. A. Mirza, S. L. Cohen, and B. T. Chait, *Anal. Chem.* **65**, 1–6 (1993).

35. C. S. Maier, M. I. Schimerlik, and M. L. Deinzer, *Biochemistry* **38**, 1136–1143 (1999).

36. M. Przybylski and M. O. Glocker, *Angew. Chem., Int. Ed. Engl.* **35**, 807–826 (1996).

37. M. Svoboda, A. Bauhofer, P. Schwind, E. Bade, I. Rasched, and M. Przybylski, *Biochem. Biophys. Acta* **1206**, 35 (1994).

38. M. O. Glocker, C. Borchers, W. Fiedler, D. Suckau, and M. Przybylski, *Bioconjugate Chem.* **5**, 583 (1994).

39. D. Suckau, M. Mak, and M. Przybylski, *Proc. Natl. Acad. Sci.* (USA) **89**, 5630 (1992).

40. A. D. Robertson and R. L. Baldwin, *Biochemistry* **30**, 9907–9914 (1991).

41. Y. Bai and S. W. Englander, *Prot. Struct. Funct. Genet.* **26**, 145–151 (1996).

42. V. Katta, and B. T. Chait, *Rapid Commun. Mass Spectrom.* **5**, 214–217 (1991).

43. V. Katta, and B. T. Chait, *J. Am. Chem. Soc.* **115**, 6317–6321 (1993).

44. G. Thevenon-Emeric, J. Kozlowski, Z. Zhang, and D. L. Smith, *Anal. Chem.* **64**, 2456–2458 (1992).

45. D. S. Wagner and R. J. Anderegg, *Anal. Chem.* **66**, 706–711 (1994).

46. C. L. Stevenson, R. J. Anderegg, and R. T. Borchardt, *J. Am. Soc. Mass Spectrom.* **4**, 646–651 (1993).

47. D. S. Wagner, L. G. Melton, Y. Yan, B. W. Erickson, and R. J. Anderegg, *Prot. Sci.* **3**, 1305–1314 (1994).

48. S. W. Englander, T. R. Sosnick, J. J. Englander, and L. Mayne, *Curr. Opin. Struct. Biol.* **6**, 18–23 (1996).

49. S. W. Englander and N. R. Kallenbach, *Quart. Rev. Biophys.* **16**, 521–655 (1984).

50. Y. Bai, J. S. Milne, L. Mayne, and S. W. Englander, *Prot. Struct. Funct. Genet.* **17**, 75–86 (1993).

51. Y. Bai, J. S. Milne, L. Mayne, and S. W. Englander, *Prot. Struct. Funct. Genet.* **20**, 4–14 (1994).

52. J. Clarke, L. S. Itzhaki, and A. R. Fersht, *Trends Biochem. Sci.* **22**, 284–287 (1997).

53. D. W. Miller and K. A. Dill, *Prot. Sci.* **4**, 1860–1873 (1995).

54. K.-S. Kim and C. Woodward, *Biochemistry* **32**, 9609–9613 (1993).

55. C. V. Robinson, in J. R. Chapman, ed., *Protein and Peptide Analysis by Mass Spectrometry*, Humana Press, Totowa, NJ, 1996, pp. 129–139.

56. A. Miranker, C. V. Robinson, S. E. Radford, R. T. Aplin, and C. M. Dobson, *Science* **262**, 896–900 (1993).

57. D. K. Heidary, L. A. Gross, M. Roy, and P. A. Jennings, *Nature Struct. Biol.* **4**, 725–731 (1997).

58. J. B. Udgaonkar and R. L Baldwin, *Nature* (Lond.) **335**, 694–699 (1988).

59. S. W. Englander and L. Mayne, *Ann. Rev. Biophys. Biomol. Struct.* **21**, 243–265 (1992).

60. A. S. Raza, K. Dharmasiri, and D. L. Smith, *J. Mass Spectrom.* **35**, 612–617 (2000).

61. Z. Zhang and D. L. Smith, *Prot. Sci.* **2**, 522–531 (1993).

62. J. J. Rosa and F. M. Richards, *J. Mol. Biol.* **133**, 399–416 (1979).

63. J. J. Englander, J. R. Rogero, and S. W. Englander, *Anal. Biochem.* **147**, 234–244 (1985).

64. Y. Deng, Z. Zhang, and D. L. Smith, *J. Am. Soc. Mass Spectrom.* **10**, 675–684 (1999).

65. Y. Deng and D. L. Smith, *Biochemistry* **37**, 6256–6262 (1998).

66. R. S. Johnson and K. A. Walsh, *Prot. Sci.* **3**, 2411–2418 (1994).

67. K. Dharmasiri and D. L. Smith, *Anal. Chem.* **68**, 2340–2344 (1996).

68. Y. Liu and D. L. Smith, *J. Am. Soc. Mass Spectrom.* **5**, 19–28 (1994).

69. Z. Zhang, C. B. Post, and D. L. Smith, *Biochemistry* **35**, 779–791 (1996).

70. Z. Zhang and D. L. Smith, *Prot. Sci.* **5**, 1282–1289 (1996).

71. H. H. Yang and D. L. Smith, *Biochemistry* **36**, 14992–14999 (1997).

72. Z. Zhang, W. Li, M. Li, T. M. Logan, S. Guan, and A. G. Marshall, *Tech. Prot. Chem. VIII* 703–713 (1997).

73. L. Konermann, B. A. Collings, and D. J. Douglas, *Biochemistry* **32**, 5554–5559 (1997).

74. V. W. S. Lee, Y.-L. Chen, and L. Konermann, *Anal. Chem.* **71**, 4154–4159 (1999).

75. L. Konermann, F. I. Rosell, A. G. Mauk, and D. J. Douglas, *Biochemistry* **32**, 6448–6454 (1997).

76. O. O. Sogbein, D. A. Simmons, and L. Konermann, *J. Am. Soc. Mass Spectrom.* **11**, 312–319 (2000).

77. J. G. Mandell, A. M. Falick, and E. A. Komives, *Anal. Chem.* **70**, 3897–3995 (1998).

78. I. D. Figueroa, O. Torres, and D. H. Russell, *Anal. Chem.* **70**, 4527–4533 (1998).

79. I. D. Figueroa and D. H. Russell, *J. Am. Soc. Mass Spectrom.* **10**, 719–731 (1999).

80. J. Buijs, C. C. Vera, E. Ayala, E. Steensma, and, P. Hakansson, *Anal. Chem.* **71**, 3219 (1999).

81. H. H. Yang, X. C. Li, M. Amft, and J. Grotemeyer, *Anal. Biochem.* **258**, 118–126 (1998).

82. R. Kriwacki, J. Wu, G. Siuzdak, and P. E. Wright, *J. Am. Chem. Soc.* **118**, 5320–5321 (1996).

83. R. Kriwacki, J. Wu, L. Tennant, P. E. Wright, and G. Siuzdak, *J. Chromtogr. A* **777**, 23–30 (1996).

84. R. W. Nelson, *Mass Spectrom. Rev.* **16**, 353–376 (1997).

85. C. Wu, W. F. Siems, G. R. Asbury, and H. H. Hill, Jr., *Anal. Chem.* **70**, 4929–4938 (1998).

86. D. E. Clemmer and M. F. Jarrold, *J. Mass Spectrom.* **32**, 577–592 (1997).

87. K. Shelimov and M. F. Jarrold, *J. Am. Chem. Soc.* **118**, 10313–10314 (1996).

88. K. Shelimov, D. E. Clemmer, R. R. Hudgins, and M. F. Jarrold, *J. Am. Chem. Soc.* **119**, 2240–2248 (1997).

89. K. Shelimov and M. F. Jarrold, *J. Am. Chem. Soc.* **119**, 2987–2994 (1997).

90. S. J. Valentine and D. E. Clemmer, *J. Am. Chem. Soc.* **119**, 3558 (1997).

91. S. J. Valentine, J. Anderson, A. E. Ellington, and D. E. Clemmer, *J. Phys. Chem.* **101**, 3891–3900 (1997).

92. S. J. Valentine, A. E. Counterman, and D. E. Clemmer *J. Am. Soc. Mass Spectrom.* **8**, 954–961 (1997).

93. R. W. Purves, D. A. Barnett, B. Ells, and R. Guevremont, *J. Am. Soc. Mass Spectrom.* **11**, 738–745 (2000).

12

INVESTIGATION OF NONCOVALENT INTERACTIONS

The noncovalent interaction of supramolecules (e.g., of proteins and oligonucleotides) is a common occurrence. For example, proteins exhibit a binding affinity toward a variety of other molecular species. Enzyme–substrate, enzyme–inhibitor, protein–protein, antibody–antigen, receptor–ligand, protein–metal, protein–DNA, and DNA-DNA associations are some of the examples of the noncovalent interactions in which biomolecules participate. These complexes are stabilized by a combination of electrostatic, hydrophobic, hydrogen bonding, and van der Waal's forces. Some interactions are involved in the transport of bioactive compounds to important locations. In other situations, these interactions are of physiological significance. For example, many cellular events are triggered via the formation of certain noncovalent complexes. Several examples are known in which a protein–protein interaction imparts an additional stability to the participating protein. An interruption of these normal interactions can lead to abnormalities and a diseased state. Therefore, study of these interactions is important from the viewpoint of understanding the complexities of human health and pathologies. These studies also have a potential of discovering potent inhibitors of enzymes or other molecules that participate in specific disease processes.

12.1. CONVENTIONAL METHODS FOR STUDY OF NONCOVALENT COMPLEXES

Several instrumental techniques have been devised for the study of noncovalent interactions of macromolecules. Some of these methods use spectroscopic techniques, such as circular dichroism (CD), light scattering, and fluoroscence [1]; others are based on separation techniques (e.g., ultracentrifugation, gel-permeation chromatography, and gel electrophoresis) [2]. Although the methods that involve the separation-based techniques can provide important stoichiometry information on protein complexes, the resulting data are sensitive to the shape and conformational changes in proteins. Advanced spectroscopy techniques that include NMR and X-ray crystallography are excellent methods in the study of three-dimensional structures of biomolecules and their complexes, but they do not provide any molecular mass information. In addition, they require large amounts of precious sample. Also, not all biomolecules can be readily crystallized for X-ray measurements. The NMR-based techniques are further limited to < 40-kDa proteins. Differential calorimetry, isothermal titration calorimetry, and surface plasmon resonance are other instrumental techniques that have been used for study of noncolvalent complexes of biomolecules, specifically, to measure their dissociation rate constants [3].

12.2. MASS SPECTROMETRY-BASED APPROACHES FOR STUDY OF NONCOVALENT INTERACTIONS

Since late 1980s, electrospray ionization (ESI) [4,5] and matrix-assisted laser desorption/ionization (MALDI) [6] have emerged as important tools for biomedical research. With the use of these techniques, several unique features (e.g., sensitivity, specificity, speed) of mass spectrometry can be exploited for the study of noncovalent complexes of supramolecules. The initial reports of the use of mass spectrometry for the detection of noncovalent complexes have emerged from the work of Ganem et al. [7] and Katta and Chait [8], who found that the gentle nature of ESI is a significant factor in this type of research. These research groups, respectively, investigated the receptor–ligand binding and the heme–globin interaction of myoglobin. Since then, a large volume of work has appeared on the applications of ESIMS to study various forms of protein interaction. The advantages of mass spectrometry methods over traditional methods include the accuracy of mass measurement, speed of analysis, and low sample amounts. Several excellent reviews have been written on the mass spectrometry characterization of noncovalent complexes of supramolecular compounds [9–13].

12.2.1. Electrospray Ionization for Study of Noncovalent Complexes

Electrospray is a delicate mode of ionization. The ionization of biomolecules by electrospray involves the formation of charged droplets by spraying their solution under the influence of a strong electric field (see Chapter 2). A combination of thermal energy, gas flow, and collision activation in the low-pressure interface region assists the desorption of molecular ions from charged droplets into the gas phase. A series of multiply charged ions of supramolecules, such as $[M + nH]^{n+}$, are formed by the ESI process.

Conditions for Observation of Noncovalent Complexes.

In order to characterize noncovalent complexes with ESIMS, it is important to recognize the presence of nonspecific complexes, which is a common occurrence in this mode of ionization [11,14]. The solution conditions and the ESI interface environment are both responsible for the formation of nonspecific complexes. Certain solution components, such as buffers, salts, surfactants, and counterions, can form solvent adducts with biomolecules. In addition, the molecules of the ESI solvents are involved in adduct formation with the molecular ions of the analyte. These nonspecific adducts must be destroyed for the unambiguous identification of noncovalent complexes. By adjusting the interface variables to provide a higher thermal and/or ion activation energy, the adduction and/or solvation of the molecular ions can be reduced. Two different designs of the ESI interface are in common use; one design employs a heated capillary, and the other uses a differentially pumped nozzle–skimmer (see Chapter 2). In the former design, the extent of thermal energy is controlled by heating the transfer capillary (usually to $> 150°C$), whereas the ion-source chamber is heated in the nozzle–skimmer interface. In both designs, a warm countercurrent gas is also circulated. By increasing the voltage difference between the capillary and the skimmer (or nozzle and skimmer), the ion activation energy can be increased. However, these harsher conditions may be detrimental to the survival of noncovalent complexes. The successful observation of such complexes requires that the complex survive the processes of droplet formation, their fission, ejection of ions into gas phase, and transfer of the ejected ions to the mass analyzer.

In addition, the solution environment should be compatible with the ESI operation. A higher sensitivity in the positive-ion ESI analysis is attained when the solution is adjusted to a pH of 2–4 and an organic cosolvent (e.g., methanol or acetonitrile) is admixed with water. However, the lowering of the solution pH, the presence or absence of buffers, and the addition of organic solvents can cause proteins to unfold (see Chapter 11), a process that will destroy noncovalent association. Therefore, a careful control of the following instrumental and chemical conditions is required for success in detection of noncovalent interactions:

- Solutions are adjusted to the physiological pH.
- Only volatile buffers (e.g., ammonium acetate) are used. Nonvolatile buffers can usually participate in the nonspecific adduct formation.
- The nonspecific adduct formation is minimized by working at low analyte concentrations.
- Organic solvents are avoided.

- The interface region temperature is kept low
- A moderate value of the capillary–skimmer (nozzle–skimmer) voltage is used.

The nature of an individual noncovalent complex is the deciding factor for a particular condition of any instrumental variable. Some complexes are stable, whereas others are more fragile. For example, the protein–DNA and protein–RNA complexes are formed via electrostatic interactions and, therefore, are highly stable. These complexes can survive high interface energies. In contrast, the complexes between acyl coenzyme A (CoA) binding protein and acyl CoA derivatives involve weak interactions and thus are detected only at low temperatures (< 80°C) [15]. In actual practice, a compromise should be made while adjusting the values of various instrumental parameters. On one hand, enough heating and collision energy is provided to ensure the absence of ion–solvent clusters and other nonspecific associations. On the other hand, these parameters are kept low to maintain the integrity of noncovalent complex.

12.2.2. Matrix-Assisted Laser Desorption/Ionization for Study of Noncovalent Complexes

Although a great majority of research on noncovalent interactions has been performed with ESIMS, reports on the use of MALDIMS have begun to appear in the literature. MALDIMS is a powerful technique for determination of the molecular mass of proteins and other biomolecules. The analysis of biomolecules by MALDI involves mixing the analyte with a suitable matrix, and bombarding the cocrystallized matrix–analyte mixture with a high intensity laser beam of an appropriate wavelength. The technique is usually combined with a TOF mass analyzer for the detection of the desorbed ions. The high-mass capability and high detection sensitivity are hallmarks of the MALDI-TOFMS combination. Biopolymers with a mass of > 300 kDa, can be analyzed by this technique. MALDIMS is at least 10 times more sensitive than ESIMS. Despite these advantages, not much success has been reported regarding its use in the study of noncovalent aggregations of proteins. One probable reason is that the energy deposition in the complex during desorption/ionization leads to its dissociation. However, the technique has shown some promise for detection of at least high-binding-energy complexes [16,17]. MALDIMS can be used to investigate nonco-valent aggregation in situations that are difficult to handle by ESIMS.

The successful detection of protein complexes by MALDIMS requires that certain variables be controlled. These variables include the protein concentration, type of matrix, matrix:analyte ratio, pH, ionic strength, and solvent. A detailed study by Jespersen et al. has demonstrated that the detection of noncovalent complexes by MALDIMS is highly dependent on the matrix used [17]. Among several matrices tested, a noncovalent complex between streptavidin and glutathione-*S*-transferases was observed when 3-hydroxypicolinic acid (3-HPA) was used as the matrix at a pH of 3.8. The dissociation of the complex is minimized with this matrix. A similar behavior is observed with detection of large single- or double-stranded oligonucleotides (see Section 12.5.5 and Chapter 15) [18].

In a thought-provoking perspective on this subject, Farmer and Caprioli have reviewed several MALDIMS approaches that have evolved to study the protein–protein complexes [19]. In the first approach, a specific matrix and laser combination is optimized. For example, the combinations of a frequency-quadrupled Nd : YAG laser (at $\lambda = 266$ nm) and nicotinic acid has been successful in the detection of the dimer, trimer, and tetramer of streptavidin [20]. This laser–matrix combination has also provided evidence for the existence of multimers of glucose isomerase [21] and jack bean (*Canavalia ensiformis*) urease [22]. A heptamer of a membrane-associated protein aerolysin was detected (at m/z 333, 850) with the nitrogen laser (at $\lambda = 337$ nm)–sinapinic acid combination [23].

The second approach makes use of the "first shot" spectra, in which only the first shots of the laser on the sample are collected and averaged [17,24]. This approach can be successful in those situations where the complexes are segregated or precipitated on the crystal surface during sample preparation. With this approach, the intact noncovalent complexes of streptavidin, alcohol dehydrogenase, and beef liver catalase (Figure 12.1) have been observed with 2,6-dihydroxyacetophenone as the matrix [25].

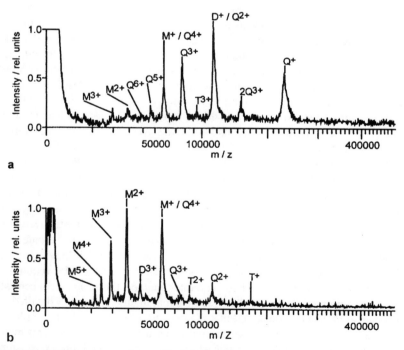

Figure 12.1. MALDI mass spectrum of beef liver catalase: (*a*) first shot only and (*b*) 11 shots on the same spot (M = monomer, D = dimer, T = trimer, Q = tetramer). [Reproduced from Ref. 25 by permission of Elsevier Science Inc. (copyright 1995 American Society for Mass Spectrometry).]

In the third approach, the pH of the matrix and/or sample mixture is manipulated by the addition of buffers. The change in pH and ionic strength can disrupt inter- and intramolecular forces, which in turn affect the stability of the complex. As a case in point, a lower pH value and a high concentration of organic solvents are both counterproductive to protein complexes. Woods et al. have observed the enzyme–substrate complexes between aminopeptidase I and growth-hormone-releasing factor (GHRF), arylamidase and GHRF, and trypsin and rat parathyroid fragment (1–34) when the solution was made more basic [26]. Ribonuclease (RNAse) S protein exists as a noncovalent complex that is formed between S-peptide and S-protein. This complex has been detected at a pH of 5.5 by MALDIMS with 6-aza-2-thiothymine (ATT) as the matrix, whereas no evidence of the complex is obtained at pH 2.0 [27]. Similarly, dimers of the leucine zipper polypeptides are observed in ATT at a basic pH only [27].

Chemical crosslinking with imido esters is another MALDIMS approach that has been successful in the study of protein complexes [19]. A typical example of this method is the investigation of complexes of bovine hemoglobin, which is a heterotetramer of two α and two β subunits (i.e., $\alpha_2\beta_2$). Without the chemical crosslinking, the heterotetramer is not detected. However, in the presence of glutaraldehyde as a crosslinking agent, the existence of the physiological dimer $\alpha\beta$ and tetramer $\alpha_2\beta_2$ is revealed by MALDIMS. The crosslinking with glutaraldehyde and subsequent MALDIMS analysis have also been applied to detect many other noncovalent complexes, which include human farnesyl protein transferase, an X complex of DNA polymerase III holoenzyme and glucose 6-phosphate dehydrogenase, yeast alcohol dehydrogenase, and bovine glutamate dehydrogenase [19].

12.2.3. Hydrogen/Deuterium Isotope Exchange for Study of Noncovalent Complexes

The technique of hydrogen/deuterium (H/D) exchange (described in more detail in Chapter 11) has achieved a high level of success in the detection of the protein denaturation, conformational states, and folding–unfolding dynamics [28–30]. The protocol can also be used to detect noncovalent complexes of proteins [31–33]. The basic premise is that the regions of the proteins that participate in molecular interactions should have a different rate of exchange relative to the regions that are more accessible to the solvent. In a typical procedure for the study of antigen–antibody complexes, the protein is first deuterated at a pH between 6 and 7, then passed through an antibody column, and deuteriums of the antigen–antibody complex are exchanged back to hydrogens (deuteriums of the bound epitope are unaccessible to H/D exchange). After quenching the H/D exchange, the antigen–antibody complex is digested with pepsin, and the peptide fragments are analyzed by LC-ESIMS. Epitopic peptides show an increase in the mass as a result of retention of the D labels.

A typical example of the use of H/D exchange and ESIMS for the investigation of noncovalent complexes is the interaction of the enzyme E. coli dihydrodipicolinate reductase with a substrate nicotinamide adenine dinucleotide (reduced form) (NADH) and an inhibitor 2,6-pyridinedicarboxylate [33]. As expected, these interactions were found to reduce the extent of deuterium exchange. The pepsin digestion and subsequent LC-ESIMS analysis of the protein digest identified four peptides with a slow deuterium exchange rate. Two of these peptides are part of the regions that are involved in the binding of the substrate and inhibitor, and the other two peptides are located at the interdomain hinge region. These two peptides are accessible to the solvent in the open, catalytically inactive conformation, but are inaccessible in the closed, catalytically active conformation that is formed on binding with NADH and 2,6-pyridinedicarboxylate. Another example of the use of this technique is the detection of the complex between the molecular chaperone GroEL and bovine α-lactalbumin [31].

12.2.4. Proteolytic Cleavage for Study of Noncovalent Complexes

Proteolytic cleavage can also provide indirect information about protein–ligand associations. In this approach, the protein is cleaved in the absence and presence of an interacting ligand, and the peptide maps that results are compared. The rationale for the use of this approach is that the sites where the binding occurs are not accessible to the proteolytic enzymes. Thus, the digestion of the protein in the presence and absence of a ligand will generate qualitatively different maps, which can be analyzed by MALDI-TOFMS. This approach has been used to characterize the protein–DNA [34] and protein–protein noncovalent complexes [35], and to identify epitopes (see Section 12.4). An example of a protein–protein interaction is the binding of a protein that represents the kinase inhibitory domain of the cell cycle regulatory protein (p^{21}-B) and cyclin-dependent kinase 2 (cdk 2) [35].

12.3. DETERMINATION OF BINDING AFFINITIES

In addition to stoichiometry, mass spectrometry can be used to obtain information about the solution binding constants of noncovalent complexes. These measurements are based on the rationale that the relative abundance of the gas-phase noncovalent complexes can be directly correlated with the measured solution-phase binding constants [36,37]. It has also been observed that the relative abundance of the products that are formed during dissociation of noncovalent complexes is a sensitive measure of the stability of a complex [38–40]. The tightly bound ligand is least readily dissociated; thus, its relative abundace will be lower. In one study, the in-source CID has been used to dissociate noncovalent complexes. This approach was applied to assess the stability of the enzyme–cofactor–inhibitor complexes of aldose reductase [39]. The ternary complexes were formed from the equilimolar mixtures of aldose reductase, coenzyme (NADP), and inhibitors (aminoSNM, imirestat, LCB3071, or IDD384), and were fragmented by increasing the nozzle–skimmer voltage (V_c). The V_c value needed to dissociate 50 % of the complex initially present was chosen as a measure of the stability of the ternary complex. Figure 12.2 shows the behavior of four different ternary complexes that were studied in this manner [39]. The data were found to correlate well with the energy of the electrostatic and hydrogen bonding interactions that were computed from the crystallographic model. Penn et al. used ESI with an FTMS instrument to determine the relative binding strengths of the complexes of four different peptides with a ligand, permethylated-β-cyclodextrin [40]. These peptide–cyclodextrin complexes were dissociated by using two different experimental approaches: heated capillary dissociation (HCD) and CID. In the HCD approach, increasing the temperature of the ES transfer capillary provided dissociation energy. The CID was accomplished in the FTMS cell. Both methods indicate the same ordering of the relative stabilities.

Tandem mass spectrometry (MS/MS) of noncovalent complexes can also be used to determine their relative stabilities. In this approach, the mass-selected noncovalent complexes are dissociated via collision activation [38]. In a typical experiment on the complexes of benzenesulfonamide-based inhibitors with zinc-bound carbonic anhydrase, the relative abundance of each released inhibitor was used as a criterion to provide the information on the relative binding constants [41].

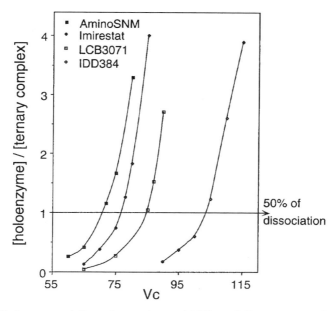

Figure 12.2. A representation of gas-phase stabilities of the enzyme–cofactor–inhibitor complexes of aldose reductase. The ternary complexes were formed from equilimolar mixtures of aldose reductase, coenzyme (NADP), and inhibitors (aminoSNM, imirestat, LCB3071, or IDD384). [Reproduced from Ref. 39 by permission of Elsevier Science Inc. (copyright 1999 American Society for Mass Spectrometry).]

12.3.1. Absolute Values of Binding Constants

From the ESIMS data, it is also feasible to derive the absolute solution-phase equilibrium constants [42]. In these measurements, it is assumed that the relative abundances of gas-phase complexes reflect solution-phase interactions. Consider the following equilibrium reaction between an analyte A (receptor, antibody, peptide, etc.) and its ligand L:

$$[A] + [L] \leftrightarrows [AL] \tag{12.1}$$

The binding constant (association K_a or dissociation K_d) is given by Eq. 12.2 or 12.3:

$$K_a = \frac{1}{K_d} = \frac{[AL]}{[A][L]} \tag{12.2}$$

$$\frac{[\text{Bound L}]}{[\text{Free L}]} = K_a[\text{Free A}] = K_a \left([\text{Total A}] - [\text{Bound L}]\right) \tag{12.3}$$

Thus, by titration of a fixed amount of A with an increasing amount of the target ligand, a Scatchard plot is obtained, the slope of which provides the value of $-K_a$ and x-intercept the stoichiometry of binding. Measurements of binding constants via Scatchard plots have been reported for the complexes between bovine serum albumin (BSA) and oligonucleotides [43], vancomycin antibiotics and target peptides [38], and pp60v-src SH2 domain protein with peptide ligands [36].

A method that can calculate the association constants of complexes of several ligands from a single measurement has been developed [44]. In this method, ESIMS measurements are made from equimolar concentrations of the analyte and ligands. This method was used for binding studies of antibiotics (vancomycin and ristocetin) and three peptide ligands. In a solution that contains an antibiotic A and three peptide ligands—L1, L2, and L3—the equilibrium concentrations can be calculated from the peak intensities by using the equation

$$[A_i] = \frac{A_i[A]_0}{A + AL1 + AL2 + AL3} \tag{12.4}$$

Here, the square brackets refer to the equilibrium concentrations, A, AL1, AL2, and AL3 are the ion intensities, and $[A]_0$ is the initial concentration of the antibiotics. $[A]_0$ can be taken as the initial concentration of each peptide because equimolar concentrations are used here. These values can be substituted into Eq. 12.2 to calculate the binding constants. For example, K_a for AL1 is given by

$$K_a = \frac{[AL1]}{[A] \times ([A] + [AL2] + [AL3])} \tag{12.5}$$

12.4. PROTEIN EPITOPE MAPPING

A monoclonal antibody (mAb) that has been raised against a particular protein binds to a specific region of that protein. This region is termed the *antigenic site* or *epitope*.

The linear (continuous) epitopes are made of a stretch of contiguous amino acids in the protein sequence, whereas the nonlinear epitopes are composed of amino acids that are distant, but proximal in terms of the protein's folded structure. The mapping of protein epitopes is of a prime significance in understanding basic immunology of the protein's immune response, and for the development of synthetic peptide vaccines.

A battery of techniques have been employed to identify these binding sites. These techniques include X-ray crystallography [45], measurement of solvent accessibility [46] and hydrophobicity [47], binding affinity of an antibody to evolutionary variants and antigen mutants, binding assays of sets of synthetic peptides that represent overlapping segments of the antigen sequence [48], and the use of bacteriophage peptide libraries [49]. In early 1990s, mass spectrometry was added to the protocol of protein epitope mapping [50–57].

In a typical mass spectrometry protocol, the antigen is complexed with an immobilized antibody, the epitopic peptides are released by limited proteolysis of the antigen–antibody complex, the antibody–peptide complex is extracted, and the epitopic peptides are identified by mass spectrometry analysis after their release from the complex [50]. ^{252}Cf-PD [50] and MALDI [51,52] have both been used to identify epitopes. Because the epitope domains are shielded from the proteolytic enzymes, suppression of the proteolysis of the antigen–antibody complex in the vicinity of an epitope can cause ambiguity. In addition, the antibody–peptide complex must survive the extraction and release steps. The following two alternative procedures have been devised to alleviate these limitations. In one variation, the antigen is digested before its complexation with the mAb [53]. The epitopic peptides are separated by immunoprecipitation with mAb, and are identified by MALDIMS analysis. In another variation, the antigen is first digested with an endoproteinase. A portion of this digest is reacted with mAb, and is analyzed by MALDIMS without removal of the antibody. A comparison of the mass spectra of the antigen–antibody reaction mixture and an unreacted antigen provides identity of the epitopic peptides [54]. Immobilization of the antibody and the extraction and dissociation of the antibody–peptide complex are not required.

Hydrogen/deuterium exchange can also be applied to protein epitope mapping. As discussed in Section 12.2.3, the protein is first deuterated at a pH of 6–7. The fully deuterated protein is complexed with the antibody, and the deuteriums of the antigen–antibody complex are exchanged back to hydrogens. After quenching the H/D exchange reaction, the antigen–antibody complex is digested with pepsin, and the peptide fragments are analyzed with LC-ESIMS. Because the deuteriums of the bound epitope are unaccessible to H/D exchange, the epitopic peptides show an increase in the mass as a result of the retention of the D labels.

12.5. APPLICATIONS

A large number of noncovalent complexes have been characterized by mass spectrometry techniques. It would be an uphill task to discuss all those applications in this chapter. However, a few representative examples of these interactions are reviewed here. For further reading, a thought-provoking review by Loo may be consulted [12].

12.5.1. Protein–Protein Complexes

The association of protein subunits into the active form of a protein structure is a common biochemical process. These subunits are held together via intermolecular electrostatic and van der Waals interactions. ESIMS and MALDIMS have evolved into viable tools to obtain information on the self-assembly process of polypeptide subunits [58–73]. An example in this category is the demonstration that leucine zipper proteins exist as dimers [27,62]. The self-assembly of alcohol dehydrogenase (ADH) has also been studied by ESIMS [73]. This protein is a metalloenzyme that participates in the interconversion of acetaldehyde and ethanol. The equine ADH exists as a dimer, whereas the active species of yeast ADH assumes a tetrameric form. Indeed, ESIMS has demonstrated that these two proteins exist as dimeric and tetrameric species with molecular masses of approximately 80 and 147 kDa, respectively (see Figure 12.3). An octamer of yeast ADH has been observed in a quadrupole ion trap at 296,719 Da [61].

The self-assembly of subunits in streptavidin is another widely studied system. This protein exists as a tetramer with a molecular mass of ∼60 kDa. Each subunit in the tetramer consists of a 159-amino-acid-long polypeptide chain. ESIMS [71] and MALDIMS [20] have both conclusively provided evidence of the tetrameric structure. The ESI mass spectrum of avidin also shows that it exists as a tetramer [65].

12.5.2. Protein–Peptide and Peptide–Peptide Complexes

ESIMS has been used to detect 1:1 noncovalent complexes between margatoxin, a 39-amino-acid-long peptide present in *Centruroides margaritatus* scorpion, and several peptide ligands [74]. These ligands represent amino acid segments that are located at the putative voltage-gated potassium channel binding site for margatoxin.

The binding of calmodulin with a peptide melittin has been reported [75,76]. These two species form a 1:1 complex, but only in the presence of calcium. Other examples of protein/peptide–peptide interactions include complexes of bovine serum albumin [77], glutathione-*S*-transferase [10], ribonuclease-*S*-protein [27,78,79], and src SH2 domain protein [36,80].

12.5.3. Protein–Small-Molecule Complexes

A large number of complexes of proteins with low-molecular-mass compounds have been characterized by mass spectrometry. The first reported example in this category is the study of a heme–globin complex [8,81]. Since then, a wide variety of complexes have been characterized. Typical examples are enzyme–inhibitor [82–

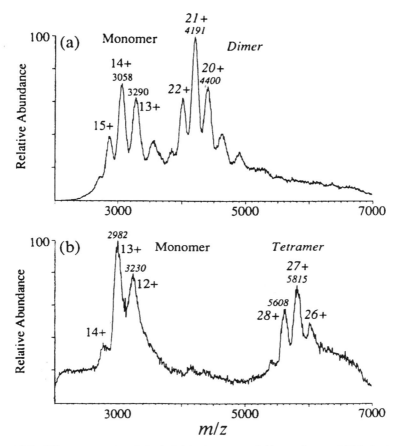

Figure 12.3. ESI mass spectra of alcohol dehydrogenase: (*a*) horse liver and (*b*) baker's yeast. [Reproduced from Ref. 73 by permission of John Wiley & Sons (copyright 1995).]

87], enzyme–cofactor [16], enzyme–substrate [88,89], receptor–ligand [8,77,89,90], and protein–ligand complexes [65,80,91–96].

Each polypeptide chain of streptavidin and avidin contains a high-affinity binding site. These proteins are known to bind four molecules of biotin. These strong noncovalent interactions have found wide applicability in the fields of biochemistry, immunochemistry, and affinity-based separations. An ESIMS study has shown that tetrameric avidin binds to four molecules of biotin (Figure 12.4) [65]. The binding behavior of the nonglycosylated analog streptavidin to biotin is also similar [91,92]. Schwartz et al. have shown that biotin and iminobiotin both interact with strepta-vidin; a stronger complex is formed with biotin than that formed with iminobiotin ($K_d \sim 10^{-15}$ M vs. $\sim 10^{-7}$ M) [92]. The noncovalent complex formation between streptavidin and a biotinylated 12-bp oligonucleotide, 5′-biotin-dACTTATCTTTCT-3′, was also reported in the same study [92].

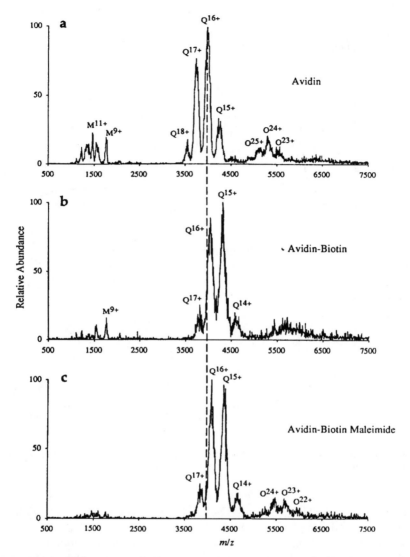

Figure 12.4. ESI mass spectra of (*a*) avidin, (*b*) the avidin–biotin complex, and (*c*) the avidin–biotinmaleimide complex. [Reproduced from Ref. 65 by permission of Elsevier Science Inc. (copyright 1994 American Society for Mass Spectrometry).]

12.5.4. Polypeptide–Metal Ion Complexes

ESIMS also has a potential to detect protein–metal ion complexes. Metal ions are known to play essential roles in the catalytic function and conformational stability of many metalloenzymes. Several mass spectrometry studies have revealed the stoichiometry of metal binding to proteins and peptides [68,97–103] and metal ion–induced conformational changes in proteins [104,105].

Vogl et al. have used MALDIMS to study the calcium-induced noncovalent association of MRP8 and MRP14 proteins [106]. These proteins are members of the S100 family of calcium binding proteins that are found in phagocytes. This study revealed that the two proteins form a tetrameric species, which is bound to eight Ca^{2+} ions.

12.5.5. Noncovalent Complexes of Oligonucleotides

Study of noncovalent interactions between strands of oligonucleotides is of prime importance in biomedical research. MALDIMS [107] and ESIMS [108–112] have both been used to detect double-stranded oligonucleotides. This term applies to true complementary pairing as well as dimers of single-stranded species. By a careful choice of the ESI interface and solution conditions, the duplexes of oligonucleotides can be observed. The first observation of the duplexes of oligonucleotides by mass spectrometry was obtained by Light-Wahl et al. [108]. The polyanionic backbone of these biomolecules facilitates their detection by negative-ion ESIMS. Light-Wahl et al. used this technique to obtain the mass spectrum of the duplex complex of two complementary oligonucleotides, 5'-dCCTTCCTCCCTCTCTCCTCC (5826.9 Da) and 5'-dGGAGGAGAGAGGGAGGAAGG-3' (6410.3 Da). Multiply charged anions with a charge distribution between −2 to −8 states are observed. Ganem et al. have also used ESIMS to detect the duplex complexes of self-complementary octanucleotides [109]. The noncovalent association of two complementary single-stranded oligonucleotides, 5'-dATGC-3' and 5'-dATGCAT-3', was studied by massive cluster impact ionization [113]. MALDIMS was successful in the detection of intact double-stranded DNA (EcoR1 adapter 12/16) when ATT was used as the matrix (Figure 12.5) [107]. The two strands of the EcoR1 adapter are 16 and 12 bases long (i.e., 5'AATTCCGTTGCTGTCG-3' and 3'GGCAACGACAGC-PO$_4$-5'). Other reports of the detection of the duplexes of self-complementary and non-self-complementary oligonucleotides include the recognition sequence of the GCN4 leucine zipper protein [10] and d(T)$_{10}$/d(A)$_{10}$ [10].

An association of drug molecules with self-complementary oligonucleotides has also been reported [112,114–116]. It was shown that distamycin A, a minor groove-binding molecule, exhibits a strong affinity for duplex oligonucleotides [112,115]. It forms 1:1 and 2:1 ligand–oligonucleotide duplexes. Similar complexes of pentamidine and Hoechst 33258 drugs have also been detected [112]. In an ESIMS study, actinomycin D has been shown to form noncovalent complexes with single-stranded oligonucleotides [116].

12.5.6. Noncovalent Complexes of Proteins and Oligonucleotides

Because of their involvement in the expression of genetic information, the noncovalent complexes of proteins with oligonucleotides have also been a focus of several studies [12,33,42,117]. Positive-ion and negative-ion ESIMS have both been

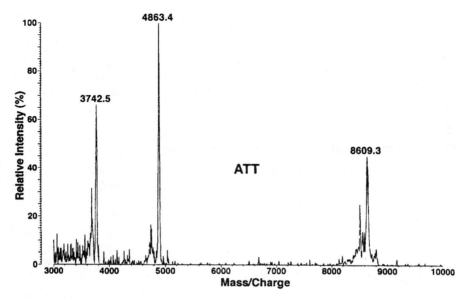

Figure 12.5. MALDI mass spectrum of EcoR1 12/16 adapter with ATT as the matrix. The ion at 8609.3 is the double-stranded complex. [Reproduced from Ref. 107 by permission of Elsevier Science Inc. (copyright 1995 American Society for Mass Spectrometry).]

successful in the characterization of such complexes. Cheng et al. have investigated the association of gene V protein with 13-mer, 16-mer, and 18-mer oligonucleotides [117]. The 13-mer d(p-T)13 was shown to form a complex with the dimer of gene V protein, whereas 16-mer and 18-mer oligonucleotides each associated with a pair of protein dimers to form a 4:1 protein:DNA complex. Loo and colleagues have characterized a noncovalent complex between Tat protein (40 residues, 4644 Da) and

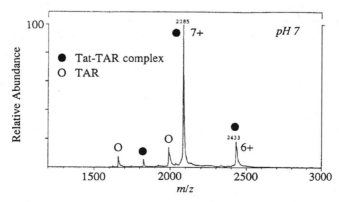

Figure 12.6. ESI mass spectrum of the 1:1 Tat peptide–TAR RNA complex (solid circles). [Reproduced from Ref. 12 by permission of John Wiley & Sons (copyright 1997).]

a 31-mer TAR RNA species (9941 Da) [12,42]. Positive-ion ESIMS revealed the formation of a 1 : 1 stochiometry complex between these molecules (Figure 12.6).

REFERENCES

1. P. Hensley, *Structure* **6**, 367–373 (1996).

2. T. M. Schuster and J. M.Toedt, *Curr. Opin. Struct. Biol.* **6**, 650–658 (1996).

3. A. Szabo, L. Stolz, and R. Granzow, *Curr. Opin. Struct. Biol.* **5**, 699–705 (1995).

4. J. B. Fenn, M. Mann, C. K. Meng, S. F. Wong, and C. M. Whitehouse, *Science* **246**, 64–71 (1989).

5. R. D. Smith, J. A. Loo, R. R. Ogorzalek Loo, M. Busman, and H. R. Udseth, *Mass Spectrom. Rev.* **10**, 359–451 (1991).

6. M. Karas and F. Hillenkamp, *Anal. Chem.* **60**, 2299–2301 (1988).

7. B. Ganem, Y. T. Li, and J. D. Henion, *J. Am. Chem. Soc.* **113**, 6294–6296 (1991).

8. V. Katta and B. T. Chait, *J. Am. Chem. Soc.* **113**, 8534–8535 (1991).

9. D. L. Smith and Z. Zhang, *Mass Spectrom. Rev.* **13**, 411–429 (1994).

10. M. Przybylski and M. O. Glocker, *Angew. Chem., Int. Ed. Engl.* **35**, 807–826 (1996).

11. R. D. Smith, X. Cheng, B. L. Schwartz, R. Chen, and S. A. Hofstadler, in A. P. Snyder, ed., *Biochemical and Biotechnological Applications of Electrospray Ionization Mass Spectrometry*, ACS, Washington, DC, 1996, pp. 294–314.

12. J. A. Loo, *Mass Spectrom. Rev.* **16**, 1–23 (1997).

13. B. N. Pramanik, P. L. Bartner, U. A. Mirza, Y.-H. Liu, and A. K. Ganguly, *J. Mass Spectrom.* **33**, 911–920 (1998).

14. R. D. Smith and K. J. Light-Wahl, *Biol. Mass Spectrom.* **22**, 493–501 (1993).

15. C. V. Robinson, E. W. Chung, B. B. Kragelund, J. Knudssen, R. T. Aplin, F. M. Poulsen, and C. M. Dobson, *J. Am. Chem. Soc.* **118**, 8646–8653 (1996).

16. B. Rosinke, K. Strupat, F. Hillenkamp, J. Rosenbusch, N. Dencher, U, Krüger, and H. J. Galla, *J. Mass Spectrom*, **30**, 1462–1468 (1995).

17. S. Jespersen, W. M. A. Niessen, U. R. Tjaden, and J. van der Greef, *J. Mass Spectrom.* **33**, 1088–1093 (1998).

18. K. J. Wu, A. Steding, and C. H. Becker, *Rapid Commun. Mass Spectrom.* **7**, 142–146 (1993).

19. T. B. Farmer and R. M. Caprioli, *J. Mass Spectrom.* **33**, 697 (1998).

20. M. Karas, U. Bahr, A. Ingendoh, E. Nordhoff, B. Stahl, K. Strupat, and F. Hillenkamp, *Anal. Chim. Acta* **241**, 175 (1990).

21. M. Karas and U. Bahr, *Trends Anal. Chem.* **9**, 321 (1990).

22. F. Hillenkamp and M. Karas, in *Proc. 37th ASMS Conf. Mass Spectrometry and Allied Topics*, Miami, 1989, Vol. 24, pp. 1168–1169.

23. M. Moniatte, F. G. van der Goot, J. T. Buckley, F. Pattus, and A. van Dorsselaer, *FEBS Lett.* **384**, 269 (1996).

24. M. Moniatte, C. Lesieur, V. Semjen, J. T. Buckley, F. Pattus, F. G. van der Goot, and A. van Dorsselaer, *I. J. Mass Spectrom. Ion Proc.* **169/170**, 179–199 (1997).

25. L. R. H. Cohen, K. Strupat, and F. Hillenkamp, *J. Am. Soc. Mass Spectrom.* **8**, 1046–1052 (1997).

26. A. S. Woods, J. C. Buchsbaum, T. A. Worrall, J. M. Berg, and R. J. Cotter, *Anal. Chem.* **67**, 4462 (1995).

27. M. O. Glocker, S. H. J. Bauer, J. Kast, J. Volz, and M. Przybylski, *J. Mass Spectrom.* **31**, 1221–1227 (1996).

28. D. L. Smith, Y. Deng, and Z. Zhang, *J. Mass Spectrom.* **32**, 135–146 (1997).

29. V. Katta, and B. T. Chait, *J. Am. Chem. Soc.* **115**, 6317–6321 (1993).

30. Z. Zhang and D. L. Smith, *Protein Sci.* **2**, 522–531 (1993).

31. C. V. Robinson, M. Gross, S. J. Eyles, J. J. Ewbank, M. Mayhew, F. U. Hartl, C. M. Dobson, and S. E. Radford, *Nature* (Lond.) **372**, 645–651 (1996).

32. R. J. Anderegg and D. S. Wagner, *J. Am. Chem. Soc.* **117**, 308–312 (1995).

33. F. Wang, J. S. Blanchard, and X.-J. Tang, *Biochemistry* **36**, 3755–3759 (1997).

34. S. L. Cohen, A. R. Ferre-D'Amare, S. K. Burley, and B. T. Chait, *Prot. Sci.* **4**, 1088–1099 (1995).

35. R. Kriwacki, J. Wu, G. Siuzdak, and P. E. Wright, *J. Am. Chem. Soc.* **118**, 5320–5321 (1996).

36. J. A. Loo, P. Hu, P. McConnell, W. T. Mueller, T. K. Sawyer, and V. Thanabal, *J. Am. Soc. Mass Spectrom.* **8**, 234–243 (1997).

37. K. X. Wan, M. L. Gross, and T. Shibue, *J. Am. Soc. Mass Spectrom.* **11**, 450–457 (2000).

38. Y.-L. Hsieh, B. Ganem, and J. D. Henion, *J. Mass Spectrom.* **30**, 708–714 (1995).

39. H. Rogniaux, A. van Dorsselaer, P. Barth, J. F. Biellmann, J. Barbanton, H. van Zandt, B. Chevrier, E. Howard, A. Mitschler, N. Potier, L. Urzhumtseva, D. Moras, and A. Podjarny, *J. Am. Soc. Mass Spectrom.* **10**, 635–647 (1999).

40. S. G. Penn, F. He, M. K. Green, and C. B. Lebrilla, *J. Am. Soc. Mass Spectrom.* **8**, 244–252 (1997).

41. X. Cheng, R. Chen, J. E. Bruce, B. L Schwartz, G.A. Anderson, S. A. Hofstadler, D. C. Gale, R. D. Smith, J. Gao, G. B. Sigal, M. Mammen, and G. M. Whitesides. *J. Am. Chem. Soc.* **117**, 8859–8860 (1995).

42. J. A. Loo and K. A. Sannes-Lowery, in B. S. Larsen and C. N. McEwen, eds., *Mass Spectrometry of Biological Materials*, Marcel Dekker, New York, 1998, pp. 345–367.

43. M. J. Greig, H. Gaus, L. L. Cummins, H. Sasmor, and R. H. Griffey, *J. Am. Chem. Soc.* **117**, 10765–10766 (1995).

44. T. J. D. Jorgensen, P. Roepstorff, and A. J. R. Heck, *Anal. Chem.* **70**, 4427–4432 (1998).

45. D. R. Davies, S. Sheriff, and E. A. Padlan, *J. Biol. Chem.* **263**, 10541–10544 (1988).

46. A. S. Kolaskar and P. C. Tongaokar, *FEBS Lett.* **276**, 172–174 (1990).

47. T. P. Hopp and K. R. Woods, *Proc. Natl. Acad. Sci.* (USA) **78**, 3824–3828 (1981).

48. H. M. Geysen, S. J. Rodda, T. J. Mason, G. Tribbick, and P. G. Schoofs, *J. Immunol. Methods* **102**, 259–274 (1987).

49. J. K. Scott and G. P. Smith, *Science* **249**, 386–390 (1990).

50. D. Sukau, J. Kohl, G. Karwath, K. Schneider, M. Casaretto, D. Bitter-Suermann, and M. Przybylski, *Proc. Natl. Acad. Sci.* (USA) **87**, 9848–9852 (1990).

51. D. I. Papac, J. Hoyes, and K. B. Tomer, *Prot. Sci.* **3**, 1485–1492 (1994).

52. C. E. Parker, D. I. Papac, S. K. Trojak, and K. B. Tomer, *J. Immunol.* **157**, 198–206 (1996).

53. Y. Zhao and B. T. Chait, *Anal. Chem.* **66**, 3723–3726 (1994).

54. M. Macht, W. Fiedler, K. Kurzinger, and M. Przybylski, *Biochemistry* **35**, 15633–15639 (1996).

55. Y. V. Lyubarskaya, Y. M. Dunayevskiy, P. Vouros, and B. L. Karger, *Anal. Chem.* **69**, 3008–3014 (1997).

56. J. G. Kiselar and K. M. Downard, *Anal. Chem.* **71**, 1792–1801 (1999).

57. L. Yu, S. J. Gaskell, and J. L. Brookman, *J. Am. Soc. Mass Spectrom.* **9**, 208–215 (1998).

58. X. Xie, T. Kokubo, S. L. Cohen, U. A. Mirza, A. Hoffmann, B. T. Chait, R. G. Roeder, Y. Nakatani, and S. K. Burley, *Nature* (Lond.) **380**, 316–322 (1996).

59. S. Jones and J. M. Thornton, *Proc. Natl. Acad. Sci.* (USA) **93**, 13–20 (1996).

60. Y.-T. Li, Y.-L. Hsieh, J. D. Henion, M. W. Senko, F. W. McLafferty, and B. Ganem, *J. Am. Chem. Soc.* **115**, 8409–8413 (1993).

61. Y. Wang, M. Schubert, A. Ingendoh, and J. Franzen, *Rapid Commun. Mass Spectrom.* **14**, 12–17 (2000).

62. S. Witte, F. Neumann, U. Krawinkel, and M. Przybylski, *J. Biol. Chem.* **271**, 18171–18175 (1996).

63. H. E. Witkowska, C. H. L. Shackleton, K. Dahlman-Wright, J. Y. Kim, and J.-A. Gustafsson, *J. Am. Chem. Soc.* **117**, 3319–3324 (1995).

64. X. J. Tang, C. F. Brewer, S. Saha, I. Chernushevich, W. Ens, and K. G. Standing, *Rapid Commun. Mass Spectrom.* **8**, 750–754 (1994).

65. B. L. Schwartz, K. J. Light-Wahl, and R. D. Smith, *J. Am. Soc. Mass Spectrom.* **5**, 201–204 (1994).

66. M. C. Fitzgerald, I. Chernushevich, K. G. Standing, C. P. Whiteman, and S. B. H. Kent, *Proc. Natl. Acad. Sci.* (USA) **93**, 6851–6856 (1996).

67. A. K. Ganguly, B. N. Pramanik, E. C. Huang, A. Tsarbopoulos, V. M. Girijavallabhan, and S. Liberles, *Tetrahedron* **49**, 7985–7996 (1993).

68. P. Hu and J. A. Loo, *J. Mass Spectrom.* **30**, 1076–1082 (1995).

69. K. J. Light-Wahl, B. E. Winger, and R. D. Smith, *J. Am. Chem. Soc.* **115**, 5809–5870 (1993).

70. K. J. Light-Wahl, B. L. Schwartz, and R. D. Smith, *J. Am. Chem. Soc.* **116**, 5271–5278 (1994).

71. B. L. Schwartz, J. E. Bruce, G. A. Anderson, S. A. Hofstadler, A. L. Rockwood, R. D. Smith, A. Chilkoti, and P. S. Stayton, *J. Am. Soc. Mass Spectrom.* **6**, 459–465 (1995).

72. R. T. Aplin, C. V. Robinson, C. J. Schofield, and N. J. Westwood, *J. Chem. Soc., Chem. Commun.* 2415–2417 (1994).

73. J. A. Loo, *J. Mass Spectrom.* **30**, 180–183 (1995).

74. R. Bakhtiar and M. A. Bednarek, *J. Am. Soc. Mass Spectrom.* **7**, 1075–1080 (1996).

75. O. Nemirovskiy, R. Ramanathan, and M. L. Gross, *J. Am. Soc. Mass Spectrom.* **8**, 809–812 (1997).

76. T. D. Veenstra, A. J. Tomlinson, L. Benson, R. Kumar, and S. Naylor, *J. Am. Soc. Mass Spectrom.* **9**, 580–584 (1998).

77. L. Baczynskyj, G. E. Bronson, and T. M. Kubiak, *Rapid Commun. Mass Spectrom.* **8**, 280–286 (1994).

78. D. R. Goodlett, R. R. Ogorzalek-Loo, J. A. Loo, J. H. Wahl, H. R. Udseth, and R. D. Smith, *J. Am. Soc. Mass Spectrom.* **5**, 614–622 (1994).

79. R. R. Ogorzalek-Loo, D. R. Goodlett, R. D. Smith, J. A. Loo, *J. Am. Chem. Soc.* **5**, 4391–4392 (1993).

80. R. J. Anderegg and D. S. Wagner, *J. Am. Chem. Soc.* **117**, 308–312 (1995).

81. J. A. Loo, A. Giordani, and H. Muenster, *Rapid Commun. Mass Spectrom.* **7**, 186–189 (1993).

82. R. Feng, A. L. Castelhano, R. Billedeau, and Z. Yuan, *J. Am. Soc. Mass Spectrom.* **6**, 1105–1111 (1995).

83. N. Potier, P. Barth, D. Tritsch, J. F. Biellmann, and A. Van Dorsselaer, *Eur. J. Biochem.* **243**, 274–282 (1997).

84. J. Gao, X. Cheng, R. Chen, G. B. Sigal, J. E. Bruce, B. L. Schwartz, S. A. Hofstadler, G. A. Anderson, R. D. Smith, and G. M. Whitesides, *J. Med. Chem. Soc.* **39**, 1949–1955 (1996).

85. M. Baca and S. B. H. Kent, *J. Am. Chem. Soc.* **114**, 3992–3993 (1992).

86. J. A. E. Kraunsoe, R. T. Aplin, B. Green, and G. Lowe, *FEBS Lett.* **396**, 108–112 (1996).

87. Q. Wu, J. Gao, D. Joseph-McCarthy, G. B. Sigal, J. E. Bruce, G. M. Whitesides, and R. D. Smith, *J. Am. Chem. Soc.* **119**, 1157–1158 (1997).

88. B. Ganem, Y. T. Li, and J. D. Henion, *J. Am. Chem. Soc.* **113**, 7818–7819 (1991).

89. B. Ganem and J. D. Henion, *Chemtrcts-Org. Chem.* **113**, 6294–6296 (1991).

90. Y.-L. Hsieh, J. Cai, Y. T. Li, and J. D. Henion, *J. Am. Soc. Mass Spectrom.* **6**, 85–90 (1995).

91. M. Hamdan, O. Curcuruto, E. Di Modugno, *Rapid Commun. Mass Spectrom.* **9**, 883–887 (1995).

92. B. L. Schwartz, D. C. Gale, R. D. Smith, A. Chilkoti, and P. S. Stayton, *J. Mass Spectrom.* **30**, 1095–1102 (1995).

93. K. Eckart and J. Spiess, *J. Am. Soc. Mass Spectrom.* **6**, 912–919 (1995).

94. R. Bakhtiar and R. A. Stearns, *Rapid Commun. Mass Spectrom.* **9**, 240–244 (1995).

95. G. Siuzdak, J. F. Kerbs, S. J. Benkovic, and H. J. Dyson, *J. Am. Chem. Soc.* **116**, 7937–7938 (1994).

96. M. Ishigai, J. I. Langridge, R. S. Bordoli, and S. J. Gaskell, *J. Am. Soc. Mass Spectrom.* **11**, 606–614 (2000).

97. C. Q. Jiao, B. S. Freiser, S. R. Carr, and C. J. Cassady, *J. Am. Soc. Mass Spectrom.* **6**, 521–524 (1995).

98. P. F. Hu, Q, Z. Ye, and J. A. Loo, *Anal. Chem.* **66**, 3858–3863 (1994).

99. T. D. Veenstra, K. L. Johnson, A. J. Tomlinson, S. Naylor, and R. Kumar, *Biochemistry* **36**, 3635–3542 (1997).

100. T. W. Hutchens, R. W. Nelson. M. H. Allen. C. M. Li, and T. T. Yip. *Biol. Mass Spectrom.* **21**, 151–159 (1992).

101. A. Loo, P. Hu, and R. D. Smith *J. Am. Soc. Mass Spectrom.* **5**, 959–965 (1994).

102. S. Moreau, A. C. Awade, D. Molle, Y. Le Graet, and G. Brule, *J. Agric. Food Chem.* **43**, 883–889 (1995).

103. Y. Pettillot, E. Forest, J. Meyer, and J. M. Moulis, *Anal. Biochem.* **228**, 56–63 (1995a).

104. F. Wang, W. Li, M. R. Emmett, A. G. Marshall, D. Corson, and B. Sykes, *J. Am. Soc. Mass Spectrom.* **10**, 703–710 (1999).

105. O. Nemirovskiy, D. Giblin, and M. L. Gross, *J. Am. Soc. Mass Spectrom.* **10**, 711–718 (1999).

106. T. Vogl, J. Roth, C. Sorg, F. Hillenkamp, and K. Strupat, *J. Am. Soc. Mass Spectrom.* **10**, 1124–1130 (1999).

107. P. Lecchi and L. K. Pannell, *J. Am. Soc. Mass Spectrom.* **6**, 972–975 (1995).

108. K. J. Light-Wahl, D. L. Springer, B. E. Winger, C. G. Edmonds, D. G. Camp, B. D. Thrall, and R. D. Smith, *J. Am. Chem. Soc.* **115**, 803–804 (1993).

109. B. Ganem, Y.-T. Li, and J. D. Henion, *Tetrahedron Lett.* **34**, 1445–1448 (1993).

110. E. Bayer, T. Bauer, T. Schmeer, K. Bleicher, M. Maler, and H. J. Gaus, *Anal. Chem.* **66**, 3858–3863 (1994).

111. X. Cheng, Q. Gao, R. D. Smith, K.-E. Jung, and C. Switzer, *J. Chem. Soc. Chem. Commun.* 746–748 (1996).

112. D. C. Gale and R. D. Smith, *J. Am. Soc. Mass Spectrom.* **6**, 1105–1111 (1995).

113. D. Fabris, Z. Wu, and C. Fenselau, *J. Mass Spectrom.* **30**, 140 (1995).

114. X. Cheng, P. E. Morin, A. C. Harms, J. E. Bruce, Y. Ben-David, and R. D. Smith. *Anal. Biochem.* **239**, 35–40 (1996).

115. D. C. Gale, D. R. Goodlett, K. J. Light-Wahl, and R. D. Smith, *J. Am. Chem. Soc.* **116**, 6027–6028 (1994).

116. Y.-L. Hsieh, Y.-T. Li, J. D. Henion, and B. Ganem, *Biol. Mass Spectrom.* **23**, 272–276 (1994).

117. X. H. Cheng, A. C. Harms, P. N. Goudreau, T. C. Terwilliger, and R. D. Smith, *Proc. Natl. Acad. Sci.* (USA) **93**, 7022–7027 (1996).

13

STRUCTURE
DETERMINATION OF LIPIDS
AND GLYCOLIPIDS

13.1. A GENERAL VIEW OF LIPID STRUCTURE

Lipids are a diverse group of compounds. They can be classified into four principal groups (Figure 13.1): fats and waxes, complex lipids, steroids, and prostaglandins and leukotrienes. They play diverse functions in the body such as in the storage of energy, cell–cell communication, cell–cell recognition, and human diseases. Fats are esters, in which the alcohol component is glycerol, and the acid components are long-chain fatty acids. Triglycerides, in which all three OH groups on the glycerol backbone (i.e., sn-1, sn-2, and sn-3) are esterified (see Figure 13.2), are the most abundant class of lipids, and are significant components of fat storage cells. Mono- and diglycerides are also not so common materials. Complex lipids are the main constituents of membranes. Two main groups of complex lipids are phospholipids and glycolipids. Phospholipids are further divided into two subgroups: glycerophospholipids (also called *phosphoglycerides*) and sphingolipids, each composed of an alcohol, fatty acids, and a phosphate group; the difference is that the alcohol in glycerophospholipids is glycerol, and in sphingolipids, it is sphingosine (Figure 13.3). Structurally, glycerophospholipids are similar to triglycerides, but one of the fatty acids of the glycerol backbone (i.e., C-3 or the sn-3 position) is replaced by a phosphate group, which, in turn, is connected to another alcohol. The most abundant of the glycerophospholipids is glycerophosphocholine (GPC; also known

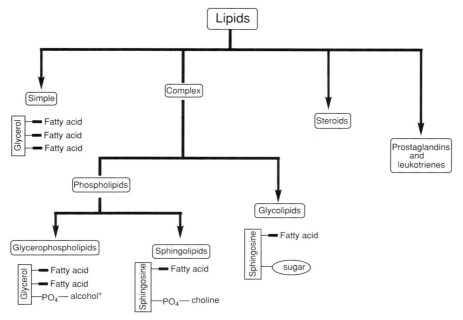

Figure 13.1. Classification of lipids.

as *phosphatidylcholine*; its common name is *lecithin*), in which the alcohol part is choline. Other less common forms of glycerophospholipids are glycerophosphoser-ine (GPS; also termed *phosphatidylserine*), glycerophosphoethanolamine (GPE; also termed *phosphatidylethanolamine*), glycerophosphoglycerol (GPG; also known as *phosphatidylglycerol*), and glycerophosphoinositol (GPI; also called *phosphatidyli-nositol*), in which the second alcohol is replaced with serine, ethanolamine, glycerol, or inositol, respectively. Glycerophosphatidic acid (GPA) has no substitution on the phosphate group. The second glycerol in GPG can also be esterified to give a form called *cardiolipin*. Sphingolipids or sphingomyelins contain a modified form of the sphingosine backbone called *ceramide*, in which a long-chain fatty acid is connected

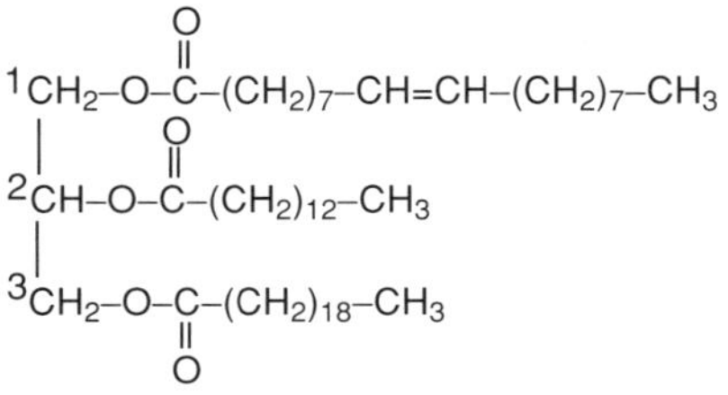

$$^{1}CH_2-O-\overset{\overset{O}{\|}}{C}-(CH_2)_7-CH=CH-(CH_2)_7-CH_3$$
$$^{2}CH-O-\overset{\overset{O}{\|}}{C}-(CH_2)_{12}-CH_3$$
$$^{3}CH_2-O-\underset{\underset{O}{\|}}{C}-(CH_2)_{18}-CH_3$$

Triglyceride

Figure 13.2. Structure of a typical triglyceride.

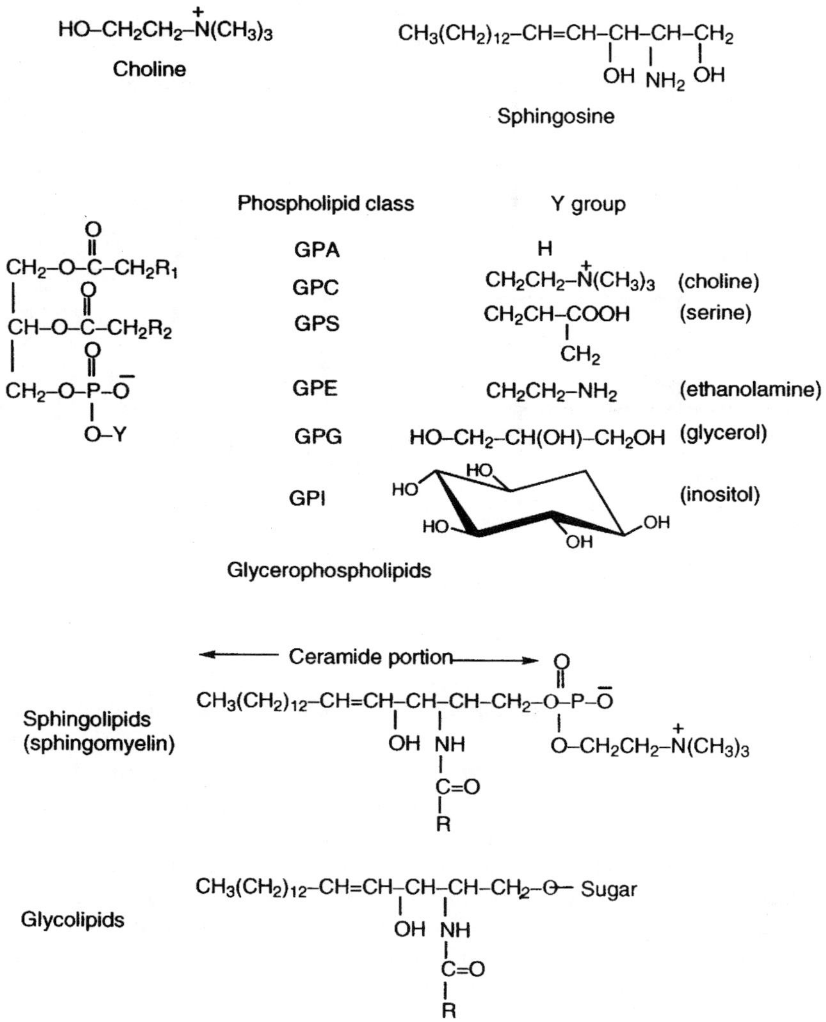

$$HO-CH_2CH_2-\overset{+}{N}(CH_3)_3$$

Choline

$$CH_3(CH_2)_{12}-CH=CH-CH-CH-CH_2$$
$$| \quad | \quad |$$
$$OH \quad NH_2 \quad OH$$

Sphingosine

	Phospholipid class	Y group	
	GPA	H	
$CH_2-O-\overset{O}{\overset{\|}{C}}-CH_2R_1$	GPC	$CH_2CH_2-\overset{+}{N}(CH_3)_3$	(choline)
$CH-O-\overset{O}{\overset{\|}{C}}-CH_2R_2$	GPS	$CH_2CH-COOH$ $\|$ CH_2	(serine)
$CH_2-O-\overset{O}{\overset{\|}{P}}-\overset{-}{O}$	GPE	$CH_2CH_2-NH_2$	(ethanolamine)
$\|$ $O-Y$	GPG	$HO-CH_2-CH(OH)-CH_2OH$	(glycerol)
	GPI		(inositol)

Glycerophospholipids

Sphingolipids
(sphingomyelin)

$$CH_3(CH_2)_{12}-CH=CH-CH-CH-CH_2-O-\overset{O}{\overset{\|}{P}}-\overset{-}{O}$$

$$\overset{\longleftarrow \text{ Ceramide portion} \longrightarrow}{}$$

$$| \quad | \qquad\qquad | \quad \overset{+}{}$$
$$OH \quad NH \qquad\qquad O-CH_2CH_2-N(CH_3)_3$$
$$|$$
$$C=O$$
$$|$$
$$R$$

Glycolipids

$$CH_3(CH_2)_{12}-CH=CH-CH-CH-CH_2-O- \text{ Sugar}$$
$$| \quad |$$
$$OH \quad NH$$
$$|$$
$$C=O$$
$$|$$
$$R$$

Figure 13.3. Structures of phospholipids, glycolipids, and related molecules.

to the amino group via an amide bond. Similar to GPC lipids, the polar head in sphingomyelins is also phosphorylcholine, which is linked to the −OH group of ceramide. Glycolipids also contain ceramide backbone, but differ from sphingolipids in that the 1-hydroxyl group of the ceramide backbone is linked to one or more sugar units. A subclass of glycolipids is gangliosides, in which the sugar portion is sialic acid.

The third major class of lipids is steroids. The most abundant steroid in the human body is cholesterol, which serves as a membrane component and also as a

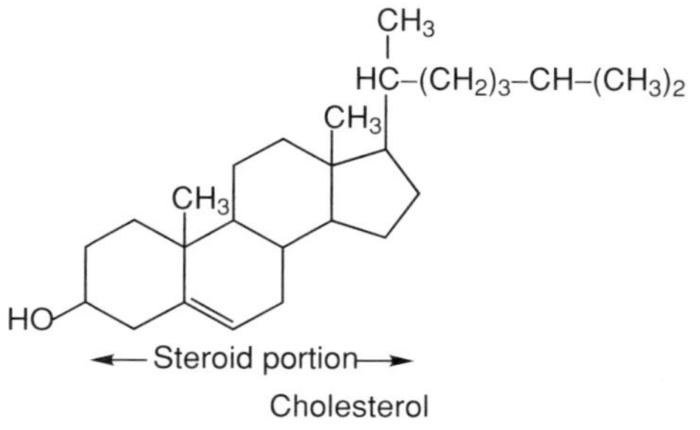

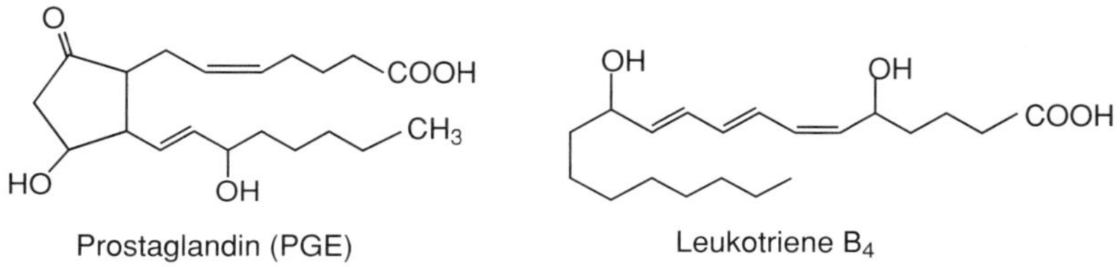

Figure 13.4. Structures of cholesterol, prostaglandins, and leukotrienes.

precursor of other steroid hormones and bile acids, which are the oxidation products of cholesterol. Prostaglandins and leukotrienes are a group of lipids that are involved in several physiologic functions. For example, they cause inflammation and fever and act as mediators of hormonal responses. Both types of molecules are synthesized in the body from arachidonic acid through the action of oxygenases, cyclooxygenases, and 5-lipooxygenase. Arachidonic acid, in turn, is released in free form from phospholipids by the action of phospholipase A_2. The structures of cholesterol, prostaglandins, and leukotrienes are shown in Figure 13.4.

13.2. MASS SPECTROMETRY ANALYSIS OF FATTY ACIDS AND GLYCERIDES

Fatty acids are a common occurrence in lipid extracts, where they exist in free as well as esterified form. Their notation includes the number of carbons followed by a number that indicates the degree of unsaturation. For example, octadecenoic acid is denoted as 18:1. Free fatty acids are isolated along with neutral lipids. Esterified fatty acids can be released via hydrolysis of lipids. Electron ionization (EI)MS had been a successful mass spectrometry approach for analysis of fatty acids, but it is applicable only after converting them to methyl, trimethylsilyl, or t-butyldimethylsilyl (TBDMS) derivatives [1]. Methyl derivatives are of special interest because they can be readily identified on the basis of a characteristic ion m/z 74, which is formed via a McLafferty rearrangement [2]. Although EIMS is successful in the structural characterization of saturated fatty acids, the location of the double bond is difficult to discern from the EI mass spectrum. Also, fragmentation is sparse in the high-mass region. Gross et al. have developed a fast-atom bombardment (FAB)-MS/MS approach to obtain the structural information of fatty acids [3–5]. This approach makes use of charge-remote fragmentation (CRF). With this approach, the site of double bonds can be readily located because the preferred site of fragmentation is the allylic C−C bond. CRF is facilitated by collision-induced dissociation (CID) of the $[M - H + cation]^+$ species. Li^+ and Na^+ are the common cations, although alkaline-earth metal ions have also been used [6]. Chang and Watson have prepared aminoethyl triphenylphosphonium bromide derivatives to structurally characterize fatty acids by FAB-MS/MS [7]. The FAB mass spectra of these derivatives are structurally more informative, and fragmentations are dominated via CRF processes. Electrospray ionization (ESI) and matrix-assisted laser desorption/ionization (MALDI) are also useful tools in structural studies of fatty acids [8,9]. Wheelan et al. have demonstrated that negative-ion ESI, coupled with low-energy CID, can provide important structure-related data for polyhydroxy unsaturated fatty acids [8]. Fragmentation is directed at the position of a hydroxy group or a double bond. As a consequence, isomer-specific CID spectra are produced. As shown in Figure 13.5, four isomeric dihydroxyeicosatetraenoic (diHETE) acids can be differentiated on the basis of their unique CID spectra.

As discussed above, triglycerides (or triacylglycerols) are fatty acid derivatives of glycerol, and are the most abundant class of lipids in nature. They are the storehouse of energy in animals and plants, and fats and oils derived from these species are composed exclusively of triglycerides. A complete structure characterization of triacylglycerols involves identification of each fatty acyl substituent and their location on the glycerol backbone. Various mass spectrometry techniques have served to characterize triacylglycerols, including EI [10,11], chemical ionization (CI) [12–14], desorption CI [15,16], gas chromatography (GC)MS [17], FAB [18], thermospray [19,20], ESI [21,22], and atmospheric pressure chemical ionization (APCI) [23,24]. EI of triacylglycerols leads to the formation of characteristic acylium ions such as $[M - RCOO]^+$, $[RCOO + 128]^+$, $[RCOO + 74]^+$, and $[RCO]^+$ ions. The carbon number and degree of unsaturation on each acyl group can be calculated from the fragmentation pattern. However, the location of double bonds and of acyl groups on the glycerol backbone is difficult to pinpoint. ESI with low-energy CID is also problematic with respect to identification of the positions of acyl groups. Cheng et al. reported the applications of high-energy CID of the ESI- and FAB-produced $[M + NH_4]^+$ and $[M + metal]^+$ adducts for the complete characterization of triacylglycerols [25]. In addition to the normal type charge-directed fragmentations, high-energy CID produces ions due to CRF. From the data, it is possible to deduce the carbon number and degree of unsaturation on each acyl group, the location of double bonds, and the position of the acyl groups on the glycerol backbone. Figure 13.6 shows the CID spectra of the FAB- and ESI-produced $[M + Na]^+$ ions of 1-palmitoyl-2-oleoyl-3-stearoylglycerol.

Mono- and diacylglycerols do not exist in appreciable amounts in biological systems. Analytic methods similar to those used for triacylglycerols can be applied for the structural characterization of these molecules.

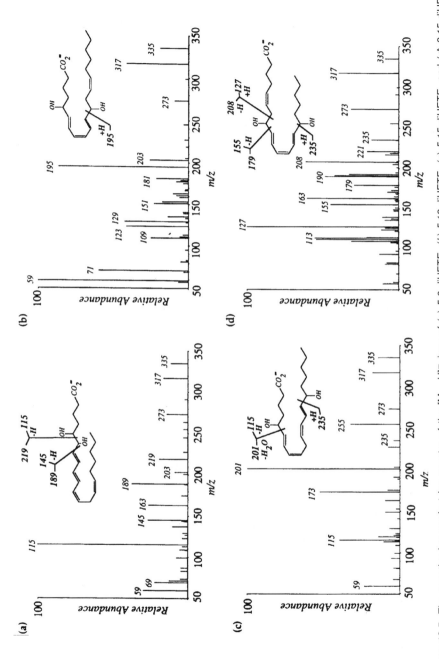

Figure 13.5. The negative-ion product ion spectra of the [M − H]⁻ ions of (a) 5,6-diHETE, (b) 5,12-diHETE, (c) 5,15-diHETE, and (d) 8,15-diHETE. [Reproduced from Ref. 8 by permission of Elsevier Science Inc. (copyright 1996 American Society for Mass Spectrometry).]

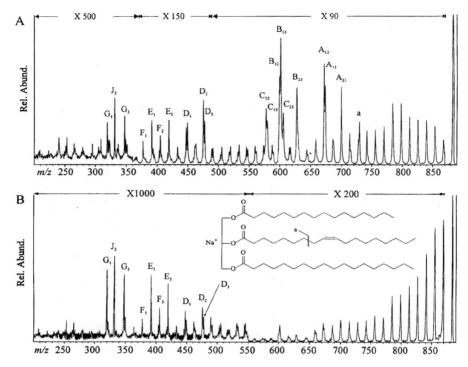

Figure 13.6. The CID spectra of the (A) FAB- and (B) ESI-produced [M + H]$^+$ ions of 1-palmitoyl-2-oleoyl-3-stearoylglycerol. [Reproduced from Ref. 25 by permission of American Chemical Society, Washington, DC (copyright 1998).]

13.3. MASS SPECTROMETRY ANALYSIS OF PHOSPHOLIPIDS

Phospholipids are amphipathic molecules; that is, they contain hydrophilic (polar head) and hydrophobic (tail) groups (Figure 13.3). They easily form the lipid bilayer that constitutes the plasma membrane of the living cells. The hydrophobic part of the bilayer forms an effective barrier to prevent the entry of ions and polar molecules into the cell. The structural diversity and unique physicochemical properties of phospholipids render these compounds difficult to characterize. A complete characterization of phospholipids involves identification of the fatty acyl groups as well as the nature of the polar head group. FAB has been the mainstay of mass spectrometry techniques for analysis of these complex compounds [26–30]. Before the advent of FAB, analysis of intact phospholipids was an impossible task. In traditional approaches, the fatty acyl profile is obtained after the release of the sn-1 and sn-2 fatty acids from phospholipids via alkaline hydrolysis, followed by analysis of the released fatty acid components by GC or GCMS [29,31]. A better approach is to use phospolipase C hydrolysis, which releases the acyl groups in the form of 1,2-diacylglycerols [32]. Field desorption [33], desorption CI [34], and liquid chromatography (LC)MS with a moving belt [35] and thermospray [36,37] interfaces were able to produce an ion signal from intact phospholipids, but with certain practical complications.

13.3.1. Fast-Atom Bombardment Mass Spectrometry Analysis of Phospholipids

FABMS has proved to be highly successful in the analysis of phospholipids because the hydrophobic tail facilitates their desorption from the FAB matrices, and the polar head group allows facile ionization. Positive- and negative-ion FAB, as well as FAB tandem mass spectrometry (MS/MS), have provided important information in the characterization of these compounds. An informative review on FABMS analysis of phospholipids has appeared in the literature [30].

Phospholipids produce very simple FAB mass spectra. Because of the built-in positive charge on the choline moiety, GPC lipids are easily ionized in positive-ion mode. In contrast, other subclasses of glycerophospholipids form negative ions more readily. The characteristic features of the positive-ion FAB spectra of glycerophospholipids are the presence of $[M + H]^+$ ions and the ions that are formed via cleavage of the phosphate ester bond. This cleavage produces the $[M + H - 98]^+$ ion (i.e., loss of H_3PO_4) in GPA, the phosphocholine ion (m/z 184) in GPC, and the $[M + H - 141]^+$ ion (loss of phosphoethanolamine) in GPE. GPC lipids are differentiated from other classes of phospholipids by the presence of the characteristic phosphocholine ion at m/z 184. The presence of a strong acid is required in the matrix to produce abundant $[M + H]^+$ ion signal.

The behavior of GPC lipids studied under negative-ion FAB has been described in detail [38]. The characteristic phosphocholine ion at m/z 184 is missing in the negative-ion FAB spectra. Instead, the carboxylate anions (R_1COO^- and R_2COO^-) that are related to the sn-1 and sn-2 fatty acyl substituents are characteristic features of these spectra. In addition, the $[M - 15]^-$, $[M - 60]^-$, and $[M - 86]^-$ ions that represent the losses of CH_3, $[CH_3 + NH(CH_3)_2]$, and $[CH_3 + CH_2=CHN(CH_3)_2]$, respectively, from the choline moiety are also formed [38,39]. In negative-ion mode, diethanolamine is the preferred matrix.

Tandem mass spectrometry of the $[M + H]^+$ ions of glycerophospholipids also yields fragment ions similar to those produced in conventional FAB. For example, CID of the $[M + H]^+$ ions of GPC and GPE lipids form only a single ion that corresponds to m/z 184 and $[M + H - 141]^+$, respectively. Whereas CID of the $[M - H]^-$ or $[M - 15]^-$ ions yields the abundant carboxylate anions as shown in Figure 13.7, which is the CID spectrum of the $[M - 15]^-$ ion of a diacyl–glycerophosphocholine (16:0a/18:1) [30]. It was demonstrated by Jensen et al. and others that in the CID spectra of $[M - 15]^-$, $[M - 60]^-$, or $[M - 86]^-$, the sn-2 carboxylate anion is always more abundant relative to the sn-1 carboxylate anion [30,38]. Huang et al. have shown that CID of the $[M - 86]^-$ ion is a better diagnostic approach for structural studies of glycerophospholipids [40]. FAB-MS/MS has the

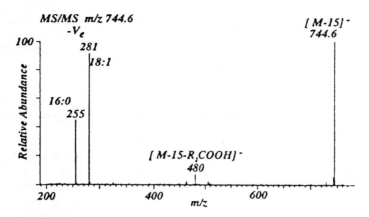

Figure 13.7. The CID spectrum of the $[M - 15]^-$ ion of a diacyl–glycerophosphocholine (16:0a/18:1). [Reproduced from Ref. 30 by permission of John Wiley & Sons (copyright 1994).]

advantage that it can provide information on the identity of these carboxylate anions [41,42].

Tandem mass spectrometry is also useful in the identification of glycero-phospholipids in complex mixtures extracted from biological samples. GPC lipids have been characterized in purified preparations of human immunodeficiency virus (HIV-1) [43], canine myocytes [44], rat kidney [45], rabbit kidney, human myocardium [46], and rat heart mitochondrial fractions [47]. In the last study, six monohydroxylated fatty acid substituents in GPC molecules were detected. The negative-ion spectrum of the mitochondrial extract is shown in Figure 13.8. The m/z 255, 279, 281, 283, 303, and 305 were attributed to 16:0, 18:2, 18:1, 18:0, 20:4, 20–3 fatty acids, respectively. The structure of one of these molecules was identified as 12-hydroxy 9-octadecenoic acid.

A few studies of real-world samples for the characterization of GPE and GPI lipids has also been reported [48,49]. Positive-ion FAB was successful in identification of ten major GPE molecular species from the isolates of canine myocardial sarcolemma [48]. Negative-ion FAB analysis of human polymorphonuclear leukocytes has revealed the presence of 8 major and 14 minor GPE molecular species that are related to arachidonic acid [49].

The behavior of sphingophospholipids under FAB and CID conditions is similar to that of GPC lipids in positive and negative ionization modes [30]. This similarity is due to the presence of phosphocholine in both groups of phospholipids. FABMS has been a useful technique for analyzing the molecular species of canine myocardial sarcolemma sphingomyelin [48]. In that study, the fatty acids were released by chemical hydrolysis to reveal the identity of fatty acyl substituents.

FABMS has also been applied to identification of phospholipid species from lipid extracts of human cerebrospinal fluid (CSF) [50]. GPC, sphingomyelin, diacylglycerol, and lysophosphatidylcholine have been identified in leukemia patients.

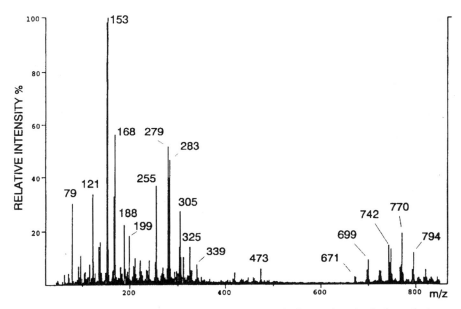

Figure 13.8. The negative-ion FAB mass spectrum of glycerophosphocholines that were derived from the mitochondrial extract. [Reproduced from Ref. 47 by permission of Elsevier Science Inc. (copyright 1996 American Society for Mass Spectrometry).]

13.3.2. Electrospray Ionization–Mass Spectrometry of Phospholipids

A remarkable attribute of ESIMS is its high level of detection sensitivity, which makes it an ideal technique for analysis of phospholipids [51–53]. Positive and negative ionization modes both provide facile means to characterize phospholipids. In positive-ion mode, the protonated and sodiated molecular ion adducts are the primary species [52]. In negative-ion mode, all classes of phospholipids, except for GPC, yield abundant $[M - H]^-$ ions. GPC lipids can be detected as $[M + Cl]^-$ ion adducts in this mode of ionization.

In order to obtain the structure-related fragment ions of phospholipids, ESI has also been coupled with tandem mass spectrometry. Han and Gross have used ESI-MS/MS to investigate in detail the fragmentation pathways of different classes of phospholipids [54]. The spectra can be acquired from low picomole amounts of samples. Compared to FAB-MS/MS [38], new product ions are observed in the product ion spectrum of ESI-generated $[M - H]^-$ ions. As an example, the product ion spectrum of the $[M + Cl]^-$ ion adduct of 1-hexadecanoyl-2-octadec-9′-enoyl-*sn*-glycero-3-phosphocholine is shown in Figure 13.9*a*. The lysoglycerophospholipid-like products, which are difficult to observe in FAB-MS/MS, are readily seen in the ESI-MS/MS spectrum. The cluster of ions around *m/z* 480 is related to these product ions. Regiospecificity of the acyl chain can also be derived from the MS/MS spectra of phospholipids. The CID behavior of the $[M + Na]^+$ ions of GPC and GPE lipids was also investigated in this study [54]. The fragmentation patterns of the

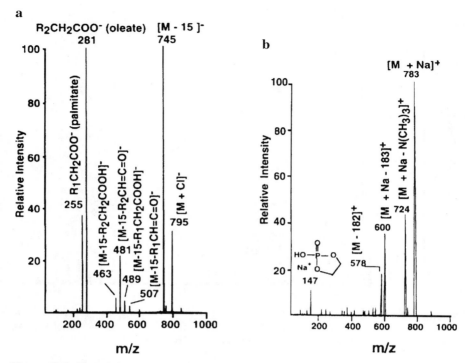

Figure 13.9. The product ion spectrum of the ESI-produced (*a*) [M + Cl]⁻ ion of 1-hexadeca-noyl-2-octadec-9′-enoyl-*sn*-glycero-3-phosphocholine and (*b*) the Na⁺ adduct of this phospho-lipid. [Reproduced from Ref. 54 by permission of Elsevier Science Inc. (copyright 1995 American Society for Mass Spectrometry).]

sodiated adducts are different from those observed for the $[M - H]^-$ ions. This observation is exemplified in Figure 13.9*b*, in which the product ions at *m/z* 724, 600, 578, and 147 correspond to the loss of trimethylamine (i.e., $[M + Na - 59]^+$) and the formation of two diglyceride-like cations ($[M + Na - 183]^+$ and $[M + Na - 05]^+$), and of a sodiated five-member cyclophosphane moiety, respectively. The behavior of sphingomyelins under positive- and negative-ion ESI-MS/MS conditions is also similar to that of GPC lipids [54].

Lithiated adducts of GPC [55] and GPE [56] lipids have also been generated under ESI conditions. The CID-MS/MS of these adducts yields ions that represent losses of fatty acid substituents. The relative abundance of these ions can reveal the identity and position of the fatty acid substituents. Parent ion and neutral loss scans can also be used to identify the head group and the fatty acid substituents. Figure 13.10 shows the usefulness of these scans in the identification of GPC species from rat liver extracts. Panels *A*, *C*, *E*, and *F* identify via parent scan MS/MS stearate-, arachidonate-, linoleate-, and palmitate-containing GPC species, respectively. Panels *B* and *D* show the neutral loss scans of 343 and 475Da, respectively. The identity of the stearate- and arachidonate-containing GPC species is revealed from these scans.

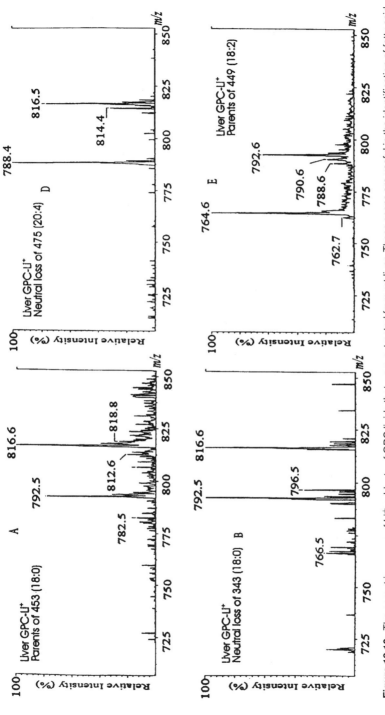

Figure 13.10. The parent ion scans of Li⁺ adducts of GPC lipids that were extracted from rat liver. These scans are useful in the identification of fatty acid substituents. Panels A, C, E, and F are parent scans, and panels B and D are neutral loss scans of 343 and 475 Da, respectively. [Reproduced from Ref. 55 by permission of Elsevier Science Inc. (copyright 1998 American Society for Mass Spectrometry).]

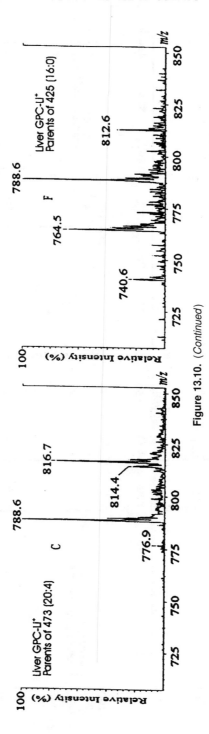

Figure 13.10. (*Continued*)

13.3.3. Matrix-Assisted Laser Desorption/Ionization–Mass Spectrometry of Phospholipids

Matrix-assisted laser desorption/ionization (MALDI)–mass spectrometry has been used sparsely for the analysis of phospholipids. Harvey has used positive- and negative-ion modes of MALDI to analyze several molecular forms of phospholipids (GPC, GPE, GPS, GPI, GPG, phosphatidic acid, and cardiolipin) [57]. Several derivatives of cinnamic acid, benzoic acid, and coumarin were tested as MALDI matrices. Of these, α-cyano-4-hydroxycinnamic acid (α-CHCA), esculetin (6,7-dihydroxycoumarin), and 2,5-dihydroxybenzoic acid (2,5-DHB) provided intense signals. In positive-ion mode, all phospholipids yield signals due to the $[M + H]^+$, $[M + Na]^+$, and $[M + K]^+$ ions. Because of the presence of a quaternary nitrogen with a fixed positive charge on the choline moiety, GPC gives the strongest M^+ ion signal. Derivatization of the phosphate group by diazomethane provides simpler molecular ion regions, and the metal ion adducts are absent. A few fragment ions are also formed by MALDI of phospholipids. A major fragmentation reaction is the cleavage of the phosphate ester bond to produce 1,2-diacylpropanediol. Fragmentation is much more pronounced with α-CHCA as the matrix, particularly when the matrix solution contains water. Additional fragmentations are observed in spectra that acquired on a magnetic sector instrument. This aspect is obvious in Figure 13.11, which shows the spectra of 1-palmitoyl-2-oleoylphosphatidylcholine that was acquired with α-CHCA as the matrix. With the exception of GPC, all phospholipids can also be analyzed in negative-ion MALDI mode, where the molecular ion region is much cleaner, and only the $[M - H]^-$ signal is produced.

Harvey also conducted a similar MALDIMS study of sphingo- and glyco-sphingolipids [58]. Similar conclusions were drawn for this set of compounds. The Na^+ adduct is the major ion in the molecular ion region. In negative-ion mode, only the $[M - H]^-$ ions are formed.

13.3.4. Liquid Chromatography–Mass Spectrometry Analysis of Phospholipids

Liquid chromatography (LC)–mass spectrometry has made a significant contribution in identification of phospholipids in complex mixtures of lipid extracts from biological samples, [19,36,52,59,60]. Several reports from Kim's group have extended the applications of LCMS to the analysis of phospholipids [19,36,52,59]. LCMS with a thermospray interface was used to separate and characterize several glycerophospholipids [19,36,59]. The fragment ion characteristics of each class are produced by the filament-on thermospray approach. These fragments can identify the head group and the constituent fatty acids. The technique was applied to identify glycerophospholipids from a whole rat brain lipid extract [59].

LCMS with a thermospray interface, however, suffers from an overall lack of sensitivity. More sensitive on-line LCMS approaches that use continuous-flow (CF)-FAB [60] or ESI as an interface have been developed [52]. In the CF-FAB approach,

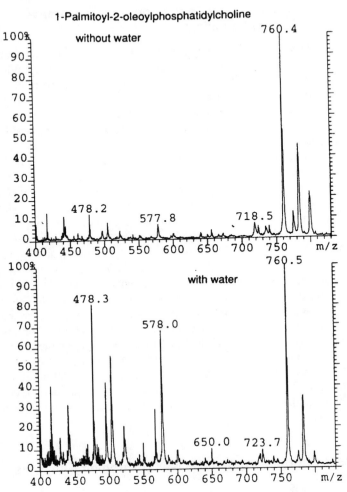

Figure 13.11. The MALDI mass spectra of 1-palmitoyl-2-oleoylphosphatidylcholine that was acquired with α-CHCA as the matrix. The top panel shows the matrix prepared without water; the bottom panel, matrix prepared in 30% water in acetonitrile. [Reproduced from Ref. 57 by permission of John Wiley & Sons (copyright 1995).]

glycerophospholipids are separated by a reversed-phase C-18 capillary column, and each separated component is characterized on the basis of the characteristic full-scan positive- and negative-ion mass spectra [60]. Each spectrum contains ions due to the molecular ion and structure-specific fragments, from which the information on the molecular mass, the polar head group, and the fatty acyl substituents can be extracted. The limit of detection of this approach is in subnanogram for each component. For GPC lipids, positive-ion detection is more sensitive than negative-ion detection. With the LC-ESIMS approach, a mixture of phospholipids was resolved on a C-18 column with 0.5% NH_4OH in a water/methanol/hexane solvent mixture [52]. The separated phospholipids were detected in positive-ion mode as $[M + H]^+$ and $[M + Na]^+$ species. The relative response depended on the nature of the polar head group and not on the fatty acyl substituents. GPC lipids exhibited the most sensitive response, followed by GPE and GPS lipids.

13.4. MASS SPECTROMETRY OF GLYCOLIPIDS

Glycolipids are ubiquitous constituents of cell membranes, and they play a major role in cell–cell recognition. Much of the biological interest in glycolipids is centered on glycosphingolipids (GSL). Structurally, GSLs are made of a ceramide portion and a sugar portion (Figure 13.3). The latter may contain one or several sugar units; the common ones are glucose, galactose, N-acetylglucosamine, N-acetylgalactosamine, fucose, sialic acid, and glucuronic acids. Extensive heterogeneity in the structure of ceramide portion also exists. A variety of fatty acids that differ in the alkyl chain length and in the presence or absence of a double bond and an OH group are attached to the sphingosine base. The characterization of glycolipids has been a challenging task because of this structural diversity and labile nature of the bonds that link the various building blocks. The complete structural characterization of GSLs involves structure elucidation of the constituents of the ceramide and sugar portions. The majority of the mass spectrometry research since the late 1980s has been accomplished with FABMS [61–63] and FAB-MS/MS techniques [64–66]. Prior to the use of FAB, glycolipids were cleaved into individual structural units, and analyzed by EIMS or GCMS after converting them to acetylated, permethylated, or permethylated-reduced derivatives [67]. EI-MS/MS of intact GSLs after permethylation–reduction steps has also been discussed [68].

Negative-ion FAB mode yields better results than does positive-ion mode. Several examples of negative-ion FAB for characterizing glycolipids in biological extracts have been reported [61–63]. Positive-ion FAB mode also provides excellent results when glycolipids are converted to permethylated, peracetylated derivatives. The negative- and positive-ion data both contain a sufficient number of fragment ions from which the structural information can be retrieved. Negative-ion FAB of native GSLs yields the $[M - H]^-$, $[Cer]^-$, and sequence-specific carbohydrate ions. Positive-ion FAB of native GSLs yields mainly the $[M + H]^+$ or $[M + cation]^+$, and Cer^+ species [62]. Positive-ion FAB of permethylated and peracetylated GSLs in addition produces sequence-specific carbohydrate ions [62].

Applications of FAB-MS/MS to the structural characterization of GSLs have been reported. This technique eliminates the contribution of the matrix and other chemical background, and improves fragment ion yield. The fragmentation behavior of glycolipids under CID conditions is well documented [64–66]. A nomenclature similar to that used to designate peptide fragment ions has also been proposed [64]. The nomenclature for fragmentation in the carbohydrate portion was discussed in Chapter 10 (shown in Figure 10.7). The designation of the ceramide fragmentation is shown in Figure 13.12.

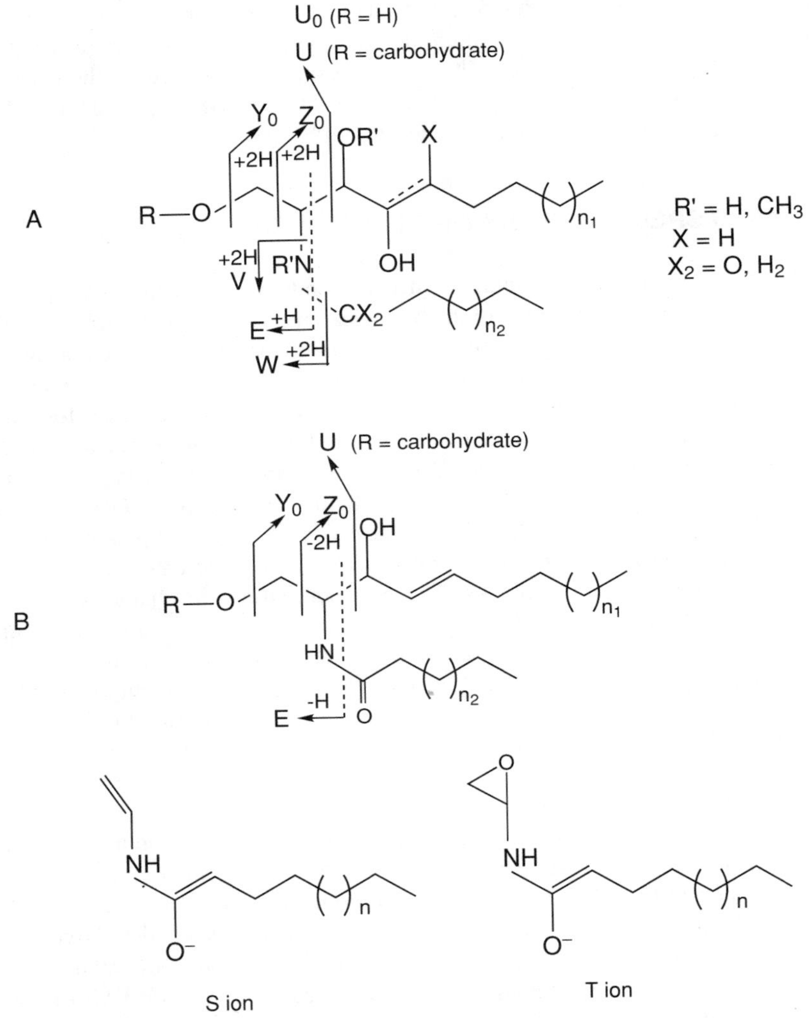

Figure 13.12. Designation of fragment ions that are formed in the ceramide portion of glycolipids: (A) positive ions and (B) negative ions.

Costello and colleagues have documented that the positive-ion CID mass spectra of the $[M + H]^+$ ion of native GSLs contain mainly the ceramide-specific fragment [64–66]. In contrast, the CID mass spectrum of the mass-selected $[M - H]^-$ ion is dominated by ions that are related to the structure of the carbohydrate moiety, with charge retention on the nonreducing end. The complementary information obtained from two modes of ionization often helps unravel the structure of unknown glycolipids. Figure 13.13 compares the positive- and negative-ion CID mass spectra of lactosyl-N-palmitoylsphinganine [64]. Often, it becomes essential to derivatize a

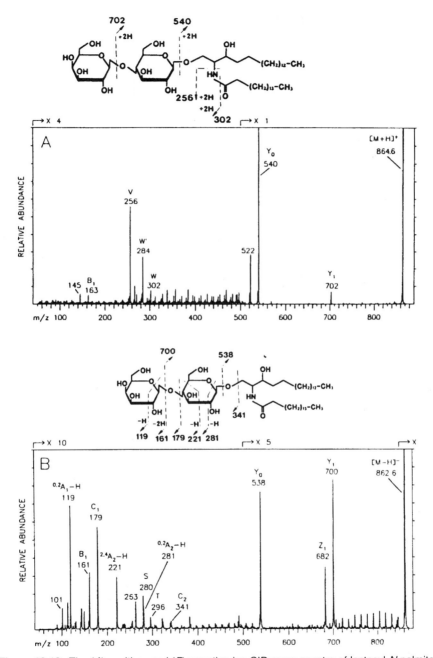

Figure 13.13. The (*A*) positive- and (*B*) negative-ion CID mass spectra of lactosyl-*N*-palmitoyl-sphinganine. [Reproduced from Ref. 64 by permission of Academic Press (copyright 1990).]

glycolipid prior to CID-MS/MS analysis, especially if it contains more sugar units. This step improves the yield of the $[M + H]^+$ and $[M - H]^-$ ions, and directs fragmentation into the useful structure-specific channels. Permethylation of amide and hydroxyl groups and borane reduction of amides are the common derivatization steps. Reduction of amides to amines creates an additional charge retention site that can induce additional structure-specific fragment ions. This aspect is demonstrated in Figure 13.14 for the reduced form of lactosyl-N-palmitoylsphinganine (compare it with the data in Figure 13.13) [64]. The structure determination of GSLs via CID of the $[M + Li]^+$ ions has also been reported by Ann and Adams [69]. These authors proposed a different nomenclature system to designate fragment ions; however, this nomenclature has not gained acceptance.

Harvey has applied MALDIMS for the analysis of GSLs [58]. 2,5-DHB, α-CHCA, and esculetin were found to be appropriate MALDI matrix materials. The molecular ion signal is produced in the form of $[M + Na]^+$ species. Fragmentation is limited and matrix-dependent; α-CHCA produces more fragmentation. The losses of water, an intact oligosaccharide moiety, and an acylamide are the common modes of fragmentation reactions that can be observed under MALDI conditions. Costello and colleagues have also recorded the MALDI mass spectra of GSLs [70,71]. They have shown that a better sensitivity for the native compounds is achieved in negative-ion mode. The dependence of fragmentation on the matrix–laser wavelength combination and on the laser power is also noted in this study. Also, as with FABMS,

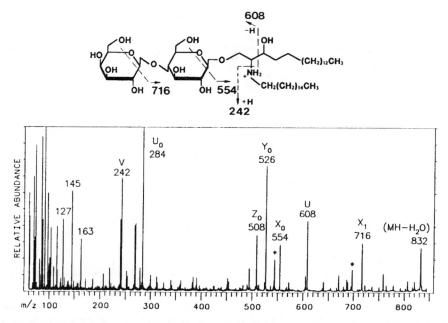

Figure 13.14. The positive-ion CID mass spectra of the reduced form of lactosyl-N-palmitoyl-sphinganine. [Reproduced from Ref. 64 by permission of Academic Press (copyright 1990).]

permethylation and peracetylation improve the sensitivity and resolution. A new matrix, 2-(4-hydroxyphenylazo)benzoic acid (HABA), was introduced by these researchers for analyzing GSLs [71].

Costello's group has developed two different approaches to MALDIMS analysis of GSLs that had been separated on thin-layer chromatographic (TLC) plates [72]. In one approach, MALDIMS was performed directly on the TLC plate. In the second approach, GSLs were heat-transferred from TLC plates onto polymer membranes. The spectral quality is better for the membrane-bound analytes. The use of ESI-MS/MS has also been explored for characterization of glycolipids [73].

13.5. MASS SPECTROMETRY OF LIPOPOLYSACCHARIDES

A complex form of glyocolipids is lipopolysaccharides (LPSs), which are major components of the outer membrane of bacterial cell walls. These species are endotoxins that are produced from Gram-negative bacteria. They are involved in a range of biological activities, and are responsible for a large number of human

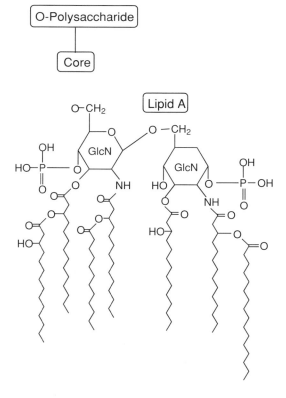

Figure 13.15. Structure of a lipopolysaccharide.

diseases (e.g., meningitis, gonorrhea, pneumonia, otitis media, chancroid). LPSs are structurally diverse, but they all share a common molecular architecture that consists of three distinct regions: the *O*-specific chain, the core oligosaccharide, and the lipid A. The first two regions are composed of the hydrophilic polysaccharide chain. Lipid A is a glucosamine disaccharide that contains *N*- and *O*-linked fatty acids, and is linked to the core region via an acidic sugar, 2-keto-3-deoxymannooctulosonic acid (KDO) (Figure 13.15). Structural heterogeneity has been observed in the lipid and glycosidic domains. The lipid A region is an endotoxic center of LPS. The structure elucidation of LPSs is important for understanding the molecular mechanism of bacterial pathology.

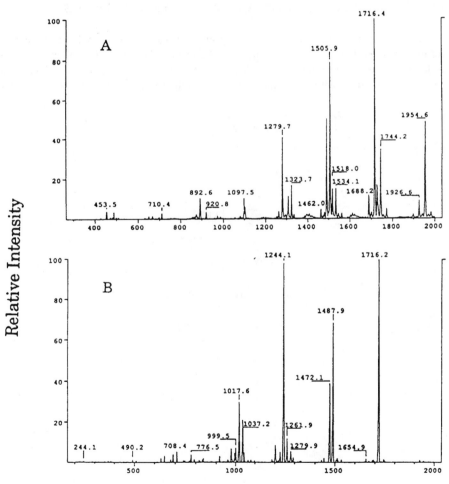

Figure 13.16. (*A*) The negative-ion ESI mass spectrum of lipid A from *Salmonella minnesota* Re95; (*B*) CID spectrum of the [M − H]⁻ ion of an hexaacyl monophosphorylated lipid A. [Reproduced from Ref. 79 by permission of Academic Press (copyright 1994).]

Mass spectrometry has played a pivotal role in characterizing LPS from different organisms. Earlier studies relied on FABMS [74] and PDMS [75–77]. Currently, interest in the use of mass spectrometry to characterize LPS has turned to ESIMS and ESI-MS/MS [78–83]. Structures of lipid A from *Salmonella typhimurium* G30/21, *Salmonella minnesota* Re95 [77], *Shigella flexneri* [77], *Enterobacter agglomerans* [80], *E. coli*, *Haemophilus influenza* 2019 [82], *Haemophilus ducreyi* 35000 [82], and *Pseudomonas aerugnosa* O6 [83] have been elucidated with these techniques. In order to obtain the structure of the lipid A moiety, it is necessary first to release it from LPS by a mild acid treatment. Improved ionization efficiency is realized in negative-ion mode, because of the presence of an acidic phosphate group on lipid A. Thus, this technique can be advantageously used to profile molecular heterogeneity in lipid A preparations. The upper panel of Figure 13.16 shows the negative-ion ESIMS spectrum of lipid A from *Salmonella minnesota* Re95 [79]. The cluster around m/z 1954 shows the alkane heterogeneity. The next four clusters represent absence of one, two, three, and four acyl groups, respectively. Finally, the remaining three clusters are a manifestation of monosaccharide structures. CID of an ion that is associated with a specific monophosphoryl lipid A species can be used to elucidate its structure. As an example, the CID spectrum of m/z 1716 is shown in the lower panel of Figure 13.16, and the fragmentation pathways that can be deduced from this spectrum are illustrated in Figure 13.17. Intact lipooligosaccharides (LOS; these compounds are the forms of LPS in which the *O*-specific chain is missing) and *O*-deacylated LOS have also been characterized by ESIMS [82].

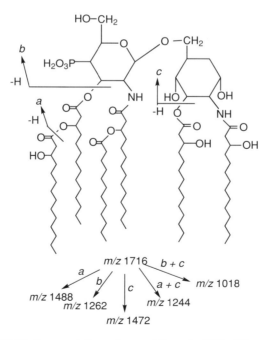

Figure 13.17. Fragmentation scheme of the m/z 1716 of Figure 13.16.

Gibson et al. have used MALDI-MS with delayed-extraction (DE)-TOFMS to characterize LOS molecules from pathogenic strains of *Haemophilus influenza*, *Haemophilus ducreyi*, and *Salmonella typhimurium* [84]. The intact and *O*-deacylated forms were both analyzed. The latter form is more amenable to MALDI analysis because of its solubility in water and the ease of cocrystalization with the 2,5-DBH matrix. Increased mass resolution and mass accuracy is realized with the DE-TOFMS analysis. Delayed-extraction and postsource decay (PSD) both provide important structure information on the phosphate and phosphoethanolamine substituents.

PDMS has been used to analyze lipid A from various sources [85]. Sufficient fragmentation was observed in negative-ion mode, from which it was feasible to localize fatty acids. Positive-ion mode also provided confirmatory data. This technique was applied to characterize lipid A from *Salmonella minnesota* Re95, *Neisseria meningitidis*, and *Yersinia enterocolitica*.

13.6. MASS SPECTROMETRY ANALYSIS OF STEROIDS

Steroids are hormones that are known to modulate DNA expression and protein expression. Adrenal hormones (e.g., aldosterone and cortisol) are known to participate in the control of metabolism and of salt and water homeostasis. The other class of steroid hormones, gonadal hormones (e.g., estradiol, progesterone, testosterone) influence mamalian sexual development and function. Several steroids also modulate neurotransmitter action. Athletes have misused anabolic steroids to enhance their physical strength and performance. A major analytic activity on steroids has been performed with GCMS and EI or CI as the ionization method [86,87]. These techniques are highly sensitive, but require multistep derivatization of these thermally labile compounds prior to mass spectrometry analysis. Alternatively, either fast GCMS with short columns [88] or a supersonic molecular beam [89] has been used to analyze underivatized steroids. LC, coupled with thermospray–MS [86,90,91], ESIMS [92–94], and APCIMS [92], has the advantage of less sample treatment. APCIMS produces the $[M + H]^+$, $[M + H - H_2O]^+$ and $[M + H - 2H_2O]^+$ ions, and the extent of H_2O loss is compound-dependent [92].

13.7. MASS SPECTROMETRY ANALYSIS OF PROSTAGLANDINS

Prostaglandins (PGs) are pharmacologically active molecules, are involved in cell–cell communication, and are oxygenated products of arachidonic acid. Like that of other simple lipid molecules, the structural analysis of these polyfunctional molecules can be carried out with EIMS, but after converting them to suitable derivatives, typically as the methyl ester trimethylsilyl ether [95]. EIMS leads to extensive fragmentation, much of which is centered around the derivative itself. The molecular

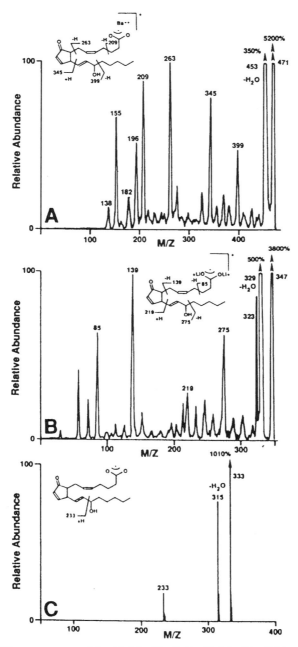

Figure 13.18. CID mass spectra of the (*A*) [M − H + Ba]$^+$, (*B*) [M − H + 2Li]$^+$, and (*C*) [M − H]$^-$ ions of PGA$_2$. These scans are useful in the identification of fatty acid substituents. [Reproduced from Ref. 96 by permission of Elsevier Science Inc. (copyright 1990 American Society for Mass Spectrometry).]

ion is quite often missing in the EI spectrum. A better approach is to use FAB coupled with MS/MS. CID of the $[M - H]^-$, $[M - H + 2Li]^+$, and $[M - H + Ba]^+$ ions of prostaglandins generates a wealth of fragment ion information that is useful for their structure analysis [96]. Several subclasses of prostaglandins such as those that belong to the PGA, PGB, PGD, PGE, and PGF series have been studied with this approach. Figure 13.18 compares the spectra of PGA_2 obtained by CID of the $[M - H]^-$, $[M - H + 2Li]^+$, and $[M - H + Ba]^+$ ions. The loss of water is a very facile process in all three spectra. Compared to CID of the carboxylate anion, CID of the metalated PGA_2 produced a greater number of fragmentations.

13.8. MASS SPECTROMETRY ANALYSIS OF LEUKOTRIENES

Leukotrienes are lipid mediators that are enzymatically derived from arachidonic acid. Mass spectrometry has played a vital role in the structural characterization of these potent bioactive species and in the identification of their biosynthetic and metabolic pathways. GCMS and EIMS of derivatized species [97–99], high- and low-energy FAB-MS/MS [100,101], and ESI-MS/MS [102] all have been used to analyze leukotrienes. Leukotriene B_4 (LTB_4), a potent neutrophil chemotactic agent, and its metabolites have been the subject of several mass spectrometry studies [100–102]. In particular, CAD of the ESI-produced $[M - H]^-$ ions has been shown to produce unique spectra for leukotriene B_4 (LTB_4) and its metabolites, and these data can be used for the complete structural characterization of these molecular species [102]. The sulfur-containing leukotrienes (LTC_4, LTD_4, and LTE_4), potent constrictors of various smooth muscles, have been characterized by EIMS, FAB, and CF-FAB of derivatized species [103].

REFERENCES

1. S. T. Weintraub, in C. N. McEwen and B. S. Larsen, eds., *Mass Spectrometry of Biological Materials*, Marcel Dekker, New York, 1990, pp. 257–286.
2. J. A. Zirrolli and R. C. Murphy, *J. Am. Soc. Mass Spectrom.* **4**, 223–229 (1993).
3. N. J. Jensen and M. L. Gross, *Lipids* **21**, 362–365 (1986).
4. N. J. Jensen and M. L. Gross, *Mass Spectrom. Rev.* **6**, 497–536 (1987).
5. K. B. Tomer, N. J. Jensen and M. L. Gross, *Anal. Chem.* **58**, 2429 (1986).
6. E. Davoli and M. L. Gross, *J. Am. Soc. Mass Spectrom.* **1**, 320–324 (1990).
7. Y.-S Chang and J. T. Watson, *J. Am. Soc. Mass Spectrom.* **3**, 769–775 (1992).
8. P. Wheelan, J. A. Zirrolli, and R. C. Murphy, *J. Am. Soc. Mass Spectrom.* **7**, 140–149 (1996).
9. F. O. Ayorinde, K. Garvin, and K. Saeed, *Rapid Commun Mass Spectrom.* **14**, 608–615 (2000).
10. R. C. Murphy, in F. Snyder, ed., *Handbook of Lipid Research 7: Mass Spectrometry of Lipids*, Plenum Press, New York, 1993, p. 213.

11. M. de Murbuker, L. G. Blomberg, N. U. Olsen, M. Bergqvist, B. G. Herslof, and F. A. Jacobs, *Lipids* **27**, 436–441 (1992).

12. P. Manninen, P. Laasko, and H. Kallio, *Lipids* **30**, 665–661 (1995).

13. R. P. Evershed, *J. Am. Soc. Mass Spectrom.* **7**, 350–361 (1996).

14. M. Cheung, A. B. Young, and A. G. Harrison, *J. Am. Soc. Mass Spectrom.* **5**, 553–557 (1994).

15. P. Laasko and H. Kallio, *Lipids* **31**, 33–42 (1996).

16. V. Stroobant, R. Rozenberg, B. el Monier, E. Deffense, and E. de Hoffmann, *J. Am. Soc. Mass Spectrom.* **6**, 498–506 (1995).

17. C. Evans, P. Traldi, M. Bambigiotti-Alberti, V. Gianelli, S. A. Corsar, and F. F. Vincieri, *Biomed. Mass Spectrom.* **20**, 351–356 (1991).

18. M. Hori, Y. Sahashi, S. Koike, R. Yamaoka, and M. Sago, *Anal. Sci.* **10**, 719–724 (1994).

19. H.-Y. Kim and N. Salem, Jr., *Anal. Chem.* **59**, 722–726 (1987).

20. P. Sundin, P. Larson, C. Wesén, and G. Odham, *Biol. Mass Spectrom.* **21**, 633–641 (1992).

21. K. L. Duffin, J. D. Henion, and J. J. Shieh, *Anal. Chem.* **63**, 1718–1788 (1991).

22. J. J. Myher, A. Kuksis, K. Geher, and P. W. Park, *Lipids* **31**, 207–215 (1996).

23. W. C. Byrdwell and W. E. Neff, *J. Liq. Chromatogr.* **18**, 2203–2225 (1996).

24. W. C. Byrdwell, E. A. Emken, and R. O. Odlof, *Lipids* **31**, 919–935 (1996).

25. C. Cheng, M. L. Gross, and E. Pittenauer, *Anal. Chem.* **70**, 4417–4426 (1998).

26. H. E. Moy and D. M. Desiderio, *Chem. Commun.* **72** (1983).

27. Y. Ohashi, *Biomed. Mass Spectrom.* **11**, 383–385 (1984).

28. S. Chen, J. Kirschner, and P. Traldi, *Anal. Biochem.* **191**, 100–105 (1990).

29. D. R. Wing, D. J. Harvey, P. La Droitte, K. Robinson, and S. J. Belcher, *J. Chromatogr.* **368**, 103–111 (1986).

30. R. C. Murphy and K. A. Harrison, *Mass Spectrom. Rev.* **13**, 57–75 (1994).

31. K. Eder, A. M. Reichlmayrlias, and M. Kirchgessner, *J. Chromatogr.* **607**, 55 (1992).

32. M. Kino, T. Matsumura, M. Gamo, and K. Saito, *Biomed. Mass Spectrom.* **9**, 363 (1982).

33. J. Sugatani, M. Kino, K. Saito, T. Matsuo, T. Matsumura, and I. Katakuse, *Biomed. Mass Spectrom.* **9**, 293–301 (1982).

34. V. N. Reinhold and S. A. Carr, *Anal. Chem.* **54**, 499–503 (1992).

35. F. B. Jungalwala, J. E. Evans, and R. H. McCluer, *J. Lipid. Res.* **25**, 738–749 (1984).

36. H.-Y. Kim and N. Salem, Jr., *Anal. Chem.* **58**, 9–14 (1986).

37. A. I. Mallet and K. Rollins, *Biomed. Environ. Mass Spectrom.* **13**, 541–543 (1986).

38. N. J. Jensen, K. B. Tomer, and M. L. Gross, *Lipids* **21**, 580–588 (1986).

39. J. A. Zirrolli, K. L. Clay, and R. C. Murphy, *Lipids* **26**, 1112–1116 (1991).

40. Z.-H. Huang, D. A. Gage, and C. C. Sweely, *J. Am. Soc. Mass Spectrom.* **3**, 71–78 (1992).

41. K. B. Tomer, N. J. Jensen, and M. L. Gross, *Anal. Chem.* **58**, 2429–2433 (1987).

42. J. Adams and M. L. Gross, *Anal. Chem.* **59**, 1576–1582 (1987).

43. D. K. Bryant, R. C. Orlando, C. Fenselau, R. C. Sowder, and L. E. Henderson, *Anal. Chem.* **63**, 1110–1114 (1991).

44. S. D. DaTorre and M. H. Creer, *J. Lipid Res.* **32**, 1159 (1991).

45. S. Chen, R. Mariot, J. Kirschner, D. Favretto, and P. Traldi, *Rapid Commun. Mass Spectrom.* **4**, 495–497 (1990).

46. S. L. Hazen, C. R. Hall, D. A. Ford, and R. W. Gross, *Clin. Invest.* **91**, 2513 (1993).

47. S. Ponchaut, K. Veitch, R. Libert, F. Van Hoof, L. Hue, and E. de Hoffmann, *J. Am. Soc. Mass Spectrom.* **7**, 50–58 (1996).

48. R. W. Gross, *Biochemistry* **23**, 158–165 (1984).

49. K. Kayganich and R. C. Murphy, *Anal. Chem.* **64**, 2965 (1992).

50. N. N. Mollova, I. M. Moore, J. Hutter, and K. H. Schram, *J. Mass Spectrom.* **30**, 1405–1420 (1995).

51. F.-F. Hsu and J. Turk, *J. Am. Soc. Mass Spectrom.* **11**, 437–439 (2000).

52. H.-Y. Kim, T.-C. L. Wang, and Y.-C. Ma, *Anal. Chem.* **66**, 3977–3982 (1994).

53. S. T. Weintraub, R. N. Pinckard, and M. Hail, *Rapid Commun. Mass Spectrom.* **5**, 309–311 (1991).

54. X. Han and R. W. Gross, *J. Am. Soc. Mass Spectrom.* **6**, 1202–1210 (1995).

55. F.-F. Hsu, A. Bohrer, and J. Turk, *J. Am. Soc. Mass Spectrom.* **9**, 516–526 (1998).

56. F.-F. Hsu and J. Turk, *J. Mass Spectrom.* **35**, 596–606 (2000).

57. D. J. Harvey, *J. Mass Spectrom.* **30**, 1333–1346 (1995).

58. D. J. Harvey, *J. Mass Spectrom.* **30**, 1311–1324 (1995).

59. Y.-C. Ma and H.-Y. Kim, *Anal. Biochem.* **226**, 293–301 (1995).

60. C. Li and A. Yergey, *J. Mass Spectrom.* **32**, 314–322 (1997).

61. H. Egge and J. Peter-Katalini, *Mass Spectrom. Rev.* **6**, 331–393 (1987).

62. J. Peter-Katalini and H. Egge, in J. A. McCloskey, ed., *Methods in Enzymology*, Academic Press, San Diego, 1990, Vol. 193, pp. 713–733.

63. J. Peter-Katalini, *Mass Spectrom. Rev.* **13**, 77–98 (1994).

64. C. E. Costello and J. E. Vath, in J. A. McCloskey, ed., *Methods in Enzymology*, Academic Press, San Diego, 1990, Vol. 193, pp. 738–768.

65. B. Dammon and C. E. Costello, *Biochemistry* **27**, 1534–1543 (1988).

66. C. E. Costello, in T. Matsuo, R. M. Caprioli, M. L. Gross, and Y. Seyama, eds., *Biological Mass Spectrometry, Present and Future*, Wiley, New York, 1994, pp. 437–462.

67. M. E. Breimer, G. C. Hansson, K.-A. Karlsson, H. Leffler, W. Pimlot, and B. E. Samuellson, *Biomed. Mass Spectrom.* **6**, 231 (1979).

68. J. M. Curtis, P. J. Derrick, J. Holgersson, B. E. Samuelsson, and M. E. Breimer, *J. Am. Soc. Mass Spectrom.* **4**, 353–359 (1993).

69. Q. Ann and J. Adams, *J. Am. Soc. Mass Spectrom.* **3**, 260–263 (1992)

70. P. Juhasz and C. E. Costello, *J. Am. Soc. Mass Spectrom.* **3**, 785–796 (1992).

71. P. Juhasz, C. E. Costello, and K. Biemann, *J. Am. Soc. Mass Spectrom.* **4**, 399–409 (1993).

72. J. Guittard, X. L. Hronowski, and C. E. Costello, *Rapid Commun. Mass Spectrom.* **13**, 1838–1849 (1999).

73. W. Wang, Z. Liu, L. Ma, C. Hao, S. Liu, V. G. Voinov, and N. I. Kalinovskaya, *Rapid Commun. Mass Spectrom.* **13**, 1189–1196 (1999).

74. N. Qureshi, K. Takayama, D. Heller, and C. Fenselau, *J. Biol. Chem.* **258**, 12947–12951 (1983).

75. N. Qureshi, K. Takayama, P. Mascagni, J. Honovich, R. Wong, and R. J. Cotter, *J. Biol. Chem.* **263**, 11971–11976 (1988).

76. R. B. Cole, L. N. Domelsmith, C. M. David, R. A. Laine, and A. J. DeLuca, *Rapid Commun. Mass Spectrom.* **6**, 616–622 (1992).

77. M. Caroff, C. Deprun, and D. Karibian, *J. Biol. Chem.* **268**, 12321–12424 (1993).

78. B. W. Gibson, W. Mclaugh, N. J. Phillips, M. A. Apicella, A. A. Campagnari, and J. M. Griffiss, *J. Bacteriol.* **175**, 2702–2712 (1993).

79. S. Chan and V. N. Reinhold, *Anal. Biochem.* **218**, 63–73 (1994).

80. S. M. Boue and R. B. Cole, *J. Mass Spectrom.* **35**, 361–368 (2000).

81. R. B. Cole, *J. Mass Spectrom.* **31**, 138–149 (1996).

82. B. W. Gibson, N. J. Phillips, W. Mclaugh, and J. J. Engstrom, in A. P. Snyder, ed., *Biochemical and Biotechnological Applications of Electrospray Ionization Mass Spectrometry*, ACS, Washington, DC, 1995, pp. 166–184.

83. S. Auriola, P. Thibault, I. Sadovskaya, E. Altman, H. Masoud, and J. C. Richards, in A. P. Snyder, ed., *Biochemical and Biotechnological Appliactions of Electrospray Ionization Mass Spectrometry*, ACS, Washington, DC, 1995, pp. 149–165.

84. B. W. Gibson, J. J. Engstrom, C. M. John, W. Hines, and A. M. Falick, *J. Am. Soc. Mass Spectrom.* **8**, 645–658 (1997).

85. D. Karibian, A. Brunelle, L. Aussel, and M. Caroff, *Rapid Commun. Mass Spectrom.* **13**, 2252–2259 (1999).

86. C. H. L. Shackleton, *J. Chromatogr. Biomed. Appl.* **379**, 91–156 (1986).

87. W. Hubbard, C. Bickel, and R. P. Schleimer, *Anal. Biochem.* **221**, 109–117 (1994).

88. S. A. Rossi, J. V. Johnson, and R. A. Yost, *Biol. Mass Spectrom.* **21**, 420–430 (1992).

89. S. Dagan and A. Amirav, *J. Am. Soc. Mass Spectrom.* **7**, 737–752, (1996)

90. N. V. Esteban and A. L. Yergey, *Steroids* **55**, 152–158 (1990).

91. J. Paulson and C. Lindberg, *J. Chromatogr.* **554**, 149–154 (1991).

92. Y.-C Ma and H.-Y Kim, *J. Am. Soc. Mass Spectrom.* **8**, 1010–1020 (1997).

93. T. M. Williams, A. J. Kind, E. Houghton, D. W. Hill, *J. Mass Spectrom.* **34**, 206–216 (1999).

94. S. Liu, J. Sjövall, and W. J. Griffiths, *Rapid Commun. Mass Spectrom.* **14**, 390–400 (2000).

95. B. S. Middleditch and D. M. Desiderio, *Anal. Biochem.* **55**, 509–526 (1973).

96. J. A. Zirrolli, E. Davoli, L. Bettazzoli, M. L. Gross, and R. C. Murphy, *J. Am. Soc. Mass Spectrom.* **1**, 325–335 (1990).

97. W. S. Powell and F. Gravelle, *J. Biol. Chem.* **264**, 5364–5369 (1989).

98. M. A. Shirley and R. C. Murphy, *J. Am. Soc. Mass Spectrom.* **3**, 762–768 (1993).

99. P. Wheelan and R. C. Murphy, *Arch. Biochem. Biophys.* **321**, 381 (1995).

100. P. Wheelan, J. A. Zirrolli, J. G. Morelli, and R. C. Murphy, *J. Biol. Chem.* **268**, 25439–25448 (1993).

101. L. J. Deterding, J. F. Curtis, and K. B. Tomer, *Biol. Mass Spectrom.* **21**, 597–609 (1992).

102. P. Wheelan, J. A. Zirrolli, and R. C. Murphy, *J. Am. Soc. Mass Spectrom.* **7**, 129–139 (1996).

103. O. A. Mamer, G. Just, C.-S. Li, P. Préville, S. Watson, R. Young, and J. A. Yergey, *J. Am. Soc. Mass Spectrom.* **5**, 292–298 (1994).

14

SCREENING
COMBINATORIAL
LIBRARIES

Combinatorial chemistry, first developed in the mid-1980s, is a fast-growing technology for synthesizing compounds on a grand scale, and testing them rapidly for desirable properties. It has radically changed the practice of preparing new compounds, and it has the potential to accelerate drug discovery and to create biologically active molecules. With this technology, it is feasible to synthesize hundreds of thousands of compounds in a time that would otherwise allow the preparation of only 50–100 compounds by conventional one-by-one synthesis, and one-by-one testing protocol. In a short period of time, combinatorial chemistry has found widespread applications in pharmaceutical and material sciences.

In a typical protocol of drug discovery with combinatorial chemistry, a large array of compounds, called *combinatorial libraries*, are produced, and are tested for high-affinity binding to enzymes, receptors, or antibodies. The compounds that provide a "hit" with the target are selected, and are further developed into "leads." Further optimization of leads results in one or two drug candidates that show significant pharmacologic activity, and can be carried on to the human trial and approval stages. The traditional approach to drug discovery has involved the synthesis, isolation, and characterization of individual compounds through a series of chemical and biochemical assays [1]. This one-at-a-time approach is expensive and time-consuming. In contrast, combinatorial chemistry has dramatically increased the number of derivatives that can be prepared and evaluated at one time for biological activity.

334

This protocol was first applied to peptide-based libraries [2–4], but it soon found applications for preparation of libraries of oligonucleotides [5,6], carbohydrates, oligocarbamate [7], and other small organic compounds [8–10].

14.1. COMBINATORIAL SYNTHETIC PROCEDURES

Combinatorial libraries are created with either a chemical synthesis or biosynthesis approach [2,11]. Several useful protocols, including parallel synthesis of arrays, the split-pool or portioning-mixing method, the biological method, and spatially addressable parallel synthesis, have been developed to prepare combinatorial libraries. Some excellent reviews on this subject can be consulted for further reading [2,11–13].

In the *parallel synthesis* procedure, an array of different compounds is simultaneously prepared. Geysen, who pioneered this technique, used multipins to synthesize a series of peptide epitopes [14]. The amino acid sequence of a peptide that is attached to a particular pin depends on the order in which those amino acids are added. The number of products synthesized is the same as the number of pins used. Several different versions of the multipin approach, such as the teabag method [1,2], Spot method (in the Spot method, compound libraries are synthesized on cellulose paper or other planar solid supports) [2,15], and the use of porous tubes [2,16], has been developed.

In the split-pool procedure, the solid support is first divided into as many equal portions as the number of amino acids in the peptide's sequence, and each portion is coupled individually to only one of the amino acids [2,3,17]. Each portion is recombined, and the whole process of splitting and combining is repeated until all amino acids have been combined. This procedure has the potential of synthesizing an exorbitantly large number of derivatives at one time. If all of the 20 naturally occurring amino acids are used in the synthesis, then the total number of the synthesized peptides can be as large as 20^n, where n is the number of amino acid residues in the peptide chains. In the mixed-reagent variation of this method, a mixture of reactants is allowed to come together in a single vessel, and is allowed to couple to a single resin support [18]. For example, combinatorial libraries can be prepared by using a mixture of amino acids in the acylation step of the synthesis protocol.

The biological method of preparing peptide libraries, in principle, is similar to the split-pool method, except that the synthesis is carried out in a host bacterium, which is infected with phages that have been prepared by inserting selected oligonucleotides in their DNA [2,19]. In the spatially addressable parallel synthesis method, the combinatorial process is achieved by controlling the addition of a chemical reagent to specified locations [2,20]. The libraries are synthesized on the surface of a glass slide, which is first functionalized with a photosensitive group via photolithography. The same photosensitive group also protects the amino acids. Synthesis is carried out only in the selected areas that are chosen by irradiating the glass slide.

14.2. GENERAL CONSIDERATIONS IN THE ANALYSIS OF COMBINATORIAL LIBRARIES

The new paradigm of combinatorial chemistry for drug discovery has radically altered the scenario of the analysis of bioactive compounds. The generation of an astronomical number of compounds in one synthesis poses substantial challenges to the screening and identification of the active compounds. A faster turn-around time and high-throughput sample analysis are the prime requirements of any analytical scheme. In addition, the analytical method must also address two other problems associated with combinatorial libraries; viz., the characterization of complex samples, and data processing speed. To this end, several new analytical approaches have been devised. Current methods include amino acid analysis [21,22], Edman degradation [23], nuclear magnetic resonance (NMR) [24,25], infrared (IR) [26], high-performance liquid chromatography (HPLC), and mass spectrometry. The last one especially stands out because of its unique advantages of low sample requirements and its ability to distinguish closely related compounds on the basis of their molecular mass and fragmentation pattern. The amount of material present on each bead makes the conventional analytical techniques (e.g., IR and NMR) less acceptable for the analysis of combinatorial libraries. Mass spectrometry has further advantage that it can be combined with high-resolution separation devices to allow the handling of complex library components. In addition, mass spectrometry-based techniques are readily applicable to peptidomimics and nonpeptide small libraries.

A conventional wisdom for verifying that a desired product has been synthesized is to analyze each component, one at a time. To increase the sample output, several parallel analyses are performed. However, time and resources are the limiting factors in analyzing one sample at a time. In order to reduce the number of samples analyzed, only a random number of library members are analyzed, and if the desired products are identified in this selection, it is assumed that no error will occur in the rest of the library [27]. An improvement over this protocol is to use deconvolution methods [28]. In one such approach, called the *mimotope* approach, the active compounds are identified via an iterative process [18]. First, rather than individually testing compounds, a pool of compounds of soluble libraries is tested. Next, a smaller pool of compounds is synthesized, and sublibraries are retested. The process is repeated until a compound with the highest activity is identified. However, this process is still laborious.

A range of mass spectrometry techniques has been developed for identification of combinatorial libraries, including those that make use of electrospray ionization (ESI)MS [29–32], HPLC-ESIMS [33–36], matrix-assisted laser desorption/ionization (MALDI)MS [37–40], and a combination of ESIMS with ion mobility-MS [41] and Fourier transform–mass spectrometry (FTMS) [42–45]. The last one has the advantage that the molecular mass of the analyte can be measured at ultrahigh resolution, which allows the separation of nominal isobaric ions. A better fingerprint of the library can be obtained at high-mass resolution. Taylor et al. have developed a system in which HPLC is coupled simultaneously to a chemiluminiscent nitrogen detector (CLND) and mass spectrometry [46]. The LC eluate is split between these detectors to provide simultaneous compound identification and quantification.

14.3. MASS SPECTROMETRY ANALYSIS OF SUPPORT-BOUND LIBRARIES

The synthesis of combinatorial libraries by the solid-phase chemistry is advantageous in many respects, especially because the reagents, solvents, and nonbound products can be easily washed away. However, manipulation of the solid support for the analysis of the products of the synthesis is problematic. Two strategies have been used to analyze the support-bound libraries by mass spectrometry methods. In one strategy, the products that are bound to the solid support are tested for the binding affinity, and the beads that exhibit activity are selected for further analysis. Those components are released from these beads, and are subsequently analyzed by MALDIMS and ESIMS or by imaging with time-of-flight (TOF)–secondary ionization mass spectrometry (SIMS). As an example, the peptides anchored via methionine are released from the SPPS resin beads by cyanogen bromide (CNBr) treatment, and are analyzed by MALDI-TOFMS [38,39]. In this study, a partial termination synthetic strategy is used. During synthesis of the library, the synthesis is terminated at each coupling step by capping the growing chain of the peptide sequence. The beads are tested via antibody–antigen interaction, and the beads that show activity are isolated for further cleavage and analysis steps. From the mass difference of the terminal products, the sequence of the full-length peptide can be deduced. Alternatively, the peptides are anchored to the resin via acid-labile and photolabile linkers, which can be released by TFA vapors [29] and UV irradiation, respectively [29].

The second approach makes use of direct in situ monitoring of the support-bound libraries. This approach is more desirable because it avoids a potentially troublesome and time-consuming chemical pretreatment step that is required for release of the support-bound products. In one protocol, the libraries are synthesized with tags to reflect the chemical history of any member of a library. The explicit information carried by these tags is encoded to provide the identity of the library components [2]. This strategy is especially useful for organic libraries. The building blocks of the encoding tags are attached to the beads in parallel with the building blocks of organic libraries. Specific sequences of peptides or oligonucleotides can serve as the identifying tags. The oligonucleotide sequences are also useful as the identifying tags for peptide libraries [47,48]. An alternate approach is binary coding, in which halobenzenes that are attached to a varying length of hydrocarbon chain are used as tags [49]. The hydrocarbon chains are attached to the beads via a cleavable spacer. The presence of the coding tag is ascertained by electron-capture gas chromatography (GC) analysis after its cleavage from the resin.

Another ingenious protocol has used a photolabile handle for analysis of the support-bound products by MALDI-TOFMS. A single laser shot cleaves the product from the SPPS resin bead, and simultaneously provides its desorption/ionization into the gas phase [50,51]. SIMS, in combination with a TOF mass analyzer, has also been used to provide in situ monitoring of the support-bound peptide intermediates [52]. The advantage of this technique is that the functionalization of the resin with a photolabile handle (for MALDIMS analysis) is not required. The polymer pins can be analyzed at any stage of the synthesis because the method is essentially nondestructive.

14.4. HIGH-THROUGHPUT SCREENING PROTOCOLS

A variety of high-throughput screening protocols have made the evaluation of combinatorial libraries practical as well as efficient [53]. A few representative protocols are described here.

Zeng and Kassel of CombiChem have designed a fully automated parallel LCMS system that allows simultaneous analysis and purification of combinatorial libraries [54]. A unique feature of this system, which is illustrated in Figure 14.1, is the use of four parallel LC columns (two each for analytic and preparative use) and a dual ESI interface. This system allows more than one sample to be analyzed at a time. In comparison with the serial LCMS systems, the parallel system provides an important increase in sample throughput, and when combined with automated data analysis and automated data processing steps, it allows up to four 96-well microtiter plates per day to be analyzed.

An automated method, based on the MALDI-FTMS technique for rapid screening of a large array of compounds that are produced by combinatorial chemistry, has been developed by Tutko et al. [55]. The library compounds are mixed with the matrix (2,5-dihydroxybenzoic acid), and are deposited on an autoindexed multiple samples disk. The matrix–sample mixture is ionized by irradiation with UV nitrogen laser (337 nm). For the mass calibration, the reference ions are generated by electron

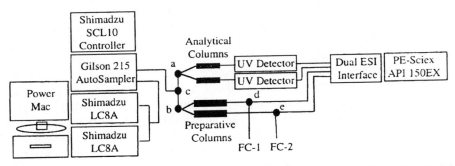

Figure 14.1. Schematic representation of the parallel LCMS system: *a*, *b*, and *c* are the switching valves; *d* and *e* the flow splitters; and FC-1 and FC-2, the fraction collectors. [Reproduced from Ref. 54 by permission of American Chemical Society, Washington, DC (copyright 1998).]

ionization of perfluorotributyl amine. This system allows the automated simultaneous analysis of 20 different compounds.

A high-throughput system that uses flow-injection analysis (FIA)MS has been developed [56,57]. The entire analysis process, including input of the sample information, adjustment of mass spectrometry analysis parameters, compound checking, and the results reporting to the customer, has been automated. A 96-well sample unit is used to manipulate the samples, and ESIMS or APCIMS is used for their analysis.

For rapid monitoring of reactions during the synthesis of combinatorial libraries, a backflush microseparation system that employs capillary columns has been reported [58]. Capillary columns have the advantage of increased sensitivity, and permit analysis of small amounts of sample, but have problems of maintaining reproducible solvent flow rates and gradients. With the recent availability of special-purpose LC pumps for capillary columns, it has become feasible to use micro-LC/MS systems on a routine basis. In particular, such a system will find increased use where the monitoring of a single bead is required for the analysis of combinatorial libraries. The system used in this report consists of two microLC pumps, an autosampler, two 6-port valves—one for injection and the other for flow switching, a short reverse-phase capillary column, and a mass spectrometer equipped with an ESI source. The sample is first loaded on the capillary column, and the trapped components are eluted into the ESI source by using a backflush flow of a high percentage of organic solvent. This system enables the in-line removal of ion-suppressing components.

A silicon chip-based nanoeletrospray (nanoES) system has been developed by Henion's group for high-throughput analysis of combinatorial libraries (for further details, see *Chemical & Engineering News*, May 15, 2000). This system contains miniature ES nozzles that are etched in the surface of a silicon wafer as circular openings, each backed with a fluid reservoir. The chips can be designed to match the wells in standard 96-, 384-, or 1536-well microtiter plates. A robotic controlled fluid-delivery probe delivers from well to well combiantorial products directly to an ESI-TOFMS instrument. Many thousands of nozzles can be arranged in parallel arrays on the surface of a single chip to provide rapid analysis of combinatorial libraries.

14.5. SCREENING OF COMBINATORIAL LIBRARIES

Screening of all the members of a library, with the intent to identify active components, is a major enterprise of combinatorial chemistry. The most popular methods are based on a specific reaction of ligands present in the library with enzymes, receptors, or antibodies. Affinity-based methods have been developed that use either immobilized receptors, immobilized ligands, or binding activity in solution [8,10,59]. A variety of solid supports, such as beads, microtiter wells, or chromatography supports, have been found convenient for immobilization of receptors. The libraries that are created on solid supports are screened directly for binding affinity of the tethered ligands to the labeled receptor. The identity of the ligand is revealed by reading the encoding tag, by its specified location, or by a structural analysis procedure. The solution-phase libraries are tested with conventional bioassays one at a time for a desired biological activity. The libraries that are synthesized on solid supports can also be tested in solution after their release from the solid support.

Several elegant screening methods that combine affinity binding with mass spectrometry have been developed for rapid analysis of compound libraries. The basic concept used in these methods is that an affinity screen selects active ligands, and mass spectrometry analysis provides their identification. A few selected methods are discussed below.

Kelly et al. have developed a technique that they call *library affinity selection–mass spectrometry* (LASMS) [59]. In this approach, the library components are reacted with the agarose gel-bound receptor in solution. After washing the beads, the peptides of increasing receptor affinity are released by treating the beads with elution buffers of different pH values, and the released peptides are identified with ESIMS or ESIMS/MS. This approach was successfully applied for analyzing the libraries that were created for the src homology 2 (SH2) domain of phosphatidylinositol 3-kinase (PI-3 kinase).

Chu et al. have developed a technique based on the combination of affinity capillary electrophoresis (ACE) with mass spectrometry for screening combinatorial libraries [60]. In this ingenious approach, the separation and characterization steps are combined into one by including the receptor in the electrophoresis buffer. When the solution of the ligands is passed through the ACE column, the ligands that have affinity for the receptor are retained, and are subsequently analyzed by ESIMS. The usefulness of this procedure was demonstrated by analyzing 100 peptide members of Fmoc-DDXX (D = Asp) library with vancomycin as the receptor.

Immunoaffinity extraction (IAE) has been combined with two HPLC columns that were coupled to mass spectrometry or tandem mass spectrometry to give on-line IAE-LC/LC-MS and IAE-LC/LC-MS/MS systems. These systems have been used for characterization of benzodiazepine libraries [61]. Benzodiazepines are first isolated from the library by trapping them in the IAE column that contains packing of antibodies specific to these molecules. A pH change in the mobile phase elutes these components onto a reverse-phase C-18 column, where benzodiazepines are separated from the antibody. The trapped benzodiazepines are backflushed onto a C-8 column for separation and identification by ESIMS or ESI-MS/MS. The use of IAE has the advantage that it can provide a simultaneous selective isolation and enrichment of the analyte. The speed of analysis is another benefit of this technique.

The same group has developed a different immunoaffinity-based approach for the characterization of small libraries [62]. This procedure differs from the one mentioned in the previous paragraph in that the affinity reaction of the ligands with the antibodies is performed in the solution, and the antibody-bound ligands are isolated with a centrifugal ultrafiltration step. The active components are released from the retained complexes by acidification, and are subsequently identified by LC-ESIMS analysis.

In another successful approach, called pulsed *ultrafiltration–mass spectrometry*, ultrafiltration has been combined on-line with ESIMS (Figure 14.2) [63]. This approach facilitates the identification of ligands via the solution-phase ligand–receptor binding reaction. The receptor is trapped in an ultrafiltration membrane. During the pulsed ultrafiltration, a pulse of library compounds is applied to the ultrafiltration cell that contains the trapped receptor. The active components are retained in the ultrafiltration cell, and the unretained components and the binding buffer are washed away to the waste. The bound components are eluted from the membrane by dissociating the complex with a flow of methanol or acidic solvent. ESIMS or ESI-MS/MS provides the identity of the liberated components. The same system can also be used to measure ligand–receptor binding constants [64]. The advantage of the solution-phase screening of receptor–ligand complexes is that their native binding interactions are fully preserved. Another strategy has used fast size-exclusion chromatography to isolate the ligand–receptor complexes that were formed by incubating the receptor with the solution of the library components [33]. The isolated ligand–receptor complexes are analyzed with ESIMS and ESI-MS/MS techniques after passing them through an on-line desalting RP cartridge.

The affinity chromatography approach to screen the active component is attractive, but has the potential of introducing artifacts. In order to minimize this problem, Smith and colleagues have developed an elegant approach that they call *bioaffinity characterization–mass spectrometry* (BACMS) [42]. Figure 14.3 is a conceptual representation of this protocol. For the screening of the active

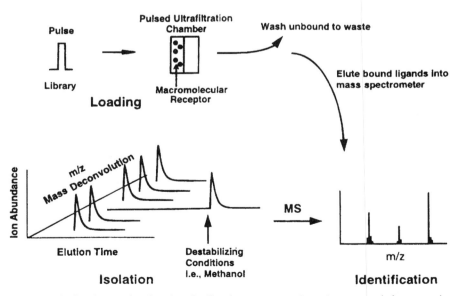

Figure 14.2. A scheme showing the ultrafiltration mass spectrometry approach for screening combinatorial libraries of those compounds that bind to a macromolecular receptor. [Reproduced from Ref. 63 by permission of American Chemical Society, Washington, DC (copyright 1997).]

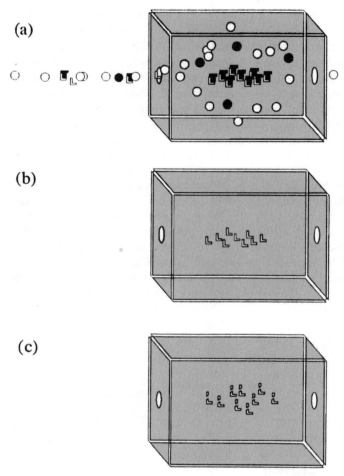

Figure 14.3. A conceptual representation of the bioaffinity characterization–mass spectrometry technique: (*a*) the complex between the ligand and target is selectively accumulated in the FTMS cell; (*b*) the complex is dissociated to liberate the ligand; (*c*) the ligand species are characterized by CAD. [Reproduced from Ref. 42 by permission of John Wiley & Sons (copyright 1995).]

components of a library, this approach also relies on formation of noncovalent complexes between the ligands and a target, but no isolation step is used prior to their mass spectrometry analysis. The ligand–target complexes formed in the solution are directly introduced into an ESI-FTMS instrument. The electrosprayed complex is first dissociated in the FTMS cell, and the free ligands are trapped in the cell; the CID spectra are acquired to characterize the free ligands. The advantages of this method over the aforementioned affinity chromatography approaches are speed of analysis, sensitivity, no contact with solid support media, and no need for interfering linker species.

REFERENCES

1. R. A. Houghton, *Proc. Natl. Acad. Sci.* (USA) **82**, 5131–5135 (1985).

2. M. A. Gallop, R. W. Barret, W. J. Dower, S. P. A. Fodor, and E. M. Gorden, *J. Med. Chem.* **37**, 1233–1251 (1994).

3. K. S. Lam, S. E. Salmon, E. M. Hersh, V. J. Hruby, and W. M. Kazmierski, *Nature (Lond.)* **354**, 82–84 (1991).

4. C. Pinnilla, J. Appel, S. Blondelle, C. Dooley, B. Dorner, J. Eichler, J. Ostresh, and R. A. Houghton, *Biopolymers* **37**, 221–240 (1995).

5. L. Gold, B. Polisky, O. Uhlenbeck, and M. Yarus, *Ann. Rev. Biochem.* **64**, 763–797 (1995).

6. J. Nielsen, S. Brenner, and K. D. Janda, *J. Am Chem. Soc.* **115**, 9812–9813 (1993).

7. C. Y. Cho, E. J. Moran, S. R. Cherry, J. C. Stephaus, S. P. A. Fodor, C. L. Adams, A. Sundaram, J. W. Jacobs, and P. G. Schultz, *Science* **261**, 1303 (1993).

8. J. A. Ellman and L. A. Thomson, *Chem. Rev.* **96**, 555–600 (1996).

9. J. A. Ellman, *Acc. Chem. Res.* **29**, 132–143 (1996).

10. B. A. Bunin, M. J. Plunkett, and J. A. Ellman *Proc. Natl. Acad. Sci.* (USA) **91**, 4708–4712 (1994).

11. E. M. Gorden, R. W. Barret, W. J. Dower, S. P. A. Fodor, and M. A. Gallop, *J. Med. Chem.* **37**, 1385–1401 (1994).

12. E. M. Gorden, M. A. Gallop, and D. V. Patel, *Acc. Chem. Res.* **29**, 144–154 (1996).

13. R. W. Armsrong, A. P. Combs, P. A. Tempest, S. D. Brown, and T. A. Keating, *Acc. Chem. Res.* **29**, 123–131 (1996).

14. H. M. Geysen, R. H. Meloen, and S. J. Barteling, *Proc. Natl. Acad. Sci.* (USA) **81**, 3998–4002 (1984).

15. R. Frank and R. Doring, *Tetrahedron* **44**, 6031–6040 (1988).

16. S. H. DeWitt, *Pharmaceut. News* **1**, 11–14 (1994).

17. A. Furka, F. Sebestyen, M. Asgedom, and G. Dibo, *Int. J. Pept. Prot. Res.* **37**, 487–493 (1991).

18. H. M. Geysen, S. J. Rodda, and T. J. Mason, *Mol. Immunol.* **23**, 709–715 (1986).

19. G. P. Smith, *Science* **228**, 1315–1317 (1985).

20. S. P. A. Fodor, J. L. Read, M. C. Pirrung, L. Stryer, A. T. Lu, and D. Solas, *Science* **251**, 767 (1991).

21. R. N. Zuckerman, J. M. Siani, and S. C. Banville, *Int. J. Pept. Prot. Res.* **40**, 497 (1992).

22. G. L. Hortin, W. D. Staatz, and S. A. Santoro, *Biochem. Int. J.* **26**, 731 (1992).

23. S. Stevanovic and G. Jung, *Anal. Biochem.* **212**, 212–220 (1992).

24. G. C. Look, C. P. Holmes, J. P. Chinn, and M. A. Gallop, *J. Org. Chem.* **59**, 7588–7590 (1994).

25. J. Chin, B. Fell, M. J. Shaprio, J. Tomesch, J. R. Wareing, and A. M. Bray, *J. Org. Chem.* **62**, 538–539 (1997).

26. W. Li. And B. Yan, *Tetrahedron Lett.* **38**, 6485–6488 (1997).

27. D. J. Ecker, T. A. Vickers, R. Hanecak, V. Driver, and K. A. Anderson, *Nucl. Acids Res.* **21**, 1853–1856 (1993).

28. J. Blake and L. Litzi-Davis, *Bioconj. Chem.* **3**, 510–513 (1992).

29. B. B. Brown, D. S. Wagner, H. M. Geysen, *Mol. Diversity.* **1**, 4 (1995).

30. Y. M. Dunayevskiy, P. Vouros, E. A. Wintner, G. W. Shipps, T. Carell, and J. Rebek, Jr., *Proc. Natl. Acad. Sci.* (USA) **93**, 6152 (1996).

31. Y. M. Dunayevskiy, P. Vouros, T. Carell, E. A. Wintner, and J. Rebek, Jr., *Anal. Chem.* **68**, 237–242 (1996).

32. C. L. Brummel, J. C. Vickerman, S. A. Carr, M. E. Hemling, G. D. Roberts, W. Johnson, J. Weinstock, D. Gaitanopouos, S. J. Benkovic, and N. Winograd, *Anal. Chem.* **67**, 2906–2915 (1995).

33. Y. M. Dunayevskiy, J.-J. Lai, C. Quinn, F. Talley, and P. Vouros, *Rapid Commun. Mass Spectrom.* **11**, 1178–1184 (1997).

34. P.-H. Lambert, J. A. Boutin, S. Bertin, J.-L. Fauchère, and J.-P Volland, *Rapid Commun. Mass Spectrom.* **11**, 1971–1976 (1997).

35. M. Ayyoub, B. Monsarrat, H. Mazarguil, and R. Gairin, *Rapid Commun. Mass Spectrom.* **12**, 557–564 (1998).

36. J. W. Metzger, K. W. Wiesmuller, V. Gnau, J. Brunjaes, and G. Jung, *Angew. Chem., Int. Ed. Engl.* **32**, 894–896 (1993).

37. C. L. Brummel, I. N. W. Lee, Y. Zhou, M. E. Hemling, and N. Winograd, *Science* **264**, 399–401 (1994).

38. R. S. Youngquist, G. R. Fuentes, M. P. Lacey, and T. Keough, *J. Am. Soc. Mass Spectrom.* **117**, 3900–3906 (1995).

39. R. S. Youngquist, G. R. Fuentes, M. P. Lacey, and T. Keough *Rapid Commun. Mass Spectrom.* **8**, 77–81 (1994).

40. J. A. Loo, D. E De John, R. R. Ogorzalek-Loo, and P. C. Andrews, *Ann. Rep. Med. Chem.* **31**, 319 (1996).

41. C. A. Srebalus, J. Li, W. S. Marshall, and D. E. Clemmer, *J. Am. Soc. Mass Spectrom.* **11**. 352–355 (2000).

42. J. E. Bruce, G. A Anderson, R. Cheng, X. Cheng, D. C. Gale, S. A. Hofstadler, B. L. Schwartz, and R. D. Smith. *Rapid Commun. Mass Spectrom.* **9**, 644–650 (1995).

43. B. E. Winger and J. E. Campana, *Rapid Commun. Mass Spectrom.* **10**, 1811–1813 (1996).

44. J. P. Nawrocki, M. Wigger, C. H. Watson, T. W. Hayes, M. W. Senko, S. A. Benner, and J. R. Eyler, *Rapid Commun. Mass Spectrom.* **10**, 1860–1864 (1996).

45. S.-A. Poulson, P. J. Gates, G. R. Cousius, and J. K. M. Sanders, *Rapid Commun. Mass Spectrom.* **14**, 44–48 (2000).

46. E. W. Taylor, M. G. Qian, and G. D. Dollinger, *Anal. Chem.* **70**, 3339–3347 (1998).

47. M. N. Needles, D. G. Jones, E. M. Tates, G. L. Heinkel, L. M. Kochersberger, W. J. Dower, R. W. Barrett, and M. A. Gallup, *Proc. Natl. Acad. Sci.* (USA) **90**, 10700–10704 (1993).

48. S. Brenner and R. A. Lerner, *Proc. Natl. Acad. Sci.* (USA) **89**, 5181–5183 (1992).

49. M. H. J. Ohlmeyer, R. N. Swanson, L. W. Dillard, J. C. Reader, G. Asouline, R. Kobayashi, and W. C. Still, *Proc. Natl. Acad. Sci.* (USA) **90**, 10922–10926 (1993).

50. M. R. Carrasco, M. C. Fitzgerald, Y. Oda, and S. B. H. Kent, *Tetrahedron Lett.* **38**, 6331–6334 (1997).

51. M. C. Fitzgerald, K. H. Harris, C. G. Shevlin, and G. Siuzdak, *Bioorg. Med. Chem. Lett.*, **6**, 979–982 (1996),

52. J. L. Aubaganac, C. Enjalbal, G. Subra, A. M. Bray, R. Combarieu, and J. Martinez, *J. Mass Spectrom.* **33**, 1094–1103 (1998).

53. J. N. Kyranos and J. C. Hogan, Jr., *Anal. Chem.* **70**, 389A–395A (1998).

54. L. Zeng and D. B. Kessel, *Anal. Chem.* **70**, 4380–4388 (1998).

55. D. C. Tutko, K. D. Henry, B. E. Winger, H. Stout, and M. E. Hemling, *Rapid Commun. Mass Spectrom.* **12**, 335–338 (1996).

56. E. Görlach, R. Richmond, and I. Lewis, *Anal. Chem.* **70**, 3227–3234 (1998).

57. G. Hegy, E. Görlach, R. Richmond, and F. Bitsch, *Rapid Commun. Mass Spectrom.* **10**, 1894–1900 (1996).

58. P. Marshall, *Rapid Commun. Mass Spectrom.* **13**, 778–781 (1999).

59. M. A. Kelly, H. Liang, I.-I. Sytwu, I. Vlattas, N. L. Lyons, B. R. Brown, and L. P. Wennogle, *Biochemistry* **35**, 11747–11755 (1996).

60. Y. H. Chu, D. P. Kirby, and B. L. Karger, *J. Am Chem. Soc.* **117**, 5419–5420 (1995).

61. M. L. Nedved, S. Habibi-Goudarzi, B. Ganem, and J. D. Henion, *Anal. Chem.* **68**, 4228–4236 (1996).

62. R. Weiboldt, J. Weigenbaum, and J. D. Henion, *Anal. Chem.* **69**, 1683–1691 (1997).

63. R. B. van Breemen, C. R. Huang, D. Nikolic, C. P. Woodbury, Y. Z. Zhao, and D. L. Venton., *Anal. Chem.* **69**, 2159–2164 (1997).

64. R. B. van Breemen, C. P. Woodbury, and D. L. Venton., in B. S. Larsen and C. N. McEwen, eds., *Mass Spectrometry of Biological Materials*, Marcel Dekker, New York, 1998, pp. 99–113.

15

CHARACTERIZATION OF OLIGONUCLEOTIDES

Deoxyribonucleic acid is one of the best known molecules; the general public knows it by its acronym, DNA. It is a storehouse of genetic information that is passed on from generation to generation. The genetic information stored in DNA is used to synthesize myriad proteins that are required to maintain essential functions of cells and organisms. Thus, it is a crucial component of all life forms. DNA testing has found widespread use in forensic identity, paternity testing, cell-line identification, and characterization of disease genes.

Structurally, DNA is a very long, thread-like macromolecule made up of large number of repeating units of nucleotides (nt) that are arranged in a doubly stranded helical structure. Each nucleotide is composed of three simpler units, a nitrogeneous base, a sugar, and a phosphate group (Figure 15.1). The nitrogenous bases are derivatives of either purine or pyrimidine. Adenine (A), guanine (G), cytosine (C), uracil (U), and thymine (T) are the five most common nitrogeneous bases, of which only adenine, guanine, cytosine, and thymine are present in DNA. The sequence of these four bases decides what genetic information a specific DNA molecule encodes. The sugar component of DNA is D-deoxyribose. The formation of the nucleotides involves the binding of the sugar to the base via the glycosidic C-1′ carbon and the N-1 of the pyrimidine or N-9 of purine. The phosphate group is linked to the sugar via the $-CH_2OH$ (i.e., at the 5′ position). The combination of only a sugar and a base is known as a *nucleoside*. Another form of a nucleic acid contains ribose as a sugar unit, hence it is named as *ribonucleic acid* (RNA). This molecule also differs from DNA in that it is formed from adenine, guanine, cytosine, and uracil as the base

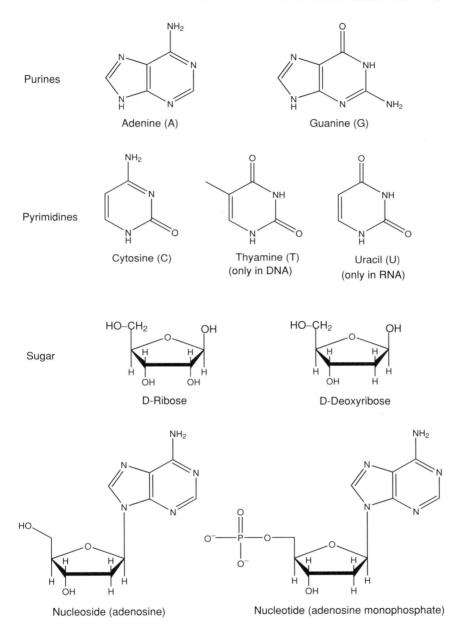

Purines

Adenine (A)

Guanine (G)

Pyrimidines

Cytosine (C)

Thyamine (T)
(only in DNA)

Uracil (U)
(only in RNA)

Sugar

D-Ribose

D-Deoxyribose

Nucleoside (adenosine)

Nucleotide (adenosine monophosphate)

Figure 15.1. The structure of nucleic acid constituents.

units (i.e., thymine in DNA is replaced by uracil in RNA). One form of the RNA, known as *messenger RNA* (mRNA), acts as a template to convey the genetic information from DNA for the synthesis of a specific protein. Two other forms of RNA are ribosomal (rRNA) and transfer RNA (tRNA). Table 15.1 lists the names of the common bases, nucleosides, and nucleotides present in DNA and RNA.

Both types of the nucleic acid are composed of chains of nucleotides. The backbone consists of sugar units that are linked by phosphodiester bridges (see Figure 15.2). The 3' carbon of one sugar and the 5' carbon of the next sugar are linked together via the phosphodiester bonds. The bases that are linked to each sugar unit constitute the side chains. Thus, the backbone of a nucleic acid terminates in either a 3'- or a 5'-OH moiety. The primary structure of nucleic acids is parallel to that of the primary structure of proteins, which are also long-chain molecules; their backbone is formed by joining 20 naturally occurring amino acids via the amide bond. By convention, the primary sequence of oligonucleotides is read from the 5' end. For example, the oligonucleotide shown in Figure 15.2 is written as 5'-d(pTGCA)-3'. An oligonucleotide chain of a specific length is referred to as an "*n*-mer" (*n* is the number of nucleotide units). Figure 15.2 also shows how to represent an oligonucleotide chain by a shorthand notation. Here, the sugar is denoted by a vertical line, the base by its letter abbreviation, and the phosphodiester bridge by the letter P within a circle.

The secondary structure of nucleic acids is composed of two strands that coil about a common axis in the form of a double helix. This observation was a landmark discovery by Watson and Crick that has important ramifications in biomedical fields. The two oligonucleotide chains of the DNA double-helix structure are the grand staircases of genetic instructions. These two chains are arranged in opposite directions, with the sugar–phosphate backbone outside and the bases directed

Table 15.1. Names of the common nucleotides and nucleosides present in DNA and RNA

Base	Nucleoside	Nucleotide	Symbol	Mass
		DNA		
Adenine, A	Deoxyadenosine	Deoxyadenosine 5'-monophosphate	dAp	313.2
Cytosine, C	Deoxycytidine	Deoxycytidine 5'-monophosphate	dCp	289.2
Thymine, T	Deoxythymidine	Deoxythymidine 5'-monophosphate	dTp	304.2
Guanine	Deoxyguanosine	Deoxyguanosine 5'-monophosphate	dGp	329.2
		RNA		
Adenine, A	Adenosine	Adenosine 5'-monophosphate	Ap	329.2
Cytosine, C	Cytidine	Cytidine 5'-monophosphate	Cp	305.2
Uracil, U	Uridine	Uridine 5'-monophosphate	Up	306.2
Guanine	Guanosine	Guanosine 5'-monophosphate	Gp	345.2

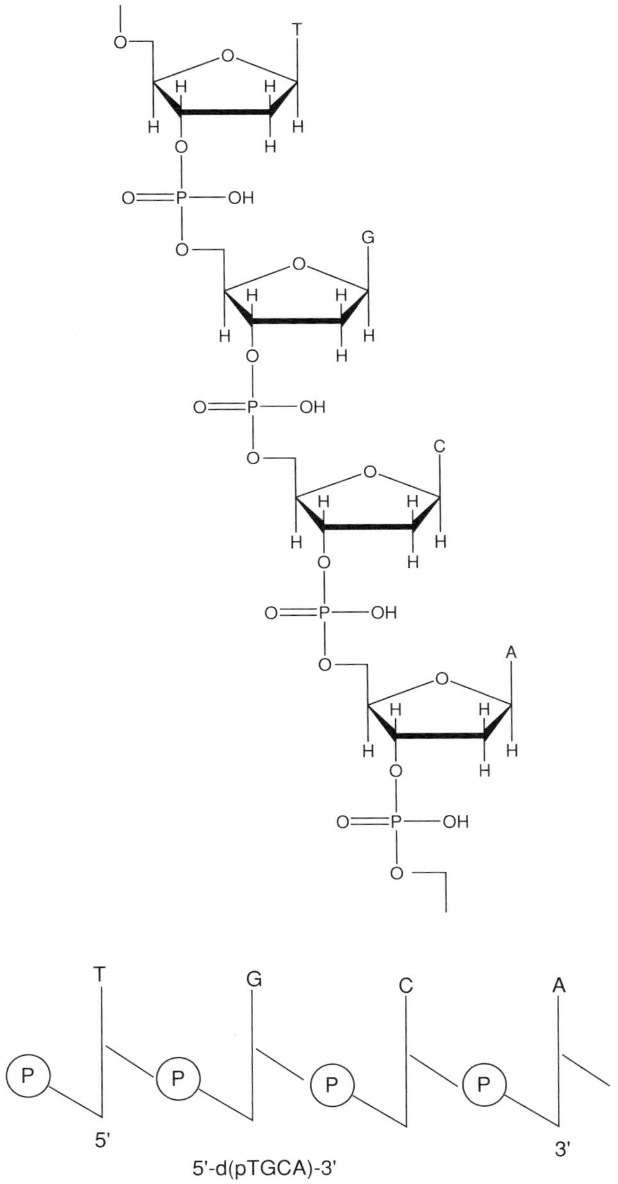

Figure 15.2. Structure of a part of DNA chain (upper part) and shorthand notation for a single-stranded DNA chain (lower part).

inside. The two oligonucleotide strands are complementary in their base sequences. Each adenine on one chain is connected via hydrogen bonds to a thymine on the other chain. Similarly, guanine and cytosine align with each other. Thus, A–T and G–C are complementary base pairs. The joining of complementary strands in DNA is termed *hybridization*. The stability of the double-stranded complexes increases with increasing length of the complementary sequence. The melting-point temperature (T_m; i.e., the temperature at which half of the number of complexes dissociates into the constituent single strands) is the measure of the stability of the double-stranded complex.

15.1. TRADITIONAL METHODS OF OLIGONUCLEOTIDE SEQUENCING

Biochemists employ a wide range of methods to sequence oligonucleotides. The most popular of these is the Sanger (or dideoxy) method [1], in which four different sets of ladders of oligonucleotides, one for each DNA base (i.e., A, C, G, and T), are synthesized by the polymerase chain reaction (PCR). A complementary single-stranded fragment is obtained from a restriction enzyme digest to serve as a primer for the synthesis. The sequence of this particular fragment is copied using DNA polymerase I (T7 polymerase and *Taq* polymerase have also been used). The synthesis is carried out by incubation with the four deoxyribose nucleosides and a $2',3'$-dideoxy analog of one of the bases (e.g., ddA for the *A* ladders). The dideoxy analogs terminate the chain, and create ladders of different lengths. The four complete sets of chain-terminated ladders are separated by gel electrophoresis, and each ladder is detected by either radiolabeling or fluorescent tagging. From the order of appearance, the sequence of the oligonucleotide is deduced. In an alternative approach, a nucleoside α-thiotriphosphate is incorporated. This reaction does not result in chain termination [2]. The four-lane sequencing ladders are produced by digestion of the full-extension products with exonuclease III, which selectively cleaves at the site of the modified base. The RNA sequencing is performed with the Sanger method either directly or after converting it to the complementary DNA. In the latter method, the reaction is catalyzed by a reverse transcriptase instead of DNA polymerase I.

The Maxam–Gilbert method makes use of base-specific chemical cleavage reactions [3]. Two steps are involved in this procedure. During the first step, the specific bases undergo chemical modifications. In the second step, the modified base is removed. Subsequent cleavage by piperidine generates a nested set of radiolabeled fragments, which are separated by gel electrophoresis. The four specific fragmentation reactions are for G, A+G, C, and C+T residues.

The sizing of the fragments that are generated by the Sanger or Maxim–Gilbert method by gel electrophoresis is time-consuming, and requires several hours. Non-gel-based methods have been developed to speed up the DNA sequencing. Multiplex capillary electrophoresis [4], miniature capillary electrophoresis chips [5], flow cytometry [6], and the use of the DNA sequencing chips [7] are some of the alternative approaches that are being pursued in this field. Since the introduction of electrospray ionization (ESI)MS [8] and matrix-assisted laser desorption/ionization (MALDI)MS [9], mass spectrometry is playing an ever-increasing role in the analysis of nucleic acids. The mass analysis of oligonucleotide fragments by mass spectrometry is much more accurate than gel-based analysis. Several excellent reviews have been written on the subject of mass spectrometry of nucleic acids [10–13].

15.2. OLDER MASS SPECTROMETRY TECHNIQUES

Although mass spectrometry has played an important role in the characterization of bases and nucleosides, it was not applicable to the direct analysis of nucleic acids and oligonucleotides that contain more than two nucleotides. The larger oligomers were degraded to constituent nucleosides; that situation is similar to what existed for proteins and peptides. The highly polar nature of nucleosides and nucleotides has posed several challenges to their analysis in the past. Before the advent of desorption and spray ionization techniques, these polar compounds had to be modified by chemical derivatization for their analysis by electron ionization (EI) or chemical ionization (CI) [14,15]. Trimethylsilylation has been the most effective approach to prepare derivatives of nucleosides [16]. Nucleic acids are usually hydrolyzed by treatment with an enzyme nuclease P1, followed by alkaline phosphatase treatment to generate the constituent nucleosides [17]. The hydrolytic mixture, after trimethyl-silylation, is screened by gas chromatography (GC)MS analysis [16,18,19]. Alternatively, high-performance liquid chromatography (HPLC) has been used to isolate individual nucleosides for further characterization by EIMS. With the availability of robust LCMS systems that used the thermospray interface, the screening of the hydrolytic mixture became a much simpler task [20]. This system also made it possible to obtain the structural information.

The introduction of field desorption (FD) [21], ^{252}Cf-plasma desorption (PD) [22], and fast-atom bombardment (FAB)–mass spectrometry [23] allowed the analysis of larger oligomers. The molecular ions of an underivatized dinuleoside monophosphate could be obtained with FDMS [24]. McNeal et al. used positive- and negative-ion PD-TOFMS for the analysis of fully protected oligonucleotides [25–28]. The potential of sequencing by this technique was demonstrated by analyzing the trichloroethyl ester of CAACCA [26]. A better success was achieved with FABMS [29-31].

15.3. DEGRADATION OF NUCLEIC ACIDS AND OLIGONUCLEOTIDES

Enzymatic hydrolysis and chemical digestion are effective means of degrading nucleic acids and oligonucleotides to provide smaller components for mass spectro-metry analysis. Enzymatic hydrolysis includes the use of base-specific endonu-cleases, nonspecific endonucleases, and exonucleases. Several base-specific endonucleases are available; for example, RNase T1 preferentially cleaves RNA at the 3′ side of all G residues to generate 3′-Gp-terminating oligonucleotides. Rnase U_2 usually cleaves at the 3′ side of all purine residues, but under specific conditions it can preferentially cleave at all A residues to generate oligonucleotides that terminate in 3′-Ap. Benzonase, nuclease P1 and phosphodiesterase I, are nonspecific endonucleases that are suitable to prepare smaller-size oligonucleotides. Exonu- claeases can cleave an oligonucleotide from either the 5′ or 3′ end. Calf spleen phosphodiestrase (CSP) is an exonuclease that cleaves DNA in a stepwise manner from the 5′ end to release the nucleoside 3′-phosphate. Snake venom phosphodies- trase (SVP) acts on the 3′ end.

15.4. MOLECULAR MASS DETERMINATION OF OLIGONUCLEOTIDES

A simple molecular mass determination can be used to provide the solution to many problems. For example, confirmation of the sequence of synthetic oligonucleotides can be readily obtained. Sizing of the DNA fragments, mapping of the DNA sequencing reactions, mass measurement of PCR-amplified products, and screening of the nucleoside oligomers are some other applications of the molecular mass determination. In addition, the base composition of oligonucleotides can be derived. With the advent of ESIMS and MALDIMS, this task is readily accomplished.

15.4.1. Electrospray Ionization for Molecular Mass Determination

Because of the multiple-charging phenomenon, ESI has become a standard techni- que for the determination of the molecular mass of macromolecules. For nucleotides, negative-ion ESIMS yields a better signal than does positive-ion mode [32]. ESI typically generates a series of ions of the type $(M - nH)^{n-}$. In addition, the cationized clusters of the type $[M - (n + 1)H + Na/K]^{n-}$ are also formed. The first use of ESIMS for the detection of oligonucleotides was reported in 1988 [33].

Sample Purification. Metal ion clusters are detrimental in ESIMS analysis because they reduce the ion signal and cause error in the molecular mass measure- ments. Therefore, removal of salts from the sample is an essential step prior to ESIMS analysis. One approach is the precipitation of nucleic acids from an ethanol solution that contains 2.5 M ammonium acetate [34]. This step removes salts, and replaces cations with ammonium ions. A significant improvement in the signal intensity is the consequence of the suppression of the metal ion adducts [34]. In this study, no Na^+ ions were detected with a 48-mer oligonucleotide, and a 77-mer contained only one Na^+ ion. During fragmentation of the ammonium phosphate ion- pairs in the gas phase, the ammonium ion transfers a proton to the phosphate group and expels an ammonia molecule, therefore, the free acid form of oligonucleotides dominates the spectrum. A further improvement in mass accuracy is realized by the addition of chelating agents and/or triethylamine (TEA) [35]. *trans*-1,2-Diamino- cyclohexane-N,N,N',N'-tetraacetic acid (DCTA) is one of the most effective chelat- ing agents for removal of certain transition metal cations and Mg^{2+} ions that are bound to RNAs. Figure 15.3 demonstrates the usefulness of this approach. This figure shows the ESI spectrum of *E. coli* 5S rRNA (120 nt). The sample of rRNA was precipitated only one time from the 2.5 M ammonium acetate solution, and 500 pmol of CDTA and 10 μL of 0.1% TEA solution were added to it before ESIMS. This procedure helped in detection of two major rRNA components. In the absence of CDTA and TEA, no discernible mass spectrum is observed. This approach has permitted molecular mass determination of nucleic acids within ±0.01% accuracy. Organic bases such as trimethylamine (TMA), TEA, or diisopropylamine can be used in place of ammonium acetate [36]. This strategy is particularly useful for phosphorothioate oligomers. Because of their high affinity for metal ions, the usual ammonium acetate precipitation and RP-HPLC protocols for removal of cations from these oligomers are not highly effective. The ammonia precipitation with 10 M ammonium acetate, however, has been recommended for these oligonucleotides [37].

Another suitable approach for removal of salts is to use RP-HPLC. An improvement has been reported in spectral quality after the RP-HPLC step [38]. This study used FT-ICRMS to measure the mass of a 50-mer DNA. A mass accuracy of better than ±0.001% can be achieved with this instrument. Ultrafiltration and size-exlusion microdevices can also be used to remove salts from the samples [39]. Muddiman et al. have developed an approach that uses microdialysis for purification

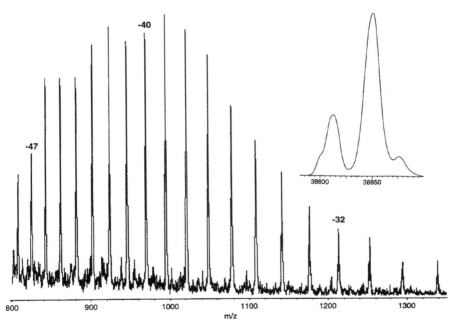

Figure 15.3. ESI mass spectrum of *E. coli* 5S rRNA (120 nt). Inset is the MaxEnt-derived molecular mass spectrum. [Reproduced from Ref. 35 by permission of Elsevier Science Inc. (copyright 1995 American Society for Mass Spectrometry).]

of the DNA fragments [40]. In this procedure, the sample is injected into a regenerated hollow cellulose fiber, and a countercurrent flow of the dialysis solvent is maintained through the annular space between a Teflon tube and the cellulose fiber.

15.4.2. MALDI for Molecular Mass Determination

Since its inception, MALDI has been a workhorse for the molecular mass determination of biopolymers. Unprecedented success has been achieved in the case of proteins. However, in the beginning, its application to the field of nucleic acids faced many challenges. The lack of an appropriate matrix, interference of sample impurities, and unexpected fragmentation of oligonucleotides under MALDI conditions were the main hurdles to the routine analysis of oligonucleotides by MALDIMS. The matrices (e.g., nicotinic acid, sinapinic acid, 2,5-dihydroxybenzoic acid) that are highly successful in the analysis of proteins provided only a limited success for the analysis of oligonucleotides. In the first application of MALDIMS, oligonucleotides with only 4–6 bases could be detected [41]. A great effort was made in the late 1990s in the search for suitable matrix materials for nucleic acid analysis. As a consequence of these efforts, a few fruitful matrix–laser wavelength combinations have emerged (see Table 2.3). Notable among these matrices is 3-hydroxypicolinic acid (HPA), which has become a standard matrix for the analysis of oligonucleotides [42]. Other matrices include picolinic acid (PA) [43], 3-aminopicolinic acid (APA) [44], 2′,4′,6′-trihydroxyacetophenone (THAP) [45], and 6-aza-2-thiothymine (ATT) [46]. All of these compounds are UV-sensitive matrices. Succinic acid and urea have found success with IR-MALDI experiments [47]. With most matrix–laser wavelength combinations, negative-ion MALDI provides a better sensitivity than does positive-ion mode of analysis. 3-HPA is an exception in this respect. It provides comparable signal intensities in both modes of ionization. Some researchers have also demonstrated the use of an ice matrix [48,49]. With laser (at 589 nm) ablation of the sample in thin ice layers, it has been possible to obtain mass spectra of a 60-base oligomer [49].

Fragmentation of oligomers during the MALDI process is a major reason for the limited mass range of the MALDI analysis of DNA. In the early days of MALDI, an intact molecular ion signal could be obtained with established matrices only for small oligonucleotides (8–10-mers). The mass spectral response was found to be a strong function of the base composition [50]. The situation has dramatically changed with the introduction of PA and its derivatives (e.g., 3-HPA) as the MALDI matrices. Fragmentation of oligomers is significantly reduced with the use of these matrices. As a consequence, large oligonucleotides with 89 bases [51] and 150 bases have been successfully detected with MALDI [43]. The use of comatrices has also improved the ion production and the stability of gas-phase oligonucleotide ions. The addition of ammonium acetate [50] and organic bases [52] has beneficial effects in terms of a reduced fragmentation of larger oligonucleotides. Simmons and Limbach have demonstrated that organic bases with a high proton affinity can serve as a proton sink, and can reduce fragmentation of oligonucleotides [53,54]. A mixture of

3-HPA with PA or diammonium citrate is currently the matrix of choice for the analysis of larger oligonucleotides. With the 3-HPA/PA combination, Tang et al. have detected DNA segments composed of as many as 500 nucleotides [55]. More recently, with an optimized matrix–laser wavelength combination, the IR-MALDI mass spectra of synthetic DNA, restriction enzyme fragments of plasmid DNA, and RNA transcripts as large as 2180 nt have been reported [56].

Sample Preparation for MALDI Analysis of Oligonucleotides. The presence of impurities is another major issue in the analysis of nucleic acids. The presence of certain impurities and additives can reduce the signal intensity. In particular, the adduction with cations, such as Na^+ and K^+, results in peak broadening and poor resolution. Some impurities may have a detrimental affect on the matrix crystallization process. As with the ESIMS analysis, these impurities must be eliminated. Several purification schemes have been developed for this purpose. Some approaches are similar to that used in ESIMS of oligonucleotides. For example, the adduction of Na^+ and K^+ ions is suppressed by the addition of ammonium salts (e.g., ammonium acetate, diammonium citrate, and diammonium tartrate) and chelating agents [45,57]. Shaler et al. have used ultrafiltration membranes to remove impurities [58]. These researchers also used a glass resin that binds the DNA molecules in the presence of a chaotropic agent. After washing of the unbound buffer components, the bound DNA is eluted from the glass beads with deionized water. Nordhoff et al. have used NH_4^+-loaded cation-exchange polymer beads to remove metal cations from the sample solution [47].

Lubman and colleagues have followed a somewhat different approach for removal of impurities from the DNA samples [59–63]. They have suggested the use of modified films of Nafian[®] and nitrocellulose as active substrates instead of normal metal sample supports for MALDIMS analysis of DNA. A Nafian[®] surface is prepared by spreading the Nafian[®] solution on the stainless-steel probe tip, followed by air drying [59]. The treatment of this film with ammonium hydroxide further improves the detection of large DNA. The active nitrocellulose substrate can also be prepared in a similar fashion by air drying the solution of Immobilon[TM]-NC pure membrane on the stainless-steel probe tip. Alternatively, the membrane can be used directly as a substrate. The use of active substrates allows on-probe purification of nucleic acid samples [60]. The use of an active nitrocellulose substrate for the analysis of a mixture of oligonucleotides is demonstrated in Figure 15.4, which shows the negative-ion spectrum of the double-stranded fragments from the DNA pBR322 after MspI digestion; fragments as large as 622 bp are visible in the spectrum [60].

15.4.3. Comparison of ESIMS and MALDIMS for Analysis of Oligonucleotides

Nordhoff et al. have compared the merits of ESI and MALDI for the detection of oligonucleotides [11]. The accessible mass range with MALDIMS is much lower than with ESIMS, primarily because of the gas-phase fragmentation of nucleic acids.

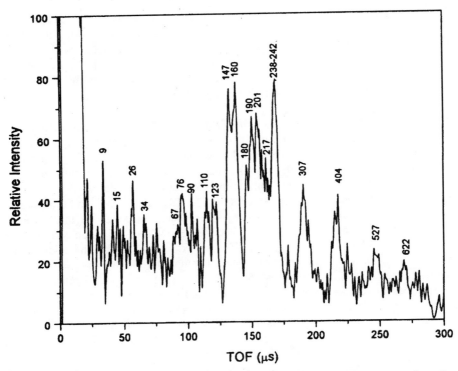

Figure 15.4. Negative-ion MALDI mass spectrum of the double-stranded fragments from the DNA pBR322 after MspI digestion on nitrocellulose substrate; matrix is a 4 : 1 mixture of 3-HPA and PA. [Reproduced from Ref. 60 by permission of American Chemical Society, Washington, DC (copyright 1995).]

The mass accuracy of ESI is also higher when coupled with FT-ICRMS. Both techniques are affected by sample impurities. For a mixture analysis, MALDI-TOFMS is superior to ESIMS for oligonucleotides that vary considerably in mass. However, ESIMS can be used for mixtures of nucleic acids that are close in mass and that contain 40–50 nt units. MALDIMS is also more sensitive than ESIMS.

15.4.4. Base Composition from Mass Measurement

One important application of the molecular mass measurement is to confirm the base composition of oligonuclotides. Mass measurement accuracy is very critical in such applications. An extensive set of calculations has shown that compositions can be derived for up to 5-mer oligonucleotides, if the mass measurement value is known within a ±0.01% accuracy [64]. The sequence confirmation can be extended to much larger oligonucleotides when certain constraints are applied. One example of such constrains is when the number of a given type of residue, which can be determined via specific chemical or enzymatic cleavage steps, is known. As a specific example, when the molecular mass is known to the nearest integer mass, base compositions can be determined up to 7-mer DNA and 8-mer RNA. If molecular mass measurement has an accuracy of ±0.01%, then base compositions up to the 14-mer level can be identified. For larger oligonucleotides, other constraints such as the chain length, compositional limits, and modification structures can be included to limit the possible compositional isomers [35]. If the chain length is known, and the molecular mass is measured within an accuracy of ±0.01%, then compositional analysis is possible up to the 25-mer level. The increased mass measurement accuracy with FT-ICRMS has been effectively used to determine the base composition of double-stranded oligonucleotides [65].

15.5. SEQUENCING OF NUCLEIC ACIDS BY MASS SPECTROMETRY TECHNIQUES

Strategies that have been developed to sequence nucleic acids depend on either gas-phase fragmentation or cleavage of nucleic acids via solution-phase chemistry followed by mass spectrometry detection of the digestion fragments.

15.5.1. Gas-Phase Sequencing Techniques

Gas-phase sequencing techniques rely on the generation and mass analysis of the sequence-specific fragment ions. These methods have the advantage of simplicity and are potentially faster than the methods based on solution-phase chemistry. A detailed account of various gas-phase sequencing methods has been provided in an excellent review [11].

Several notations have been used to designate gas-phase fragment ions. In early studies, fragment ions were designated as 5′- and 3′-sequence ions on the basis of the location of the terminal phosphate group [29,30]. Viari et al. called these fragments Y- and X-sequence ions, respectively [66]. Later, this system was modified by Nordhoff et al. to include additional fragment ions [67]. McLuckey and colleagues proposed a more systematic approach (depicted in Figure 15.5), similar to that used for peptides [68]. The four possible fragments generated by cleavage of the

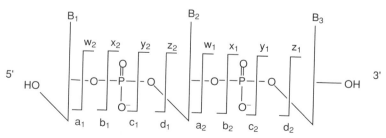

Figure 15.5. The nomenclature of fragment ions in a single-stranded DNA chain.

phosphoester and phosphorus–oxygen bonds of the sugar–phosphate backbone from the 5′ terminus are designated as *a*, *b*, *c*, and *d*. The corresponding 3′-terminus fragments are called *x*, *y*, *z*, and *w* ions. A base is indicated as B_n, where *n* is the base position from the 5′ terminus. The loss of a specific base from a fragment ion is indicated in parentheses. For example, $[a_3 - B_2(G)]$ indicates an *a*-type ion at the 3 position from which guanidine at the 2 position has been lost.

Fragmentation Characteristics of Oligonucleotides. Several ion activation techniques and instrumentation types have been employed to delineate the fragmentation characteristics of oligonucleotides. Earlier studies used FAB ionization and high- [69,70] and low-energy CID-MS/MS to characterize oligonucleotides [71– 73]. McLuckey and colleagues conducted pioneering MS/MS studies on the fragmentation of ESI-formed ions of oligonucleotides with a quadrupole ion-trap (QIT) instrument [68,74–76]. These studies have provided a great wealth of information on the fragmentation mechanisms of multiply charged oligonucleotides. According to McLuckey and colleagues, the loss of a base is a prominent reaction, and occurs via a 1,2-elimination reaction. If the charge resides on the base, then the loss of that base as an anion is preferred; otherwise, a neutral base is lost. Following loss of the base, the 3′ C–O bond is cleaved to yield complementary $(a - B_n)$- and *w*-type ions. For the 5′ base, the neutral and charged base losses both occur, whereas for the 3′ base, neutral base loss is favored. The preferrential loss of the base as an anion follows the order A > T > G > C. In the spectra obtained with a triple-sector quadrupole, no preference for loss of any base is observed, and first-generation fragment ions usually are higher in abundance [77,78]. It has also been concluded that the amount of structural information derived is greatest from this type of instrument [78]. The differences in fragmentation characteristics between the QIT and triple-sector quadrupole experiments have been explained on the grounds that a QIT samples the lowest energy pathways. In contrast, the collision activation process in a triple-sector quadrupole is competitive, and may allow access to higher-energy dissociation pathways as well [79]. Figure 15.6 shows the mechanism of

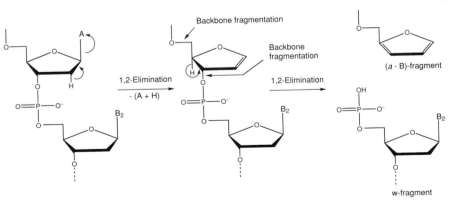

Figure 15.6. Fragmentation of DNA in the gas phase.

fragmentation of oligonucleotides. It involves protonation of the base and cleavage of the N-glycosidic bond, which lead to the loss of the base. Subsequently, phosphodiester backbone cleavage occurs at either the 5′ or the 3′-side of the sugar.

Several studies have also investigated the fragmentation mechanism of oligonucleotides under MALDI conditions [80–83]. These studies have confirmed the hypothesis that all fragment ions result from the process that is initiated by protonation of an appropriate base, followed by loss of this base as a neutral molecule or ion and further backbone cleavage.

Sequencing of Oligonucleotides by Electrospray Ionization. By adjusting the nozzle–skimmer voltage, fragmentation can be induced in the ESI source interface region via collisions of translationally excited ions with excess buffer gas molecules. This approach can be used to sequence oligonucleotides with a single-stage mass spectrometer. However, because no mass selection is done, each charge state of the analyte contributes to the fragment ion yield to produce a complex spectrum. Little et al. have demonstrated that in-source fragmentation is an efficient way to sequence oligonucleotides with < 15 bases, but requires sufficient resolution and mass accuracy to identify the resulting fragments [38]. This requirement can be satisfied with FT-ICRMS instrument, the use of which provided a mass accuracy of better than 50 ppm and a mass resolution of 10^5 for 25-mer oligonucleotides [38]. The same researchers found that the use of this technique alone could not generate complete sequence coverage for larger oligonucleotides (> 50-mers). However, by combining the data from the nozzle–skimmer dissociation and infrared multiphoton dissociation (IRMPD) experiments, nearly complete sequence coverage for these larger oligonucleotides can be obtained [84]. These two dissociation techniques yield complementary sequence ions. For example, IRMPD mainly produces the w- and $(a − B)$-type sequence ions, whereas the nozzle–skimmer dissociation generates the b-, c-, and d-type ions.

Sequencing of Oligonucleotides by MALDI-TOFMS. Several studies have reported the use of MALDI for sequencing oligonucleotides [85–92]. The molecular ions formed via MALDI are less stable than those produced by the ESI process, and they fragment at a faster timescale. Nordhoff et al. have categorized four different timescales of fragmentation of the MALDI-generated ions in TOFMS instrument: prompt, fast, fast metastable, and metastable (discussed in detail in Section 9.6.1) [11]. Fragment ions produced in any of the timescales can be used to provide sequence ion information for oligonucleotides. Prompt dissociations, which occur within the timeframe of the desorption event, have been used with linear and reflectron TOF instruments for sequencing oligonucleotides up to 21-mer in size [67]. An IR laser with succinic acid as the matrix was used in this study. Prompt dissociations have also been observed with the UV laser/2,5-DHBA matrix combination [81]. Fast dissociations occur after the desorption event, but before the ion acceleration step. The ions produced during fast dissociations can be discriminated with the delayed-extraction (DE)-TOFMS technique [85,86]. The sequencing of 11-mer oligonucleotide 5′-d(CACACGCCAGT)-3′ was achieved

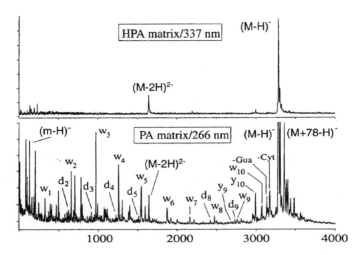

Figure 15.7. Sequencing of 11-mer oligonucleotide 5'-d(CACACGCCAGT)-3' with DE-TOFMS. PA was used as a matrix. [Reproduced from Ref. 87 by permission of American Chemical Society, Washington, DC (copyright 1996).]

using this technique, with PA as the matrix and a laser wavelength of 266 nm (see Figure 15.7) [87].

An increase of the laser pulse energy above the threshold for molecular ion production is critical for the detection of a sufficient number of fast dissociation ions. The fast metastable fragmentation reactions have a decay-time constant on the order of the acceleration event. These reactions are considered a nuisance for the MALDI measurement of large nucleic acids. In contrast, the metastable reactions that occur in the field-free region are of great utility in sequencing peptides by a technique known as *postsource decay* (PSD) analysis [88]. A few studies have extended this technique to sequence small oligonucleotides [89]. The differentiation of isomeric photomodified oligonucleotides via PSD and ESI-MS/MS in a QIT has been discussed [90]. The MALDI-induced fragmentation reactions have also been analyzed with FT-ICRMS to sequence oligonucleotides [91,92]

Sequencing of Oligonucleotides by Tandem Mass Spectrometry. Tandem mass spectrometry (MS/MS) has played a critical role in the structure elucidation of a wide variety of organic and biological compounds. The technique is discussed in more detail in Chapter 4. Several studies have reported its use to sequence oligonucleotides. In the past, FAB was used to produce molecular ions of dinucleotides for subsequent MS/MS studies with high-energy CID [69,70]. FAB-MS/MS with low-energy CID [71] and FT-ICR-MS/MS [72,73] have also been used to characterize small oligonucleotides. CID-MS/MS of the ESI-produced multiply charged ions has been performed in a QIT instrument [68,74–76,93]. Figure 15.8 demonstrates the usefulness of LC/ESI-MS coupled to a QIT to sequence a phosphorothioate oligonucleotide GCTGGCATCCGT [94]. Several

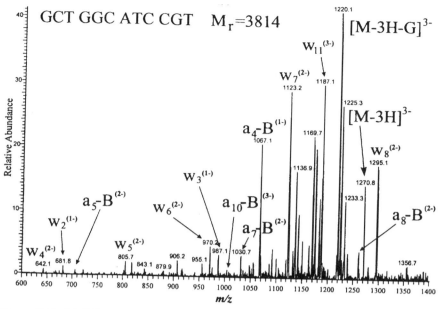

Figure 15.8. LC/ESI-MS/MS spectrum of a phosphorothioate oligonucleotide GCTGGCATCCGT. The [M+3H]$^{3+}$ ion at m/z 1270.8 was mass-selected for the MS/MS spectrum. [Reproduced from Ref. 94 by permission of John Wiley & Sons (copyright 1997).]

research groups have used ESI in combination with a triple-sector quadrupole [76–78,95,96] and FT-ICRMS [97,98] to obtain the sequence information of oligonucleotides. Labeled oligonucleotides have been sequenced with ESI-MS/MS, following their digestion with exonucleases [99].

15.5.2. Sequencing Methods Based on Solution-Phase Reactions

Mass spectrometry–based solution-phase methods for sequencing nucleic acids have been reviewed [12]. Several different approaches can be used, and are discussed below.

Ladder Sequencing. Ladder sequencing is a well-established approach in the field of protein and peptide sequencing (see Chapters 8 and 9) [100]. In essence, this approach uses consecutive cleavage reactions to generate a ladder that contains a nested set of oligonucleotide fragments. Mass spectrometry is used to obtain the mass of each member of the ladder, and the sequence is deduced from the mass difference between successive peaks in the spectrum. MALDIMS and ESIMS are both highly useful in determining the mass of the ladders. In the case of oligonucleotides, a cleavage reaction with an exonuclease sequentially removes nucleotides from either the 5′ or 3′ terminus of the molecule. The mass difference of

289.18, 304.20, 313.21, and 329.21 Da indicates the presence of C, T, A, and G, respectively. With oligoribonucleotides, the mass difference of 305.18, 306.20, 329.21, and 345.21 Da will be observed, and will correspond to C, U, A, and G, respectively. The concept of the ladder sequencing approach is depicted in Figure 15.9 with respect to an oligonucleotide 5'-AGTCACG-3'. A partial digestion of the oligonucleotide with CSP (a 3'-acting nuclease) and snake venom phosphodiesterase (a 5'-acting nuclease) can generate 5' and 3' ladders, respectively.

Pieles et al. were the first to use this approach with MALDIMS for the sequence determination of oligonucleotides [45]. A mixed-base single-stranded oligonucleotide was cleaved by digestion with 5'- and 3'-exonucleases to generate ladders from the 3' and 5' ends, respectively. Since then, several other researchers have used this approach [80,101–106]. The utility of the exonuclease-based sequencing approach can be considerably enhanced when combined with MALDI/DE-TOFMS [86]. High resolution, increased mass accuracy, and increased detection sensitivity are the benefits of this combination. As a consequence, compared to the standard MALDI-TOFMS analysis, a much longer length DNA can be sequenced with MALDI/DE-TOFMS. An example is shown in Figure 15.10, which is the negative-ion MALDI/DE-TOFMS spectrum of a 33-mer, d(GCCAGGGTTTTCCCAGT-CACGATGCAGAATTCA), following its digestion with snake venom phosphodiesterase [105].

ESIMS has also been used to measure the masses of the ladders. The use of ESIMS was demonstrated for the analysis of a 10-mer oligonucleotide [107]. The digestion mixture was continuously infused into the ESI source. The applications of

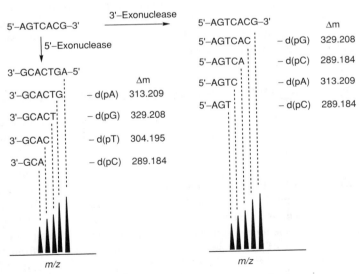

Figure 15.9. Concept of the ladder sequencing protocol. The 5' and 3' ladders can be generated with partial digestion of the oligonucleotide with CSP (a 3'-acting nuclease) and snake venom phosphodiesterase (a 5'-acting nuclease), respectively.

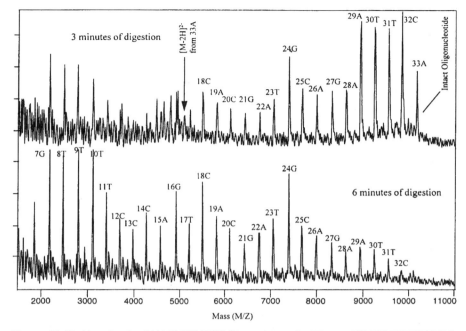

Figure 15.10. Negative-ion MALDI/DE-TOFMS spectrum of a 33-mer d(GCCAGGGTTTTCC-CAGTCACGATGCAGAATTCA) following its digestion with snake venom phosphodiesterase. [Reproduced from Ref. 105 by permission of Academic Press, San Diego (copyright 1996).]

an off-line HPLC separation and ESIMS detection of the digestion products have also been discussed [108]. This approach has the advantage that interfering salts can be removed; however, the sample amount required to sequence an oligonucleotide is larger than that required by other methods.

Sanger Sequencing and Mass Spectrometry. The Sanger dideoxy proce-dure has become the most popular method for the DNA sequencing because of its ease of application [1]. In order to speed up sequencing, the method has been combined with mass spectrometry. In this new approach, the nested sets of four sequencing reaction mixtures, one each for A, C, G, and T, are generated as usual, but in place of the gel separation and detection of the radiolabeled or photoaffinity-tagged products, mass spectrometry provides the mass analysis [12,109]. A philosophy similar to that used in gel separation is applied for readout of the oligonucleotide sequence. Mass spectra are acquired of each set of the terminated ladders by MALDI-TOFMS. The four spectra are overlaid. However, the mass measurement by mass spectrometry for the detection of the sequencing reactions is much more specific than the relative migration parameter in gel electrophoresis. The sensitivity of the mass spectrometry–based approach is poorer than the gel-based approach. Compared to the exonuclease-based sequencing approach, the Sanger sequencing is applicable to larger DNA fragments.

Smith's group first demonstrated the feasibility of the Sanger sequencing/MALDI-TOFMS combination with a mock sequencing of a potential DNA fragment that could be generated from a 35-mer primer [110]. The same group has developed a more practical approach, in which the primer extension products are generated from a vector DNA as the template; this is a common practice in the conventional Sanger sequencing method [111]. In this study, a single-stranded M13-mp bacteriophage template was obtained from infected *E. coli* cells. The results of their experiment are shown in Figure 15.11. Shaler et al. have used Sanger sequencing with the MALDI-TOFMS detection to sequence a synthetic template composed of 45 nucleotides [58]. Sequencing information could be read to 19 bases past the primer when a 12-mer primer was used, but a 21-mer primer could identify nearly the entire sequence of the template. The Sanger sequencing approach has also benefited from the aforementioned advantages of DE-TOFMS [112]. Analysis of the primer extension products from the Sanger reaction of a 50-base template with MALDI-DE-TOFMS provided the sequence information up to 32 bases past a 13-mer primer. The combination of the Sanger sequencing and mass spectrometry was applied to identify a mutation of the cystic fibrosis gene [113].

Chemical Cleavage and Mass Spectrometry. Sequence ladders can also be generated by cleavage of oligonucleotides by chemical means such as acid hydrolysis, base hydrolysis, and alkylation [12]. Chemical digestion reactions, in general, are nonspecific. The glycosidic linkage in oligonucleotides is susceptible to cleavage under acid hydrolysis conditions. This reaction is more specific for DNA, and in combination with mass spectrometry, it has a potential to sequence oligonucleotides. Nordhoff et al. have used this reaction to hydrolyze polyuridylic acids, and to analyze the sequence ladders that resulted from this reaction with MALDI-TOFMS [114].

Base hydrolysis is more specific for RNA and methylphosphonate-linked oligonucleotides. Keough et al. have used this technique to generate mass ladders of methylphosphonate oligonucleotides for their sequence determination [101,115].

Alkylation is more specific for phosphorothioate-modified oligonucleotides. Alkylation of the phosphorothioate group renders these oligonucleotides susceptible to cleavage at the site of modification. The mass ladders can be readily analyzed by MALDI-TOF for the sequence determination of phosphorothioate-linked oligonucleotides. Polo et al. applied the strategy of alkylation in combination with negative-ion MALDI/DE-TOFMS to obtain the sequence information of a phosphorothioate oligonucleotide 10-mer, d(GCATGACTCA) [116].

For specific chemical cleavage reactions, an approach similar to that used in the Maxam–Gilbert protocol has been developed. Isola et al. applied this method to provide sequence information for 30- and 60-mer DNAs [117]. In order to adopt this procedure for mass spectrometry detection, DNAs are labeled with biotin rather than with radiolabels, which is a common practice in the conventional Maxam–Gilbert approach. The DNAs are subjected to the usual cleavage protocol (see Section 15.1). The biotin-containing fragments are separated from other fragments by capturing them with streptavidin-coupled magnetic beads. The fragments are released from the

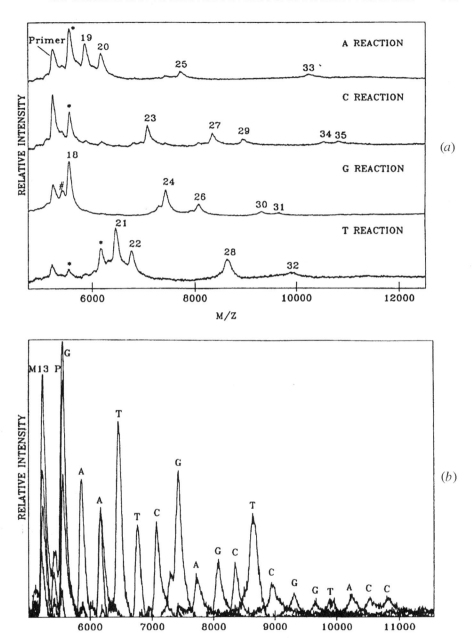

Figure 15.11. Mapping of DNA with the Sanger sequencing–mass spectrometry approach. (*a*) Negative-ion mass spectra of four sequencing reactions, A, C, G, and T. The numbers on the peaks represent the length of DNA in base pairs. (*b*) Overlay of the four mass spectra of (*a*). The letters indicate the sequence. [Reproduced from Ref. 111 with the permission of John Wiley & Sons (copyright 1996).]

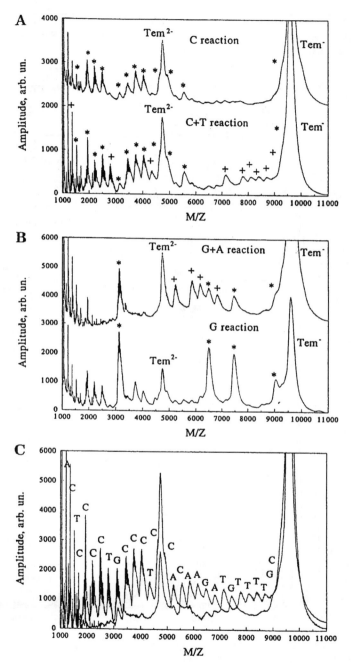

Figure 15.12. Sequencing of DNA with the Maxam–Gilbert/mass spectrometry approach. MALDI spectra of the sequencing reactions are shown in (*A*) and (*B*) and the composite spectra of the 30-mer DNA in (*C*). [Reproduced from Ref. 117 by permission of American Chemical Society, Washington, DC (copyright 1999).]

beads by hot ammonia treatment, and are analyzed by MALDI-TOFMS with a 3-HPA/PA/ammonium chloride (9 : 1 : 1, molar ratio) matrix combination. Figure 15.12 shows the MALDI spectra of the respective sequencing reactions of this protocol (Panels A and B) and the composite spectra of the 30-mer DNA (Panel C).

Analysis of PCR Products. PCR is a useful procedure to produce informative nucleic acid fragments from small quantities of DNA or RNA. PCR analysis involves the isolation of the template DNA from the sample matrix, amplification of the selected region of DNA, and determination of the size of the amplified products. Gel electrophoresis is the standard approach to separate and measure the size of the PCR products. A growing number of applications using mass spectrometry to determine the mass of the PCR products have been reported [42,71,112,113,118–123].

A key aspect in the success of mass spectrometry is the purification of PCR products. Liu et al. have used nitrocellulose films for the MALDIMS detection of several PCR products derived from the human genome [60]. Muddiman et al. have purified PCR products via microdialysis [40]. Ross and Belgrader have used affinity-

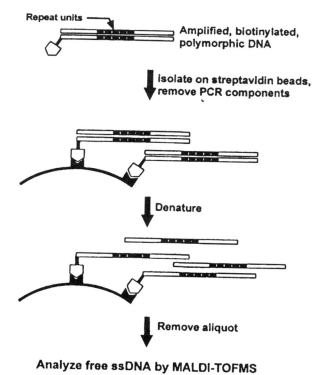

Figure 15.13. Schematic of affinity-capture purification of PCR products before MALDI-TOFMS analysis. [Reproduced from Ref. 118 by permission of American Chemical Society, Washington, DC (copyright 1997).]

capture to purify PCR products before MALDI-TOFMS analysis [118]. This procedure was applied to analyze short tandem repeat polymorphism in human DNA [118,119]. The protocol for this analysis is illustrated in Figure 15.13. Biotinylated PCR-amplified products are captured by streptavidin-coated magnetic particles [118]. After the beads have been separated and washed, the double-stranded DNA is heat-denatured, and directly analyzed by MALDI-TOFMS. Analysis of the DNA fragments from PCR devices is improved with the use of MALDI/DE-TOFMS [119]. Smith and colleagues have used ESI coupled to FT-ICRMS for analyzing double-stranded PCR products from the *Bacillus cerus* group [120–122].

Differentiation of nucleotide compositions of PCR products requires a high degree of precision in their mass measurement. Chen et al. have developed a stable-isotope mass labeling technique for this purpose [123]. During PCR amplifications of the target DNA sequences, $^{13}C/^{15}N$-labeled nucleotides are incorporated. The mass shift due to the labeling of a single type of nucleotide can reveal the number of that type of nucleotide in a given DNA. With this approach, the nucleotide composition of a mass-tagged PCR product up to 70-mer can be determined accurately from its molecular mass.

A valuable contribution of this technique is the detection of DNA polymorphism in various genes, especially those associated with genetically inherited human diseases.

REFERENCES

1. F. Sanger, S. Nicklen, and A. R. Coulson, *Proc. Natl. Acad. Sci.* (USA) **74**, 5463–5467 (1977).

2. D. B. Olsen, G. Wunderlich, G. A. Uy, and F. Eckstein, *Meth. Enzymol.* **218**, 79–92 (1993).

3. A. M. Maxam and W. Gilbert, *Meth. Enzymol.* **218**, 79–92 (1993).

4. E. N. Fung and E. S. Young, *Anal. Chem.* **67**, 1913–1919 (1995).

5. A. T. Wolley and A. R. Mathies, *Anal. Chem.* **67**, 3676–3680 (1995).

6. J. T. Petty, M. E. Johnson, P. M. Goodwin, J. C. Martin, J. H. Jett, and R. A. Keller, *Anal. Chem.* **67**, 1755–1761 (1995).

7. A. C. Pease, D. Solas, E. J. Sullivan, M. T. Cronin, C. P. Holmes, and S. P. A. Fodder, *Proc. Natl. Acad. Sci.* (USA) **91**, 5022–5026 (1994).

8. J. B. Fenn, M. Mann, C. K. Meng, S. F. Wong, and C. M. Whitehouse, *Science* **246**, 64–71 (1989).

9. M. Karas and F. Hillenkamp, *Anal. Chem.* **60**, 2299–2301 (1988).

10. P. F. Crain, *Mass Spectrom. Rev.* **9**, 505–554 (1990).

11. E. Nordhoff, F. Kirpekar, and P. Roepstorff, *Mass Spectrom. Rev.* **15**, 67–138 (1996).

12. P. A. Limbach, *Mass Spectrom. Rev.* **15**, 297–336 (1996).

13. K. K. Murray, *J. Mass Spectrom.* **31**, 1203–1215 (1996).

14. J. A. McCloskey, *Meth. Enzymol.* **193**, 771–781 (1990).

15. J. A. McCloskey, *Meth. Enzymol.* **193**, 825–842 (1990).

16. K. H. Schram, *Meth. Enzymol.* **193**, 790–796 (1990).

17. P. F. Crain, *Meth. Enzymol.* **193**, 782–790 (1990).

18. M. Buck, J. A. McCloskey, B. Basile, and B. N. Ames, *Nucl. Acids Res.* **10**, 5649 (1982).

19. M. Dizdaroglu, *Meth. Enzymol.* **193**, 842–856 (1990).

20. J. A. McCloskey, *Meth. Enzymol.* **193**, 796–824 (1990).

21. H. D. Beckey, *Principles of Field Ionization and Field Desorption Mass Spectrometry,* Pergamon, Oxford, England, 1977.

22. B. Sundqvist and R. D. Macfarlane, *Mass Spectrom. Rev.* **2**, 421–460 (1985).

23. M. Barber, R. S. Bordoli, R. D. Sedgwick, and A. N. Tyler. *J. Chem. Soc. Chem. Commun.* 325–327 (1981).

24. H.-R. Schulten and H. M. Schiebel, *Z. Anal. Chem.* **280**, 139 (1976).

25. C. J. McNeal, S. A. Narang, R. D. Macfarlane, H. M. Hsiung, and R. Brousseau, *Proc. Natl. Acad. Sci. (USA)* **77**, 735–739 (1980).

26. C. J. McNeal, C. J. Oglivie, M. Y. Theriault, and M. J. Nemer, *J. Am. Chem. Soc.* **104**, 976–980 (1982).

27. C. J. McNeal, C. J. Oglivie, M. Y. Theriault, and M. J. Nemer, *J. Am. Chem. Soc.* **104**, 981–984 (1982).

28. A. Viari, J. P. Ballini, P. Meleard, P. Vigny, P. Dousset, C. Blonski, and D. Shire, *Biomed. Environ. Mass Spectrom.* **16**, 225 (1988).

29. L. Grotjahn, R. Frank, and H. Bloecker, *Nucleic Acids Res.* **10**, 4671–4678 (1982).

30. L. Grotjahn, R. Frank, and H. Bloecker, *Biomed. Mass Spectrom.* **12**, 514–524 (1985).

31. A. M. Hogg, J. G. Kelland, J. C. Vederas, and C. Tamm, *Helv. Chim. Acta* **69**, 908–917 (1986).

32. D. M. Reddy, R. A. Reiger, M. C. Torres, and C. R. Iden, *Anal. Biochem.* **220**, 200–207 (1994).

33. T. R. Covey, R. F. Bonner, B. I. Shushan, and J. D. Henion, *Rapid Commun. Mass Spectrom.* **2**, 249–256 (1988).

34. J. T. Stults and J. C. Marsters, *Rapid Commun. Mass Spectrom.* **5**, 359–363 (1991).

35. P. A. Limbach, P. F. Crain, and J. A. McCloskey, *J. Am. Soc. Mass Spectrom.* **6**, 27–39 (1995).

36. N. Potier, A. van Dorsselaer, Y. Cordier, O. Roch, and R. Bischoff, *Nucl. Acids Res.* **22**, 3895 (1994).

37. M. Greig and R. H. Griffey, *Rapid Commun. Mass Spectrom.* **9**, 97–102 (1995).

38. D. P. Little, T. W. Thannhauser, and F. W. McLafferty, *Proc. Natl. Acad. Sci. (USA)* **92**, 2318–2322 (1995).

39. P. F. Crain, *Meth. Enzymol.* **193**, 857–865 (1990).

40. D. C. Muddiman, D. S. Wunschel, C. Liu, L. Pasa-Tolic, K. F. Fox, A. Fox, G. A. Anderson, and R. D. Smith, *Anal. Chem.* **68**, 3705–3712 (1996).

41. B. Spengler, Y. Pan, R. J. Cotter, and L.-S. Kan, *Rapid Commun. Mass Spectrom.* **4**, 99–102 (1990).

42. K. J. Wu, A. Steding, and C. H. Becker, *Rapid Commun. Mass Spectrom.* **7**, 142–146 (1993).

43. K. Tang, N. I. Taranenko, S. L. Allman, C. H. Chen, L. Y. Ch'ang, and K. B. Jacobson, *Rapid Commun. Mass Spectrom.* **8**, 673–677 (1994).

44. N. I. Taranenko, K. Tang, S. L. Allman, L. Y. Ch'ang, and C. H. Chen, *Rapid Commun. Mass Spectrom.* **8**, 1001–1006 (1994).

45. U. Pieles, W. Zürcher, M. Schär, and H. E. Moser, *Nucl. Acids Res.* **21**, 3191–3196 (1993).

46. P. Lecchi, H. M. T. Le, and L. K. Pannell, *Nucl. Acids Res.* **23**, 1276–1277 (1995).

47. E. Nordhoff, A. Ingendoh, R. Cramer, A. Overberg, B. Stahl, M. Karas, and F. Hillenkamp, *Rapid Commun. Mass Spectrom.* **6**, 771–776 (1992).

48. P. Williams, *Int. J. Mass Spectrom. Ion Proc.* **131**, 335–344 (1994).

49. R. W. Nelson, R. M. Thomas, and P. Williams, *Rapid Commun. Mass Spectrom.* **4**, 348–351 (1990).

50. G. J. Currie and J. R. Yates, III, *J. Am. Soc. Mass Spectrom.* **4**, 955–963 (1994).

51. K. J. Wu, T. A. Shaler, and C. H. Becker, *Anal. Chem.* **66**, 1637–1645 (1994).

52. T. A. Simmons and P. A. Limbach, *Rapid Commun. Mass Spectrom.* **11**, 567–572 (1997).

53. T. A. Simmons and P. A. Limbach, *J. Am. Soc. Mass Spectrom.* **9**, 668–675 (1998).

54. P. A. Limbach and T. A. Simmons, *Rapid Commun. Mass Spectrom.* **13**, 567–572 (1999).

55. K. Tang, N. I. Taranenko, S. L. Allman, L. Y. Ch'ang, and C. H. Chen, *Rapid Commun. Mass Spectrom.* **8**, 727–737 (1994).

56. S. Berkenkamp, F. Kirpekar, and F. Hillenkamp, *Science* **281**, 260–262 (1998).

57. Y. F. Zhu, N. I. Taranenko, K. Tang, S. L. Allman, S. A. Martin, L. Huff, and C. H. Chen, *Rapid Commun. Mass Spectrom.* **10**, 1591–1596 (1996).

58. T. A. Shaler, Y. Tan, J. N. Wickham, K. J. Wu, A. and C. H. Becker, *Rapid Commun. Mass Spectrom.* **9**, 942–947 (1995).

59. J. Bai, Y.-H. Liu, X. Liang, Y. Zhu, and D. M. Lubman, *Rapid Commun. Mass Spectrom.* **9**, 1172–1176 (1995).

60. Y.-H. Liu, J. Bai, X. Liang, D. M. Lubman, and P. J. Venta, *Anal. Chem.* **67**, 3482–3490 (1995).

61. Y.-H. Liu, J. Bai, Y. Zhu, X. Liang, D. Siemieniak, P. J. Venta, and D. M. Lubman, *Rapid Commun. Mass Spectrom.* **9**, 735–743 (1995).

62. J. Bai, Y.-H. Liu, D. M. Lubman, and D. Siemieniak, *Rapid Commun. Mass Spectrom.* **8**, 687–691 (1994).

63. D. M. Lubman, J. Bai, Y.-H. Liu, J. R. Srinivasa, Y. Zhu, D. Siemieniak, P. J. Venta, in B. S. Larsen and C. N. McEwen, eds., *Mass Spectrometry Biological Materials*, Marcel Dekker, 1998, pp. 405–434.

64. S. C. Pomerantz, J. A. Kowalack, and J. A. McCloskey, *J. Am. Soc. Mass Spectrom.* **4**, 203–209 (1993).

65. D. A. Aaserud, N. L. Kelleher, D. P. Little, and F. W. McLafferty, *J. Am. Soc. Mass Spectrom.* **7**, 1266–1269 (1996).

66. A. Viari, J. P. Ballini, P. Vigny, D. Shire, and P. Dousset, *Biomed. Environ. Mass Spectrom.* **14**, 83–90 (1987).

67. E. Nordhoff, M. Karas, R. Cramer, S. Hahner, F. Hillenkamp, A. Lezius, F. Kirpekar, K. Kristiansen, J. Muth, C. Meier, and J. W. Engels, *J. Mass Spectrom.* **30**, 99–112 (1995).

68. S. A. McLuckey, G. J. Van Berkel, and G. L. Glish, *J. Am. Soc. Mass Spectrom.* **3**, 60–70 (1992).

69. R. L. Cerny, M. L. Gross, and L. Grotjahn, *Anal. Biochem.* **156**, 424–435 (1986).

70. R. L. Cerny, K. B. Tomer, M. L. Gross, and L. Grotjahn, *Anal. Biochem.* **165**, 175–182 (1987).

71. D. R. Phillips and J. A. McCloskey, *Int. J. Mass Spectrom. Ion Proc.* **128**, 61–82 (1993).

72. M. T. Rodgers, S. Cambell, E. M. Marzluff, and J. L. Beauchamp, *Int. J. Mass Spectrom. Ion Proc.* **148**, 1–23 (1995).

73. M. T. Rodgers, S. Cambell, E. M. Marzluff, and J. L. Beauchamp, *Int. J. Mass Spectrom. Ion Proc.* **137**, 121–149 (1994).

74. S. Habibi-Goudarzi and S. A. McLuckey, *J. Am. Soc. Mass Spectrom.* **6**, 102–113 (1995).

75. S. A. McLuckey and S. Habibi-Goudarzi, *J. Am. Chem. Soc.* **115**, 12085–12095 (1995).

76. S. A. McLuckey, G. Viadyanathan, and S. Habibi-Goudarzi, *J. Mass Spectrom.* **30**, 122–1229 (1995).

77. P. F. Crain, J. M. Gregson, J. A. McCloskey, C. C. Nelson, J. M. Peltier, D. R. Phillips, S. C. Pomerantz, and D. M. Reddy, in A. L. Burlingame and S. A. Carr, eds., *Mass Spectrometry in the Biological Sciences*, Humana Press, Totowa, NJ, 1996, pp. 497–517.

78. J. Boschenok and M. M. Sheil, *Rapid Commun. Mass Spectrom.* **10**, 144–149 (1996).

79. J. Ni, S. C. Pomerantz, J. Rozenski, Y. Zhang, and J. A. McCloskey, *Anal. Chem.* **68**, 1989–1999 (1996).

80. F. Kirpekar, E. Nordhoff, K. Kristiansen, P. Roepstorff, S. Hahner, and F. Hillenkamp, *Rapid Commun. Mass Spectrom.* **9**, 525–531 (1995).

81. L. Zhu, G. R. Parr, M. C. Fitzgerald, C. M. Nelson, and L. M. Smith, *J. Am. Chem. Soc.* **117**, 6048–6056 (1995).

82. J. Gross, A. Leisner, F. Hillenkamp, S. Hahner, M. Karas, J. Schäfer, F. Lützenkirchen, and E. Nordhoff, *J. Am. Soc. Mass Spectrom.* **9**, 866–878 (1998).

83. J. Krause, M. Scalf, and L. M. Smith, *J. Am. Soc. Mass Spectrom.* **10**, 423–429, (1999).

84. D. P. Little and F. W. McLafferty, *J. Am. Soc. Mass Spectrom.* **7**, 1266–1269 (1996).

85. R. S. Brown and J. J. Lennon, *Anal. Chem.* **67**, 1998–2003 (1995).

86. M. L. Vestal, P. Juhasz, and S. A. Martin, *Rapid Commun. Mass Spectrom.* **9**, 1044–1050 (1995).

87. P. Juhasz, M. T. Roskey, I. P. Smirnov, L. A. Haff, M. L. Vestal, and S. A. Martin, *Anal. Chem.* **68**, 941–946 (1996).

88. B. Spengler, *J. Mass Spectrom.* **32**, 1019–1036 (1997).

89. G. Talbo and M. Mann, *Rapid Commun. Mass Spectrom.* **10**, 100–103 (1996).

90. Y. Wang, J.-S Taylor, and M. L. Gross, *J. Am. Soc. Mass Spectrom.* **10**, 329–338 (1999).

91. A. Meyer, M. Spinelli, J.-L. Imbach, and J.-J. Vasseur, *Rapid Commun. Mass Spectrom.* **14**, 234–242 (2000).

92. E. A. Stemmler, M. V. Bucahnan, G. B. Hurst, and R. L. Hettich, *Anal. Chem.* **67**, 2924–2930 (1995).

93. S. Habibi-Goudarzi and S. A. McLuckey, *J. Am. Soc. Mass Spectrom.* **5**, 740–747 (1994).

94. R. H. Griffey, M. J. Greig, H. J. Gaus, K. Liu, D. Monteith, M. Winniman, and L. L. Cummins, *J. Mass Spectrom.* **32**, 305–313 (1997).

95. J. P. Barry, P. Vouros, A van Schepdael, and S.-J. Law, *J. Mass Spectrom.* **30**, 993–1006 (1995).

96. J. Ni, M. A. A. Mathews, and J. A. McCloskey, *Rapid Commun. Mass Spectrom.* **11**, 535–540 (1997).

97. D. P. Little, R. A. Chorush, J. P. Spier, M. W. Senko, N. L. Kelleher, and F. W. McLafferty, *J. Am. Chem. Soc.* **116**, 4893–4897 (1996).

98. D. P. Little, D. J. Aaserud, G. A. Valaskovic, and F. W. McLafferty, *J. Am. Chem. Soc.* **118**, 9352–9359 (1996).

99. H. Wu, R. L. Morgan, and H. Aboleneen, *J. Am. Soc. Mass Spectrom.* **9**, 660–667 (1998).

100. B. T. Chait, R. Wang, R. C. Beavis, and S. B. H. Kent, *Science* **262**, 89–92 (1993).

101. T. Keough, J. D. Shaffer, M. P. Lacy, T. A. Riley, W. B. Marvin, M. A. Scurria, J. A. Hasselfield, and E. P. Hesselberth, *Anal. Chem.* **68**, 3405–3412 (1996).

102. C. M. Bentzley, M. V. Johnston, B. S. Larsen, and S. Gutteridge, *Anal. Chem.* **68**, 2141–2146 (1996).

103. C. M. Bentzley, M. V. Johnston, and B. S. Larsen, *Anal. Biochem.* **258**, 31–37 (1998).

104. W. P. Bartoloni, C. M. Bentzley, M. V. Johnston, and B. S. Larsen, *J. Am. Soc. Mass Spectrom.* **10**, 521–528 (1999).

105. I. P. Smirnov, M. T. Roskey, P. Juhasz, E. J. Takach, S. A. Martin, and L. A. Haff, *Anal. Biochem.* **238**, 19–25 (1996).

106. C.-W. Chou, S. E. Bingham, and P. Williams, *Rapid Commun. Mass Spectrom.* **10**, 1410–1414 (1996).

107. P. A. Limbach, P. F. Crain, and J. A. McCloskey, *Nucl. Acids Res. Symp. Ser.* **31**, 127–128 (1994).

108. R. P. Glover, G. M. A. Sweetman, P. B. Framer, and G. C. K. Roberts, *Rapid Commun. Mass Spectrom.* **9**, 97–102 (1995).

109. L. M. Smith, *Science* **262**, 530–532 (1993).

110. M. C. Fitzgerald, L. Zhu, and L. M. Smith, *Rapid Commun. Mass Spectrom.* **7**, 895–897 (1993).

111. S. Mouradian, D. R. Rank, and L. M. Smith, *Rapid Commun. Mass Spectrom.* **10**, 1475–1478 (1996).

112. M. T. Roskey, P. Juhasz, I. P. Smirnov, E. J. Takach, S. A. Martin, and L. A. Haff, *Proc. Natl. Acad. Sci.* (USA) **93**, 4724–4729 (1996).

113. L. Y. Ch'ang, K. Tang, M. Schell, C. Ringelberg, K. G. Matteson, S. L. Allman, L. and C. H. Chen, *Rapid Commun. Mass Spectrom.* **9**, 772–774 (1995).

114. E. Nordhoff, R. Cramer, M. Karas, F. Hillenkamp, F. Kirpekar, K. Kristiansen, and P. Roepstorff, *Nucl. Acids Res.* **21**, 3347–3357 (1993).

115. T. Keough, T. R. Baker, R. L. M. Dobson, M. P. Lacy, T. A. Riley, M. A. Scurria, J. A. Hasselfield, and E. P. Hesselberth, *Rapid Commun. Mass Spectrom.* **7**, 195–200 (1993).

116. L. M. Polo, T. D. McCarley, and P. A. Limbach, *Anal. Chem.* **69**, 1107–1112 (1997).

117. M. R. Isola, S. L. Allman, V. V. Golovlov, and C.-H. Chen, *Anal. Chem.* **71**, 2266–2269 (1999).

118. P. L. Ross and P. Belgrader, *Anal. Chem.* **69**, 3966–3972 (1997).

119. P. L. Ross, P. A. Davis, and P. Belgrader, *Anal. Chem.* **70**, 2067–2073 (1998).

120. D. S. Wunschel, K. F. Fox, A. Fox, J. E. Bruce, D. C. Muddiman, and R. D. Smith, *Rapid Commun. Mass Spectrom.* **10**, 29–35 (1996).

121. D. C. Muddiman, G. A. Anderson, S. A. Hofstadler, and R. D. Smith, *Anal. Chem.* **69**, 1543–1549 (1997).

122. D. Wunschel, L. Pasa-Tolic, B. Feng, and R. D. Smith, *J. Am. Soc. Mass Spectrom.* **11**, 333–337 (2000).

123. X. Chen, Z. Fei, L. M. Smith, E. M. Bradbury, and V. Majidi, *Anal. Chem.* **71**, 3118–3125 (1999).

16

APPLICATIONS TO REAL-WORLD PROBLEMS

In previous chapters, the basic principles of mass spectrometry instrumentation and appropriate mass spectrometry protocols that are used for the analysis of a variety of biomolecules were discussed in detail. In developing a method for a specific analyte, analytical chemists rely on the analysis of a set of synthetic model compounds. Excellent detection sensitivity and molecular specificity have been demonstrated with mass spectrometry–based protocols. However, the worthiness of a specific method can be judged only when it can be applied to the analysis of real-world samples. In particular, samples of biological origin are heterogeneous mixtures of an astronomically large number of compounds and contain salts as impurities, and the analyte of interest may be present as a trace component. In addition, degradation with time of biological compounds is also a major concern. Therefore, it is a challenging task to obtain the required analytical information from real-world biological samples. Current advances in the field of biological mass spectrometry are meeting this challenge. This chapter highlights some of the unique applications of mass spectrometry and demonstrates its capability in solving real-world problems.

16.1. IMMUNOLOGIC STUDIES

Mass spectrometry has emerged as a viable tool for research in immunology [1]. In particular, it has facilitated identification of the peptide antigens that are associated with class I and II major histocompatibility complexes (MHCs) [1–8]. An important role of cytotoxic T lymphocytes (CTL) is to recognize potential new antigens, and consequently to stimulate an immune response to kill the infected cells. The antigens that are recognized by T-cell receptors are peptides of 8–12 residues, and are bound to MHC class I molecules on the surface of the host cells. These peptides are derived from cytosolic proteins by the action of proteases, and are transported to endoplasmic reticulum, where the class I molecules preferentially select the peptides with a certain binding motif. In contrast, the class II peptides are longer in size (10–30 residues), and are generated from proteins that are present in vesicles of the endosomal pathway. They stimulate cytokine secretion by helper T cells. Each class of MHC molecules contains a large array of peptide antigens. Therefore, identification of the MHC-bound antigens seemingly is a major undertaking. Hunt's group has developed a comprehensive mass spectrometry–based strategy to characterize antigens presented by several class I alleles [1–6]. This protocol includes a multistep reverse-phase (RP) high-performance liquid chromatography (HPLC) fractionation of peptides, CTL-based assay to identify the active fraction, and microLC/electrospray ionization (ESI)-MS/MS to characterize the suspected peptides. This research group has characterized antigens that are presented by several class I alleles: HLA-A1, HLA-A2.1, HLA-A3, HLA-A11, HLA-A24, and HLA-B7 [1–6]. As a typical example, characterization of an antigen from the MHC class I HLA-A2.1 involved the following steps [1]:

1. The peptide–MHC complexes were isolated by passing the supernatant of the lysed cells through an HLA-A2.1-specific monoclonal antibody column.
2. The liberated peptides were separated from high-molecular-mass contaminants by ultrafiltration, and fractionated by RP-HPLC.
3. Each fraction was analyzed with CTL-based assay by using melanoma-specific human CTL cell lines.
4. The fraction that exhibited bioactivity was further fractionated with a second RP-HPLC separation step.
5. Finally, each fraction was analyzed by microLC/ESI-MS/MS, in which the flow was split to collect fractions for CTL bioassay.

In a similar approach, Yates and colleagues have combined microLC/ESI-MS/MS with database searching [7,8]. With this approach, they have characterized the peptides bound to the class II MHC alleles that are associated with rheumatoid arthritis, subcellular fractions of murine B lymphoma cells, and mutant cells [7].

16.2. DIRECT ANALYSIS OF BIOACTIVE COMPOUNDS FROM BIOLOGICAL TISSUES

Direct analysis of bioactive compounds from biological tissues and at the single-cell level is a worthy task, but nevertheless a formidable challenge. A normal procedure involves extraction of these bioactive compounds, followed by several separation steps and subsequent analysis by mass spectrometry (mass measurement, sequencing, etc.). Although an appropriate combination of chromatography and mass spectrometry steps can improve the efficiency of the procedure, it still requires the pooling of a large number of cells or tissue materials. Several advances in mass spectrometry in the late 1990s in particular improved sample preparation in matrix-assisted laser desportion/ionization (MALDI), have made it feasible to analyze directly endogenous compounds from intact cells, dissected tissue samples, and even single organelles within a cell. Some typical examples are presented below.

16.2.1. Direct Analysis of Peptides in Single Neurons

Although the classical transmitters, such as dopamine, 5-hydroxytryptamine, and epinephrine, can be conveniently analyzed with microcolumn LC or capillary electrophoresis (CE) with electrochemical detection [9], the analysis of neuropeptides requires a different methodology. Because of high structural complexities of various chemical constituents of a cell, an analytic scheme must provide high-resolution separation and unequivocal identification, and must be extremely sensitive to deal with the low sample volumes that are encountered in the analysis of single neurons. Hsieh et al. have developed a novel experimental system for characterization of neuropeptides from single neurons from the brain of the snail *Lymnaea stagnalis* [10]. The important components of this system are a microcolumn LC for the separation of neuropeptides, MALDI-TOFMS for their characterization, and a MALDI sample preparation system. The last one, along with the microcolumn LC system, is illustrated in Figure 16.1. With the ingenious arrangement of flow splitters, LC eluents emerge from the capillary column at a flow rate of 85 nL/min. The solution of MALDI matrix is delivered at a flow rate of 60 nL/min through a pressurized container. The two flows are mixed together, and deposited every 60 s onto a fresh spot of a polished stainless-steel target. This system was used to analyze neuropeptides from visceral 1 (VD1) dorsal neuron, which is located in the visceral ganglion of the snail *Lymnaea stagnalis*, and is known to play an important role in cardiorespiratory mechanisms of the snail *Lymnaea*. Figure 16.2 shows a 3D plot of peptide signals from a single VD1 neuron. Several peptides in the mass range between 1000 and 7000 Da were identified, and include δ peptide (1159 Da; in early fractions), β peptide (6375 Da; fractions 33 and 34), α1 (2401 Da) and α2 peptides (2996 Da; both in fractions 24 and 25), three modified forms (molecules C, D, and E) of α2 peptide (in fractions 23–25), two small cardioactive peptides a and b, peptide F (6025 Da), peptide G (6455 Da), and peptide I (3970 Da). This research also reported direct analysis of the single-cell neuron by MALDI-TOFMS. After dissection from the brain, the cell was placed directly on the MALDI target, and mixed with the matrix. The same set of peptides that were observed following the microcolumn LC separation step are also detected in this analysis (Figure 16.3).

16.2.2. Direct Analysis of Peptides and Proteins from Biological Tissues

Li et al. have also demonstrated direct analysis of peptides from biological tissues and single cells with MALDI–postsource decay (PSD)-TOFMS analysis [11]. This

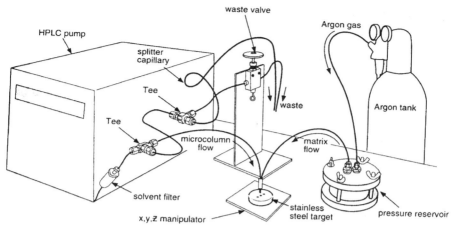

Figure 16.1. Schematic diagram of the MALDI sample preparation system directly from microLC eluents. [Reproduced from Ref. 10 by permission of American Chemical Society, Washington, DC (copyright 1998).]

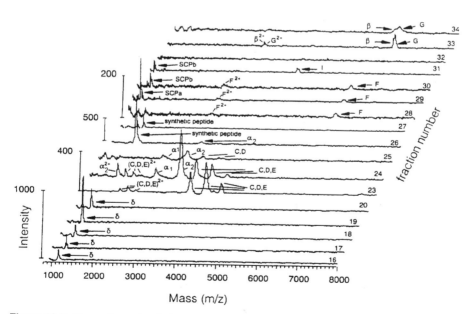

Figure 16.2. Three-dimensional plot of the data obtained by the analysis of a single VD1 neuron that was analyzed with the combination of the system shown in Figure 16.1. [Reproduced from Ref. 10 by permission of American Chemical Society, Washington, DC (copyright 1998).]

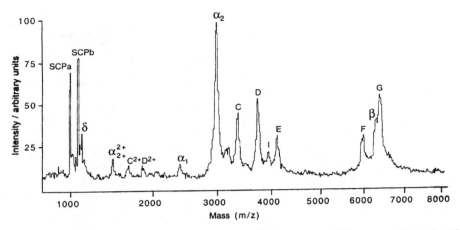

Figure 16.3. Result of the direct MALDIMS analysis of the single VD1 neuron. [Reproduced from Ref. 10 by permission of American Chemical Society, Washington, DC (copyright 1998).]

technique was applied to characterize peptides in tissue samples and single neurons from the invertebrate mould *Aplysia california*. In situ sequencing and database search could provide the identity of some proteins.

In a similar effort, Caprioli's research group has also obtained direct profiling of proteins that are present in tissue sections of several mouse organs with MALDI-TOFMS [12]. Fresh tissue sections were blotted on conductive polyurethane membranes, and the dried blot was coated with the matrix before mass spectrometry analysis. Well over 100 peptide and protein signals in the 2000–30000 Da range were observed.

In another study, MALDI-TOFMS was used to identify the products derived from the processing of a precursor protein, proopiomelanocortin (POMC), in melanotrope cells of the pars intermedia of the amphibian *Xenopus laevis* [13]. The dissected tissue pieces of the pituitary were directly subjected to MALDI delayed-extraction (DE)/TOFMS analysis [14] after mixing with the 2,5-dihydroxybenzoic acid (DHB) matrix solution. Figure 16.4 shows the mass profile of neuropeptides that were detected from this tissue. Several mass peaks have been assigned to suspected neuropeptides that include m/z 1050.46 (CYIQNCPRG-NH$_2$), 1108.49 (CYIQN-CPRGG), 1236.57 (CYIQNCPRGK), 1392.72 (CYIQNCPRGKR), 1606.83 (desacetyl-α-MSH-NH$_2$), 1655.90 (Lys-γ1-MSH-NH$_2$), 2098.98 (β-MSH), 2521.13 (CLIP POMV$_B$), and 2537.12 (CLIP POMC$_A$). The m/z 1050.46 and 1392.72 were the unidentified peaks. These peptides were identified with the PSD sequencing approach [15].

16.2.3. Direct Analysis of Single Mast Cells

Cells are the fundamental units of living organisms. Each type of cell has a complex mixture of biomolecule—some as diverse in nature as nucleic acids, proteins, lipids,

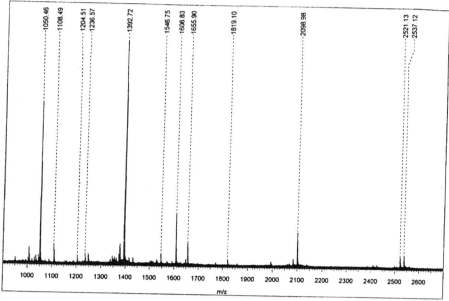

Figure 16.4. Mass profile of peptides that are detected in pars intermediate tissue from the amphibian *Xenopus laevis* with direct MALDI/DE-TOFMS analysis. [Reproduced from Ref. 13 by permission of American Chemical Society, Washington, DC (copyright 1999).]

carbohydrates, amines, and so on. Analysis of cells can provide answers to some interesting questions related to cellular metabolism, signal transduction, and differentiation. Direct analysis of cells is desirable because of the limited amounts of samples and time constraints. The most popular method for direct analysis of single cells is *flow cytometry*, in which a flowing stream of cells is first arranged in a single file with the aid of a sheath liquid [16]. The analytes of interest are labeled with fluorescent dyes. The fluorescence signal is measured with laser fluorescence spectrometry. Although this approach is fast, the analysis of an analyte is subject to the availability of an appropriate fluorescence probe that display binding specificity with the target analytes.

Fung and Yeung have combined flow cytometry with mass spectrometry detection for direct analysis of single rat peritoneal mast cells (RPMCs) [17]. A laser vaporization–ionization interface [18] is used to couple flow cytometry with a time-of-flight (TOF) mass spectrometer. The RPMCs are suspended in physiological buffer that contains 2 mM ammonium chloride and 0.1 mM $CuCl_2$. The solution is passed through a 50-cm capillary (20 μm i.d.), the end of which is inserted into the mass spectrometer to allow entry of the solution at a rate of 1 nL/s. A high-energy beam of a 480-nm laser is focused at the tip of the capillary to vaporize and ionize the eluents. This system was used to analyze serotonin and histamine—the important biogenic amines that are involved in the signal transduction mechanism of cells, in individual cells. A detection limit of 20 amol of serotonin standard with one laser shot was demonstrated. Hundreds of RPMCs were analyzed, and the average amount of serotonin and histamine per cell were found to be 0.11 ± 0.06 and 0.75 ± 0.33 fmol, respectively. This system does not require any pre- or postcolumn staining or derivatization, and is much faster. Hundreds of cells can be analyzed within 5 min. In addition, mass spectrometry provides the molecular mass, from which the suspected (and often unknown) analytes can be identified directly. Such a system can serve as an effective disease-screening method.

16.3. CLINICAL APPLICATIONS OF MASS SPECTROMETRY

Several mass spectrometry procedures are gaining importance for a routine diagnosis of a variety of clinical disorders. A few typical examples are discussed below.

16.3.1. Detection of Mutation of the Cystic Fibrosis Gene

The ease of sample preparation, high-mass analysis capability, and high sensitivity of MALDIMS have opened the opportunity to apply it in the molecular diagnosis of mutant alleles. A typical application is the detection of the cystic fibrosis (CF) gene mutation. CF is a common form of an autosomal recessive genetic disease among the Caucasian population. The mutant gene is devoid of 3 base pairs (bp) in exon 10. This change results in the loss of a Phe residue at codon 508 (ΔF5087). In a MALDIMS procedure, the DNA from exon 10 of the gene from carrier and normal patients is amplified by a polymerase chain reaction (PCR), and cleaved with MseI enzyme to produce a 59-bp fragment from the normal CF gene and a 56-bp fragment from the mutant gene [19]. These fragments are analyzed in negative-ion mode with 3-hydroxypicolinic acid (HPA) as the matrix. Figure 16.5 shows that the 56-bp fragment is present in carrier and CF patients. Thus, the detection of this fragment can form the basis of a diagnostic test of CF.

16.3.2. Detection of Transthyretin Mutants

Familial transthyretin amyloidosis (ATTR) is a hereditary degenerative disease. The single amino acid substitutions in transthyretin have been linked to ATTR. Therefore, the detection and identification of transthyretin (TTR) variants can provide a definitive diagnosis of ATTR. A mass spectrometry procedure has been developed to detect transthyretin mutants in serum [20]. This procedure uses HPLC-ESIMS to identify the mutants, and MALDIMS of the digests of the isolated HPLC fractions to unequivocally detect the site of mutation. The analysis of serum from ATTR patients successfully detected variants, Val30 $\rightarrow$ Met, Val22 $\rightarrow$ Ile, and Ser23 $\rightarrow$ Asn. The same research group has also used MALDI bioretive probe for detection of TTR variants [21].

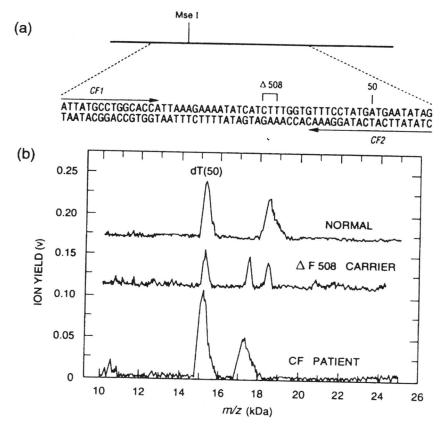

Figure 16.5. MALDIMS diagnosis of individuals with the normal and ΔF508 alleles. (*a*) The nucleotide sequence of the 59-bp oligonucleotide in exon 10. The deletion sequence is also indicated. (*b*) The negative-ion MALDIMS spectra of the amplified DNA from a normal, a ΔF508 carrier, and a CF patient. [Reproduced from Ref. 19 by permission of John Wiley & Sons (copyright 1995).]

16.3.3. Characterization of Hemoglobin Variants

Normal adult hemoglobin is a tetrameric protein that is composed of two α chains and two β chains. Structural variations in these hemoglobin chains are responsible for hemoglobin disorders, which can manifest themselves in the form of a number of disease states such as sickle cell anemia. The importance of an early detection of the health risks that are caused by the pathogenic hemoglobin variants has been fully realized; many states have implemented newborn screening programs. Because of the enormity of the task and the possibility of a large number of hemoglobin variants (over 700 are known), the need exists for a rapid and cost-effective diagnostic method. Since the early 1980s, mass spectrometry–based methods have provided accurate and rapid means to detect and characterize the structural changes that occur in the hemoglobin chains. Matsuo et al. detected sickle cell disease in 1981 with field desorption (FD)MS [22]. Since then, a number of more reliable mass spectrometry approaches have evolved [23–29]. Most of these methods rely on a mapping of tryptic digests to detect mutations. Trypsin digestion, however, is the rate-limiting step of the protocol. In addition, this step can introduce artifact peptides due to trypsin autolysis. Also, analysis of peptide fragments that are generated from the hydrophobic core region of hemoglobin can pose problem. A few innovative approaches have emerged to speed up tryptic mapping and to reduce complexity of the analysis. These methods are discussed here.

Whittal et al. [27] have combined CE with ESI and ion-trap reflectron time-of-flight (TOF)MS to identify mutants of hemoglobin. The combination provides a convenient means of separation and mass analysis of the peptide fragments of tryptic digests. A comparison of the peptide maps of the normal and mutant proteins allows detection of the site mutation.

McComb et al. have reported a new MALDI-TOFMS procedure for the detection of hemoglobin variants [28]. A critical aspect of this procedure is the use of a nonporous polyurethane membrane as a collection device, and a transportation medium of a blood sample for analysis. Less than 1 µL of the blood sample is collected with a simple lancet device, and is placed on a polyurethane membrane. As an optional step, 2 µL of methanol is added to the blood sample to disrupt coagulation, enhance cell lysis, and effect protein binding onto the membrane. The dried blood sample is washed twice with water, and the matrix solution is added to it. After drying of the matrix, the sample is analyzed by the MALDI/DE-TOFMS technique. If desired, the tryptic digestion can be performed directly on the polyurethane membrane. Figure 16.6 shows a comparison of the analysis of normal and Hb Shepherds Bush hemoglobin variant b74(E18)GlY → Asp. This variant is known to enhance oxygen affinity, and expresses itself as mild hemolytic anemia in patients. The variant b74(E18)GlY → Asp is indicated by a mass shift of 58 Da in the β chain of hemoglobin. The mapping of the tryptic fragments indicated that the sequence 67–82 of the β chain is involved in the mutation. The advantage of this experimental protocol is that the blood sample can be collected in any setting outside the laboratory by an inexperienced person with a minimal invasive procedure, and can be transported without any particular precaution. The analysis is fast and sensitive, and requires minimal sample preparation.

Houston and Reilly have used trypsin-activated bioreactive MALDI probes to generate peptide maps from hemoglobin samples [29]. These samples are obtained directly from whole human blood without further purification. The digestion is carried out by depositing a few microliters of diluted (1000 times) blood on the active probe, and heating the probe at 37°C for 15 min. This technique is rapid and allows the detection of traditionally elusive tryptic peptides. For example, the problematic tryptic peptides αT12, αT13, and βT12 could be detected with this approach.

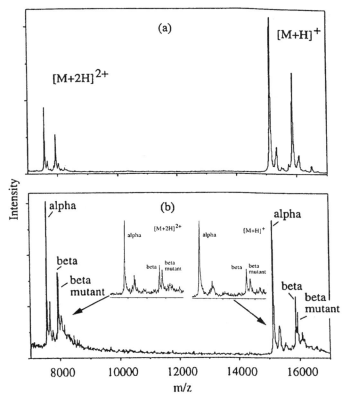

Figure 16.6. A comparison of the analysis of (*a*) normal and (*b*) Hb Shepherds Bush hemoglobin variant b74(E18)GlY→Asp. The insets are the expanded region of the spectra. The whole-blood sample was deposited on polyurethane membrane and analyzed by MALDIMS. [Reproduced from Ref. 28 by permission of American Chemical Society, Washington, DC (copyright 1998).]

16.3.4. Detection of Enzyme-Deficiency Diseases

Gerber et al. have developed a simple technique to detect rare diseases that are caused by a deficiency in specific enzymes [30]. This strategy involves assaying the enzyme with its synthetic substrate conjugate and ESIMS. If the enzyme is active in a cell culture, it can cleave the sugars to provide a product that can be identified by its characteristic mass. Biotin is used as a handle in the synthesis of conjugates. The technique was applied to diagnose glycogen storage disease, which is caused by the β-galactosidase deficiency. In a typical procedure, the substrate conjugates are incubated with cultured fibroblasts from normal and affected subjects, labeled internal standards (deuterated conjugates that lacked the sugar moiety) are added, and biotinylated components are separated from the crude mixture with biotin-binding streptavidin. After multiple washing, the biotinylated components are released and analyzed by ESIMS. This study demonstrated that very little enzyme product was present in cells from patients with β-galactosidase deficiency compared to controls.

16.3.5. Mass Spectrometry in Detection of Cachexia

An off-line combination of CE and MALDI-TOFMS has been utilized for the clinical diagnosis of cachexia in several cancer patients [31]. These patients typically suffer from accelerated protein breakdown that can cause reduction of skeletal muscle. Proteolysis-inducing factor (PIF), a sulfated glycoprotein, has been implicated with advanced muscle proteolysis [32]. Therefore, the development of a clinical assay is aimed at analysis of PIF. In this study, CE is used to analyze the urine samples of healthy volunteers and cachexia cancer patients. A comparison of CE profiles reveals the presence of additional peaks in electropherograms of the cancer patients that are associated with cachexia. However, a CE profile alone does not provide a clear indication of whether a patient suffers from cachexia. The unambiguous indication of PIF is obtained by the molecular mass analysis of the CE-fractionated samples with MALDI-TOFMS.

16.3.6. Rapid Screening of Protein Profiles of Human Breast Cancer Cell Lines

Rapid screening of cancer-related proteins can be potentially used as a diagnostic tool for detection of cancer. The basis of this approach is the fact that mutations in the DNA may manifest themselves in the cellular protein expression. Methods based on 2D gel, RP-HPLC, and CE separation are used commonly to study changes in cellular protein expression [33–36]. Chong et al. have developed an approach in which nonporous RP-HPLC is used to rapidly generate protein profiles of whole-cell lysates of human breast cancer cell lines [37]. This approach is used to search for elevated levels of proteins or for the appearance of new proteins in the cancerous cells. The fractions that correspond to the highly expressed proteins are analyzed by MALDI/DE-TOFMS to obtain the molecular mass information. Proteins are further subjected to peptide mapping and MALDI/DE-TOFMS analysis steps. The expressed proteins are identified using a database search. The utility of this approach as a diagnostic tool for cancer detection was demonstrated by analyzing the MCF-10-based breast epithelial cell lines. Several key proteins such as tumor suppressor p53 and oncogenes c-src, c-myc, and h-ras are found to be at highly elevated levels in malignant cells.

16.3.7. Analysis of Proteins in Human Cerebrospinal Fluid

Cerebrospinal fluid (CSF) plays a vital role in protecting the brain, maintaining ventricular pressure, and balancing electrolytes and proteins. Proteins and peptides can move freely between the brain and CSF. Therefore, CSF is a reflection of metabolic events that occur in the brain, and it can be utilized to screen certain metabolic diseases. Westman et al. have demonstrated that MALDIMS is an effective method of detecting proteins that are linked to human diseases such as multiple sclerosis, Alzheimer's disease, and stroke [38]. Although two proteins, cystatin C (m/z 13,344) and β_2-microglobulin (m/z 11730), were detected from the analysis of neat CSF, the authors felt the need for microHPLC separation of the CSF sample prior to MALDIMS analysis to detect those proteins that were present at low concentrations (< 200 fmol/μL). Apart from cystatin C and β_2-microglobulin, an additional protein, transthyretin, was detected. A database search, using peptide mapping data, confirmed the presence of these proteins.

Beranova-Giorgianni and Desiderio have analyzed proteomes of human pituitary [43]. The pituitary proteins were separated by 2D gel electrophoresis. After in gel digestion of the protein spots of interest, the masses of the tryptic peptides were determined by MALDI-TOFMS. The database search identified somatotropin, prolactin, lutropin β-chain, α and β chains of hemoglobin, gutathione S-transferase P, ubiquitin thiolesterase L1, and glyceraldehyde 3-phosphate dehydrogenase.

16.4. GENDER IDENTIFICATION OF HUMAN DNA SAMPLES

The DNA structure of a human being is unique and nearly identical throughout all body tissues of that individual. This fact has been utilized for the purpose of forensic analysis, disease diagnosis, paternity tests, and archeology (wildlife protection). The gender determination of human DNA samples is one important aspect of these tests,

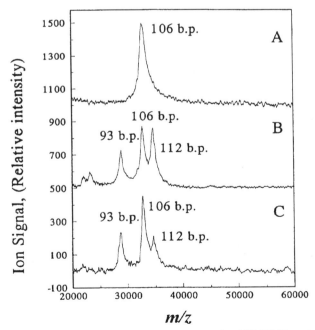

Figure 16.7. Geneder differentiation through the negative-ion MALDIMS analysis of the DNA samples. Spectra from the (*A*) female, (*B*) male, and (*C*) female/male samples. [Reproduced from Ref. 40 by permission of American Chemical Society, Washington, DC (copyright 1999).]

as well as in prenatal diagnosis of X-linked inherited diseases. In a general method, the DNA samples are subjected to coamplification of the amelogenin loci (AMEL) on the X and Y chromosomes with the sex-determining region Y (SRY) locus via PCR [39]. The PCR products are analyzed with gel electrophoresis, which is a time-consuming procedure. A new MALDI-TOFMS-based method has been validated for the sex determination of DNA samples [40]. The primer pairs used in PCR amplification are 5'-CCCTGGGCTCTGTAAAGAATAGTG-3' and 5'-ATCA-GAGCTTAAACTGGGAAGCTG-3' for the amelogenin locus, and 5'-ATAAG-TATCGACCTCGTCGGAA-3' and 5'-GCACTTCGCTGCAGAGTACCGA-3' for the SRY locus. PCR-amplified products are analyzed with a mixture of 3-hydro-xypicolinic acid, picolinic acid, and ammonium fluorate (9 : 1 : 1 molar ratio) as a MALDI matrix. The gender determination is based on the fact that the coamplification of the male DNA generates three products of 93 bp (28737 Da), 106 bp (32754 Da), and 112 bp (34608 Da), whereas the female DNA produces only one product of 106 bp. Thus, as shown in Figure 16.7, genotyping of the male and female DNA samples is achieved unambiguously. This protocol is more sensitive, rapid, and reliable than the conventional gel-electrophoresis approach. The dynamic range of this procedure is also superior to gel electrophoresis. As little as 0.1% of the male-specific DNA could be detected in the female DNA sample.

16.5. PROTEOMIC ANALYSIS

The analysis of proteome has become an active area of study now that the sequence of proteins can be determined with an integrated mass spectrometry–database search approach (see Section 8.10). With proteome analysis, an essential chemical definition of biological systems can be obtained. Chen et al. have provided a practical demonstration of this approach to identify proteins from human erythroleukemia cells [42]. In this example, the proteins from human erythroleukemia cell lines are separated with 2D gel electrophoresis. The protein spots are subjected to tryptic digestion, and the tryptic peptide fragments are separated and analyzed by a combination of capillary HPLC on-line with ESIMS analysis with an ion-trap (IT) storage reflectron–TOF instrument. A 2D topography display of the on-line data provides the mass identification of each peptide fragment. These mass values are used to search for the identity of a protein against a database. An extensive coverage of the protein sequence can be obtained with this approach. If an additional parameter such as the cell species, the approximate molecular mass of the protein, and its pI value is entered in the search procedure, a unique match can be identified. Several proteins such as hucha (hsp-60), tropomyosin α chain, stathmin (op18), tropomyosin (tm30-nm), vimentin, troponin T, Martin 3, and microfibrillar-associated protein 1 could be identified from human erythroleukemia cell lines.

16.6. ANALYSIS OF MICROORGANISMS

Identification of viruses and other microorganisms is important from the viewpoint of public health and for protection of crops. Several forms of viral infection can cause death or significant debilitation. An accurate identification of microorganisms can be a powerful tool for diagnosing diseases, monitoring potential contamination in foods, and recognizing biological and environmental hazards. The analysis of certain biomarkers (chemotaxonomy) forms the basis of microbial identification. Chemotaxonomy not only can distinguish among related organisms but can also identify species and strains of related organisms. Any of the components of the cells, such as lipids, oligosaccharides, proteins, and DNA, can serve as biomarkers for the classification of microorganisms. In contrast to the conventional chromatography methods, mass spectrometry can provide rapid and accurate information to bacterial taxonomy. In the past, FAB, laser desorption, PD, and pyrolysis GCMS were used for analysis of bacteria [44–46]. In those studies, the ions due to phospholipids or other small molecules were considered as biomarkers. Currently, MALDIMS has demonstrated great success in microbial analysis. The advantages of this technique are the direct analysis of bacterial cells, the ability to analyze large-sized molecules such as proteins, and low sample amounts. The selection of proteins as biomarkers has the advantage that they are less prone to interference from other low-mass species, which usually are more abundant in the bacterial cells. A few examples of the identification of bacteria and viruses are discussed in this section.

16.6.1. Bacterial Analysis

Several studies have reported the use of MALDIMS for the identification of bacterial samples. In some studies, the extracts of bacteria were analyzed [47–50]. For example, Krishnamurthy et al. analyzed proteins that were isolated from the whole cells of a series of bacterial species [47]. Several ion characteristics of each bacteria type were observed in the mass range of 2400–10,000 Da. Several pathogenic bacterial strains, such as *Bacillus anthracis*, *Yersinia pestis*, and *Brucella melitensis*, were distinguished from nonpathogenic species. Distinct mass spectrometry data were also obtained for various *Bacillus* species (e.g., *anthracis*, *subtilis*, *cerus*, and *thuringiensis*). Holland et al. have demonstrated that the identification of whole bacteria, rather than the extract, is also feasible with MALDIMS [51]. Since then, several studies have reported results with intact cells [33,52–55]. The laboratory-to-laboratory reproducibility of the spectra has been a concern. The bacterial spectra depend upon a number of instrumental factors (sample preparation method, solvents, matrix composition, etc.) and microbial variables (culture medium and cell growth time). Wang et al. have shown that, despite these variations, a number of peaks are conserved even when spectra are acquired under different experimental conditions [53]. These conserved peaks can serve as potential biomarkers.

Welham et al. have characterized a variety of intact Gram-positive (*Bacillus subtilis* and *Staphylococcus aureus*) and Gram-negative (*Escherichia coli*, *Klebsiella aerogenes*, *Proteus mirabilis*, and *Salmonella typhimurium*) bacterial cells [52]. Many of these microorganisms are waterborne, and thus are a great threat to public health. With a better sample preparation technique, this group of researchers was able to detect discrete peaks in the mass range of 3–40 kDa. Their sample preparation technique involves centrifugation of the cell suspension at 12,000 rpm for 30 min, resuspension of the cell pellet in 0.1% TFA, vortexing it for 30 s, and immediately mixing with an excess of the matrix (1 : 9) solution (2[4′-hydroxyphenylazo]benzoic acid, 10 mg/mL). An aliquot (2 μL) of this mixture is applied to the sample target, dried, and washed with cold water for 30 s. Winkler et al. have found that an improved sample stability is achieved when the cell culture is diluted in 50% methanol rather than in 0.1% TFA [55]. The greater stability is attributed to the fact that this solvent mixture may fix the bacteria and reduce its degradation due to proteases. With this modification, they were able to detect taxonomic marker ions in the mass range of 2–62 kDa. They used this technique to identify *Helicobacter* and *Campylobacter* species. A biomarker for *Helicobacter pylori* was found at m/z 58,268 and for *Helicobacter mustelae* at m/z 49,608 and 57,231. *Campylobacter* species did not show ions above m/z 45,000. However, characteristic biomarkers were observed at m/z 10,074 and 25,478 for *Campylobacter coli*, at m/z 10,285 and 12,901 for *Campylobacter jejuni*, and at m/z 10,726 and 11,289 for *Campylobacter fetus*. Thus, *Campylobacters* could be readily distinguished from *Helicobacters*. Nilsson was able to determine strain-specific biomarkers for six different strains of *Helicobacter pylori* with MALDIMS [49]. Nilsson and co-workers have also characterized cell-surface from *Helicobacter pylori* [56]. Forty different proteins were identified from a detergent-solubilized *Helicobacter pylori* preparation. This finding is of potential value in vaccine design against this class of bacteria.

The use of fatty acid profiling has been demonstrated to distinguish among various bacterial species (*Francisella tularensis*, *Brucella melitensis*, *Yersinia pestis*, *Bacillus anthracis*, and *Bacillus cerus*) [57]. Three different ionization techniques, EI, positive-ion CI, and negative-ion CI, were investigated. Although all ionization modes, in conjunction with principal-component analysis (PCA), could distinguish bacteria, the positive-ion CI mode produced the greatest amount of differentiation between the four genera of bacteria (see Figure 16.8). ESI-MS/MS has also been applied as a tool to search for biomarkers with the aim to identify microorganisms [58].

16.6.2. Analysis of Viruses

Viruses are made up of molecules of nucleic acids that are encased in an envelope of proteins, called *capsid proteins*. The envelope proteins are also referred to as *coat proteins* when they are part of the outermost shell. Usually, these proteins exist as a noncovalent association of protein subunits, and are responsible for an array of functions that include cell attachment, cell entry, and RNA release. The study of viruses can shed light on their molecular biology and pathology, and on their possible interactions with antibodies. Several research groups have recognized the utility of mass spectrometry for the characterization of viruses [59–63]. Because viral proteins have a mass of < 10,000 Da and are easily accessible, they are good candidate as biomarkers in the characterization of viruses. Siuzdak's tutorial on the subject of probing viruses with mass spectrometry is worth reading [64].

Despeyroux et al. have characterized the capsid proteins of cricket paralysis virus (CrPV) with ESIMS [60]. They demonstrated that viral proteins can be identified

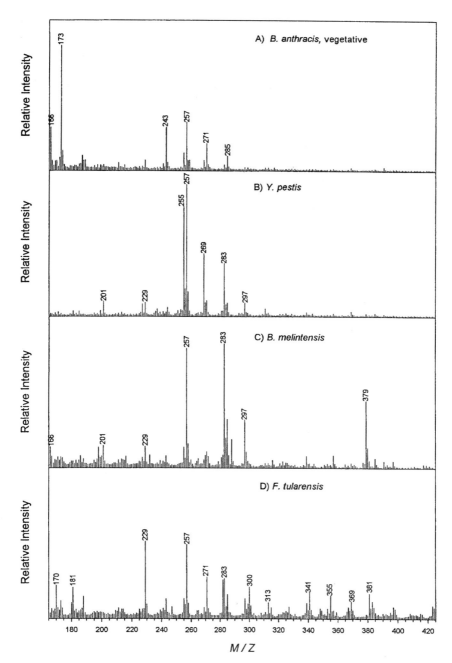

Figure 16.8. The positive-ion CI spectra of whole bacteria. [Reproduced from Ref. 57 by permission of Elsevier Science Inc. (copyright 1999 American Society for Mass Spectrometry).]

from intact viruses without subjecting the sample to any preliminary extraction or disruption step. The nondisrupted sample of the CrPV was directly injected into the flowing ESI solvent. Four capsid proteins, VP1, VP2, VP3, and VP4, were detected from the ESIMS data. They also demonstrated that the nondisrupted sample of viruses could also be characterized via LC-ESIMS. By subjecting the common cold virus to gradual enzymatic digestion of the viral surface, it has been possible to detect peptides that are derived from the outermost capsid protein (VP1) as well as from the innermost capsid protein (VP4) [63]. This observation has suggested that a virus actually breathes to expose the inner surface.

Thomas et al. used MALDIMS to characterize tobacco mosaic virus U2, MS2 bacteriophage, and Venezuelan equine encephalitis (VEE) virus [61]. The aliquots of virus suspension were mixed with acetic acid, and were sandwiched between matrix (α-cyano-4-hydroxycinnamic acid) layers on the sample holder. The addition of acetic acid (with a final concentration of 10–60%) had the effect of dissociating the coat proteins from the virus. The molecular masses of the coat proteins are highly characteristic of species and strains. Figure 16.9 shows a mass spectrum of an infected piece of a tobacco leaf that was analyzed directly with MALDI. The mass of the viral coat was determined with an accuracy of 0.02%.

MALDI-TOFMS has also been used to analyze chemically modified recombinant hepatitis B surface antigen (HBsAg) [62]. The HBsAg is a very large molecule (its mass may be over one million Da), and is composed of multiple copies of a protein subunit called S antigen, the glycosylated form of this subunit, and a lipid membrane. In hepatitis B virus (HBV) vaccines, recombinant HBsAg is used as an antigen to produce antibodies against it. Biotinylated HBsAg is used in the

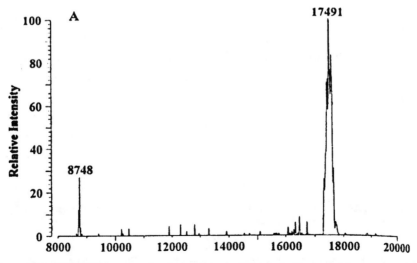

Figure 16.9. MALDIMS spectra of a piece of tobacco (*A*) infected leaf with tobacco mosaic virus (*B*) uninfected leaf. [Reproduced from Ref. 61 by permission of American Chemical Society, Washington, DC (copyright 1998).]

enzyme-based diagnostic tests to measure antibodies to HBsAg. The determination of the purity and extent of biotinylation of HBsAg is important for the success of the diagnostic test. Because of its large size and the insolubility of its subunits, the analysis of the intact HBsAg is a daunting task. Winkler et al. have developed a simple procedure, in which the HBsAg samples were dialyzed to remove salts, and reduced with DTT [62]. The MALDIMS analysis was able to distinguish between protein subunits (S antigen) and biotin-labeled S antigen.

16.7. DETECTION OF PHOSPHORYLATED PEPTIDES IN BOVINE ADRENAL MEDULLA

Li and Dass have combined iron(III)-immobilized-metal affinity chromatography [Fe(III)-IMAC] and mass spectrometry (MALDIMS and LC-ESIMS) to isolate and detect phosphopeptides in body tissues [65]. Analysis of the bovine adrenal medulla showed the presence of several phosphopeptides (Figure 16.10) [66]. Confirmatory evidence of the presence of those peptides was obtained by treating the Fe(III)-IMAC-retained fraction with phosphatase and reanalyzing those fractions with MALDIMS and LC-ESIMS. Several peaks disappeared and new peaks appeared in those spectra, which were separated by a multiple of 80 Da.

16.8. QUANTITATIVE ANALYSIS OF NEUROPEPTIDES IN PITUITARY TUMORS

Desiderio and colleagues have developed a comprehensive analytic scheme to quantify methionine enkephalin and β-endorphin in body tissues and fluids [67–69]. The important steps of this scheme are tissue homogenization, isolation of the peptide-rich fraction via a solid-phase extraction step, RP-HPLC fractionation of methionine enkephalin and β-endorphin, further purification of each fraction with isocratic RP-HPLC, and mass analysis with the multiple reaction monitoring (MRM)FABMS technique [67,68]. Because β-endorphin is a large-sized peptide, it is first tryptically cleaved into smaller segments, and the peptide NAIIK is used as a surrogate analyte for its quantification [69]. Deuterium-labeled internal standards of methionine enkephalin and β-endorphin are added at the tissue homogenization stage. This scheme was applied to monitor the levels of these endogenous peptides in prolactin-secreting tumors [67]. The methionine enkephalin level was 1032 ± 278 pmol/mg protein (control 76.6 ± 6.2) and the β-endorphin level was 450 ± 253 (594 ± 121) pmol/mg protein. Those data demonstrate that the proenkephalin A (precursor protein of methionine enkephalin) system is regulated differentially in prolactin-secreting tumors.

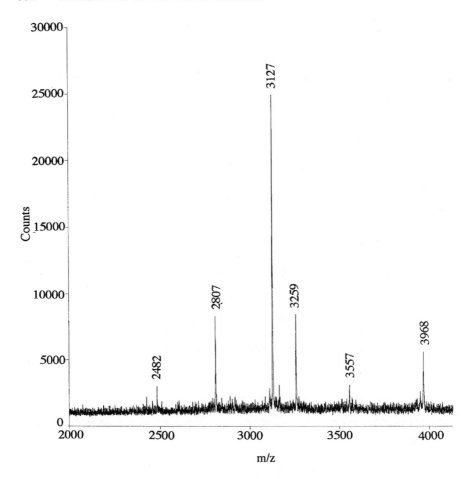

Figure 16.10. MALDIT-OF mass spectrum of the Fe(III)-retained fraction from the extract of bovine adrenal medula. These signals are due to phoshpopeptides.

16.9. HUMAN GENOME PROJECT

The goal of the human genome project (HGP) is to sequence the intact human genome. This project is a challenging and tremendous undertaking, considering the fact that the human genome contains approximately 3.3×10^9 bp that are held together to form more than 100,000 genes. The existing methods are cumbersome, time-consuming, labor-intensive, and expensive to be relied for a large-scale DNA sequencing project. Mass spectrometry has begun to play a significant role in the HGP. In its role, mass spectrometry can act as an efficient and sensitive detector for the CE-separated oligonucleotide fragments, it can be combined with the Sanger sequencing technique (see Chapter 15), or it can provide the sequence of the oligonucleotide fragments. In particular, FT-ICRMS allows the accurate mass measurement of the ESI-produced single molecular ions of single-stranded DNA. The ion can be trapped within the ICR cell, and subjected to sequential photo-dissociation steps. The sequence can be derived from the masses of the fragments from each nucleotide cleavage step.

A large number of examples presented in this chapter provide an ample proof that mass spectrometry is an indispensable component of biomedical research. In the future, it is destined to play a much expanded role in solving real-world problems in the field of health sciences.

REFERENCES

1. K. M. Downard, *J. Mass. Spectrom.* **35**, 493–503 (2000).
2. A. L. Cox, J. Skipper, Y. Chen, R. A. Henderson, T. L. Darrow, J. Shabanowitz, V. H. Engelhard, D. F. Hunt, and C. L. Slinguff, *Science* **264**, 716–719 (1994).
3. W. Wang, L. R. Meadows, J. M. M. den Haan, N. E. Sherman, Y. Chen, E. Blokland, J. Shabanowitz, A. L. Agulnik, C. E. Bishop, D. F. Hunt, and E. Goulmy, *Science* **269**, 1588–14590 (1995).
4. A. Y. C. Huang, P. H. Gulden, A. S. Woods, M. C. Thomas, C. D. Tong, W. Wang, V. H. Engelhard, G. Pasternanck, R. Cotter, D. F. Hunt, D. M. Pardoll, and E. M. Jaffe, *Proc. Natl. Acad. Sci.* (USA) **93**, 9730–9735 (1996).
5. L. R. Meadows, W. Wang, J. M. M. den Haan, E. Blokland, C. Reinhardus, J. W. Drijfhout, J. Shabanowitz, R. Pĺerce, A. L. Agulnik, C. E. Bishop, D. F. Hunt, E. Goulmy, and V. H. Engelhard, *Immunity* **6**, 273–281 (1997).
6. C. R. Bazemore Walker, N. E. Sherman, J. Shabanowitz, and D. F. Hunt, in B. S. Larsen and C. N. McEwen, eds., *Mass Spectrometry of Biological Materials*, Marcel Dekker, New York, 1998, pp. 115–135.
7. T. Monji, A. L. McCormack, J. R. Yates, III, and D. Pious, *J. Immunol.* **153**, 4468 (1994).
8. A. L. McCormack, J. K. Eng, P. C. DeRoos, A. Y. Rudensky, and J. R. Yates, III, in A. P. Snyder, ed., *Biochemical and Biotechnological Appliactions of Electrospray Ionization Mass Spectrometry*, ACS, Washington, DC, 1995, pp. 207–225.
9. K. Pihel, S. Hsieh, J. W. Jorgenson, and R. M. Whiteman, *Anal. Chem.* **67**, 4514–4521 (1995).
10. S. Hsieh, K. Dreiseuerd, R. C. van der Schors, C. R. Jimnez, J. Stahl-Zeng, F. Hillenkamp, J. W. Jorgenson, W. P. M. Geraerts, and K. W. Li, *Anal. Chem.* **70**, 1847–1852 (1998).
11. L. Li, R. W. Garden, E. V. Romanova, and J. V. Sweedler, *Anal. Chem.* **71**, 5451–5458 (1999).
12. P. Chaurand, M. Steckli, and R. M. Caprioli, *Anal. Chem.* **71**, 5263–5270 (1999).
13. S. Jespersen, P. Chaurand, F. J. C. van Strien, B. Spengler, and J. van der Greef, *Anal. Chem.* **71**, 660–666 (1999).
14. R. S. Brown and J. J. Lennon, *Anal. Chem.* **67**, 3990–3999 (1995).
15. R. Kaufmann, B. Spengler, and F. Lutzenkirchen, *Rapid Commun. Mass Spectrom.* **7**, 902–910 (1993).
16. H. M. Shapiro, *Practical Flow Cytometry*, 3rd ed., Wiley, New York, 1995.

17. E. N. Fung and E. S. Yeung, *Anal. Chem.* **70**, 3206–3211 (1998).

18. Y. Chang and E. S. Yeung, *Anal. Chem.* **69**, 2251–2259 (1997).

19. L. Y. Ch'ang, K. Tang, M. Schell, C. Ringelberg, K. G. Matteson, S. L. Allman, L. and C. H. Chen, *Rapid Commun. Mass Spectrom.* **9**, 772–774 (1995).

20. R. Théberge, L. Conners, M. Skinner, J. S. Kare, and C. E. Costello, *Anal. Chem.* **71**, 452–459 (1999).

21. R. Théberge, L. H. Connors, M. Skinner, and C. E. Costello, *J. Am. Soc. Mass Spectrom.* **11**, 172–175 (2000).

22. Y. Wada, A. Hayashi, T. Fujita, T. Matsuo, I. Katakuse, and H. Matsuda, *Biochm. Biophys. Acta.* **667**, 233–241 (1981).

23. Y. Wada, T. Matsuo, and T. Sakurai, *Mass Spectrom. Rev.* **8**, 379–434 (1989).

24. H. E. Witkowska and C. H. L. Shackleton, *Anal. Chem.* **68**, 29A–33A (1996).

25. K. J. Light-Wahl, J. A. Loo, C. G. Edmonds, R. D. Smith, H. E. Witkowska, and C. H. L. Shackleton, *Biol. Mass Spectrom.* **22**, 112–120 (1993).

26. T. Nakanishi, A. Miyazaki, M. Kishikawa, A. Shimuzu, and T. Yonezawa, *J. Am. Soc. Mass Spectrom.* **7**, 1040–1049 (1996).

27. R. M. Whittal, B. O. Kelly and L. Li, *Anal. Chem.* **70**, 5344–5347 (1998).

28. M. E. McComb, R. D. Oleschuk, A. Chow, W. Ens, K. G. Standing, H. Perreault, and M. Smith, *Anal. Chem.* **70**, 5142–5149 (1998).

29. C. T. Houston and J. P. Reilly, *Anal. Chem.* **71**, 3397–3404 (1999).

30. S. A. Gerber, C. R. Scott, F. Turecek, and M. H. Gelb, *J. Am. Chem. Soc.* **121**, 1102–1103 (1999).

31. G. Choudhary, J. Chakel, W. Hancock, A. Torres-Duarte, G. McMahon, and I. Wainer, *Anal. Chem.* **71**, 855–859 (1999).

32. P. T. Todorov, M. Deacon, and M. J. Tisdale, *J. Biol. Chem.* **272**, 12279–12288 (1997).

33. P. H. O'Farrel, *J. Biol. Chem.* **250**, 4007–4021 (1975).

34. T. Krishnamurthy, M. T. Davis, D. C. Stahl, and T. D. Lee, *Rapid Commun. Mass Spectrom.* **13**, 39–49 (1999).

35. P. J. Jensen, L. Pasa-Tolic, G. A. Anderson, J. A. Horner, M. S. Lipton, J. E. Bruce, and R. D. Smith, *Anal. Chem.* **71**, 2076–2084 (1999).

36. Y. Dai, L. Li, D. C. Roser, and S. R. Long, *Rapid Commun. Mass Spectrom.* **13**, 73–78 (1999).

37. B. E. Chong, D. M. Lubman, F. R. Miller, and A. J. Rosenspire, *Rapid Commun. Mass Spectrom.* **13**, 1808–1812 (1999).

38. A. Westman, C. L. Nilsson, and R. Ekamn, *Rapid Commun. Mass Spectrom.* **12**, 1092–1098 (1998).

39. K. M. Sullivan, A. Mannucci, C. D. Kimpton, and P. Gill, *Biotechniques* **15**, 636–641 (1993).

40. N. Taranenko, N. T. Potter, S. L. Allman, V. V. Golovlev, and C. H. Chen, *Anal. Chem.* **71**, 3974–3976 (1999).

41. J. R. Yates, III, *J. Mass Spectrom.* **33**, 1–19 (1998).

42. Y. Chen, X. Jin, D. Misek, R. Hinderer, S. M. Hanash, and D. M. Lubman, *Rapid Commun. Mass Spectrom.* **13**, 1907–1916 (1999).

43. S. Beranova-Giorgianni and D. M. Desiderio, *Rapid Commun. Mass Spectrom.* **14**, 161–167 (2000).

44. C. Fenselau, ed., *Mass Spectrometry for the Characterization of Microorganisms*, ACS Symp. Series, Vol. 541, 1994.

45. C. Fenselau and R. J. Cotter, *Chem. Rev.* **87**, 501–512 (1987).

46. S. DeLuca, E. W. Sarver, P. D. Harrington, and K. J. Voorhees, *Anal. Chem.* **62**, 1465–1472 (1990).

47. T. Krishnamurthy, P. L. Ross, and U. Rajamani, *Rapid Commun. Mass Spectrom.* **10**, 883–887 (1996).

48. B. E. Chong, D. B. Wall, D. M. Lubman, and S. J. Flyn, *Rapid Commun. Mass Spectrom.* **11**, 1919–1908 (1997).

49. C. L. Nilsson, *Rapid Commun. Mass Spectrom.* **13**, 1067–1071 (1999).

50. M. A. Domin, K. J. Welham, and D. S. Ashton, *Rapid Commun. Mass Spectrom.* **13**, 222–226 (1999).

51. R. D. Holland, J. G. Wilkes, F. Rafii, J. B. Sutherland, C. E. Voorhees, and J. O. Lay, Jr., *Rapid Commun. Mass Spectrom.* **10**, 1227–1232 (1996).

52. K. J. Welham, M. A. Domin, D. E. Scannell, and E. Cohen, and D. S. Ashton, *Rapid Commun. Mass Spectrom.* **12**, 176–180 (1998).

53. Z. Wang, L. Russon, L. Li, D. C. Roser, and S. R. Long, *Rapid Commun. Mass Spectrom.* **12**, 456–464 (1998).

54. E. C. Lynn, M.-C Chung, W.-C. Tsai, and C.-C. Han, *Rapid Commun. Mass Spectrom.* **13**, 2022–2027 (1999).

55. M. A. Winkler, J. Uher, and S. Cepa, *Anal. Chem* **71**, 3416–3419 (1999).

56. C. L. Nilsson, T. Larsson, E. Gustafsson, K.-A. Karlsson, and P. Davidson, *Anal. Chem.* **72**, 2148–2153 (2000).

57. M. B. Beverly, F. Basile, K. J. Voorhees, and T. D. Hadfield, *J. Am. Mass Spectrom.* **10**, 747–758 (1999).

58. F. Xiang, G. A. Anderson, T. D. Veenstra, M. S. Lipton, and R. D. Smith, *Anal. Chem.* **72**, 2475–2481 (2000).

59. B. Bothner, X. F. Dong, L. Bibbs, J. E. Johnson, and G. J. Siuzdak, *J. Biol. Chem.* **273**, 673–676 (1998).

60. D. Despeyroux, R. Phillpotts, and P. Watts, *Rapid Commun. Mass Spectrom.* **10**, 937–941 (1996).

61. J. J. Thomas, B. Falk, C. Fenselau, J. Jackman, and J. Ezzell, *Anal. Chem* **70**, 3863–3867 (1998).

62. M. A. Winkler, N. Xu, H. Wu, and H. Aboleneen, *Anal. Chem.* **71**, 3416–3419 (1999).

63. J. J. Smith, J. K. Lewis, B. Bothner, and G. J. Siuzdak, *Proc. Natl. Acad. Sci.* (USA) **95**, 6774 (1998).

64. G. Siuzdak, *J. Mass Spectrom.* **33**, 203–211 (1998). — *virus tutorial*

65. S. Li and C. Dass, *Anal. Biochem.* **270**, 9–14 (1999).

66. S. Li and C. Dass, *Eur. Mass Spectrom.* **5**, 279–284 (1999).

67. D. M. Desiderio and X. Zhu, *J. Chromatogr. A* **794**, 85–96 (1998).

68. C. Dass, J. J. Kusmierz, and D. M. Desiderio, *Biol. Mass Spectrom.* **20**, 130–138 (1991).

69. C. Dass, G. H. Fridland, P. W. Tinsley, J. T. Killmar, and D. M. Desiderio, *Int. J. Pept. Prot. Res.* **34**, 81–87 (1989).

INDEX